Cybersecurity Analytics

CHAPMAN & HALL/CRC DATA SCIENCE SERIES

Reflecting the interdisciplinary nature of the field, this book series brings together researchers, practitioners, and instructors from statistics, computer science, machine learning, and analytics. The series will publish cutting-edge research, industry applications, and textbooks in data science.

The inclusion of concrete examples, applications, and methods is highly encouraged. The scope of the series includes titles in the areas of machine learning, pattern recognition, predictive analytics, business analytics, Big Data, visualization, programming, software, learning analytics, data wrangling, interactive graphics, and reproducible research.

Published Titles

Feature Engineering and Selection: A Practical Approach for Predictive Models
Max Kuhn and Kjell Johnson

Probability and Statistics for Data Science: Math + R + Data
Norman Matloff

Introduction to Data Science: Data Analysis and Prediction Algorithms with R
Rafael A. Irizarry

Cybersecurity Analytics
Rakesh M. Verma and David J. Marchette

Cybersecurity Analytics

Rakesh M. Verma
University of Houston, Houston, TX
David J. Marchette
Naval Surface Warfare Center, Dahlgren, VA

CRC Press is an imprint of the
Taylor & Francis Group, an **informa** business
A CHAPMAN & HALL BOOK

Chapman & Hall/CRC Press
Taylor & Francis Group
6000 Broken Sound Parkway NW, Suite 300
Boca Raton, FL 33487-2742

First issued in paperback 2022

ISBN 13: 978-1-03-240100-3 (pbk)
ISBN 13: 978-0-367-34601-0 (hbk)
ISBN 13: 978-0-429-32681-3 (ebk)

DOI: 10.1201/9780429326813

Publisher's Note
The publisher has gone to great lengths to ensure the quality of this reprint but points out that some imperfections in the original copies may be apparent.

Library of Congress Cataloging-in-Publication Data

Names: Verma, Rakesh M., author. | Marchette, David J., author.
Title: Cybersecurity analytics / Rakesh M. Verma and David Marchette.
Description: Boca Raton, FL : CRC Press, 2020. | Series: Chapman & Hall/CRC data science series | Includes bibliographical references and indexes. | Summary: "This book organizes in one place the mathematics, probability, statistics and machine learning information that is required for a practitioner of cybersecurity analytics, as well as the basics of cybersecurity needed for a practitioner"-- Provided by publisher.
Identifiers: LCCN 2019029587 (print) | LCCN 2019029588 (ebook) | ISBN 9780367346010 (hardback) | ISBN 9780367346027 (paperback) | ISBN 9780429326813 (pdf)
Subjects: LCSH: Computer security.
Classification: LCC QA76.9.A25 V456 2020 (print) | LCC QA76.9.A25 (ebook) | DDC 005.8--dc23
LC record available at https://lccn.loc.gov/2019029587
LC ebook record available at https://lccn.loc.gov/2019029588

Visit the Taylor & Francis Web site at
http://www.taylorandfrancis.com

and the CRC Press Web site at
http://www.crcpress.com

In loving memory of my mother, Usharani Verma

Rakesh M. Verma

To my wife, Susan

David J. Marchette

Contents

Preface

This book arose from a panel discussion [71] at the ACM International Workshop on Security and Privacy Analytics (2015) [449] that led to the paper, [455]. Part of the impetus for this book also came from the Security Analytics course that the author, Rakesh M. Verma, has been teaching at the University of Houston since 2015.

It is an attempt to organize in one place the mathematics, probability, statistics and machine learning information that is required for a practitioner of cybersecurity analytics, as well as the basics of cybersecurity needed for a practitioner. We feel that any collaboration between a data scientist and a cybersecurity expert cannot be fully productive, unless there is some shared vocabulary and mutual understanding, which is our objective in this book. We aim to provide the cybersecurity analyst with the tools to understand the mathematics and machine learning methods that are used for most cybersecurity analytics.

Such an undertaking cannot be comprehensive; however, we aim to give some background on the main areas of knowledge necessary for cybersecurity analytics, and provide a basic understanding of the ideas, with citations for further reference. It is assumed that a reader has a basic understanding of mathematics at the college level. We attempt to give enough background to understand the examples and applications to cybersecurity, and provide references that can fill in the gaps. We intend the book to provide cybersecurity analysts an introduction to some of the advanced mathematics that applies to the problem domain.

In *The Art of War* ([442]), it is said:

If you know the enemy and know yourself, you need not fear the result of a hundred battles. If you know yourself but not the enemy, for every victory gained you will also suffer a defeat. If you know neither the enemy nor yourself, you will succumb in every battle.

The main purpose of the techniques discussed in this book is to provide methods to both "know thyself" and "know thy enemy". The former involves methods to understand the environment in which the systems are utilized, the data that one observes under normal conditions, and the ways to protect the systems from attack. The latter involves understanding what attacks "look like" when they are mounted against the system, design methods to detect attacks, and methods to detect unusual events that may correspond to novel, previously unseen attacks. While one also wishes to "know thy enemy" in the sense of knowing who the attackers are and what their motivations and capabilities are, this is mostly beyond the scope of this book, and is touched on only briefly in Chapter 3.

The book is organized as follows. The introduction gives some insight into our philosophy and some of the main ideas we wish the reader to take away from the book. The chapter on data anaytics outlines the basic techniques of data processing and data analysis necessary for cybersecurity applications. It describes many of the important issues and gives an introduction to some of the important concepts. Basic cybersecurity concepts are introduced. The basic mechanisms, threats and challenges are described from a data analytics point of view.

The chapter on probability and statistics gives an overview of the key concepts, and provides some discussion of practical applications. This is followed by a chapter on data mining. In some circles this term has become passé, and replaced by data analytics or data science. In this book it is used to refer to the unsupervised aspects of machine learning, including clustering, and anomaly detection, but also association rules – "if you liked this movie, you might like these" – and manifold learning, which is a method for reducing the dimensionality of a data set by modeling the "manifold structure" of the data.

The data mining chapter is followed by a chapter on machine learning, which in this book refers primarily to supervised classification. The distinction between supervised and unsupervised methodologies is important, which is why the chapters are split, but the naming convention of data mining versus machine learning for the two approaches is nonstandard. The machine learning also contains a section on topological data analysis, which is an unsupervised method for finding structure in data that is related to manifold learning, and is becoming more and more a standard tool in machine learning. Also, we give a very brief introduction to neural networks in this chapter, which have both supervised and unsupervised variants.

The next two chapters apply the previous methods to the analysis of textual data. Examples of these are documents, emails, log files, etc. Finally, the book ends with a short chapter on big data techniques and processing at scale.

There are appendices in which the basic linear algebra and graph theory required by this book are discussed. The book also has a fairly extensive Bibliography, and an effort has been made to provide references to a wide range of papers and books relevant to the field, as well as background material and some material of historical interest.

In writing a book, there is always the question of what material to present, and what areas to leave undiscussed. This is a very personal decision, based on the preferences, expertise, and interests of the authors, and the level of experience and knowledge expected of the reader. There are many things missing from this book that another author might choose to present. There are also several choices that were made, consciously and subconsciously, on how to present the material and the depth and detail of the presentations.

Some of the organization of the book was dictated by the decision to write a book jointly, and to set an ambitious deadline for the book, and some of the organization decisions were made when we wrote the paper [455]. The deadline forced us to devote the necessary time, even when other constraints on our time pressed us for attention. The fact that we were geographically dispersed was mitigated through the use of Overleaf,[1] which was considerable help in working on this project and keeping things organized.

The deadline also helped to scope down some of our ambitions. There are clearly topics that we would like to devote more discussion to. There are more examples we would like to provide. There are topics that we would like to include. We hope that the decisions on what to include and exclude is mitigated somewhat by the Bibliography, and the level of detail and examples provided is sufficient to get the important points across and give the readers the tools to explore further, and apply the techniques to their data and problems.

A book of this scope and length is a major undertaking, and would not be possible without significant help and sacrifices. With deep gratitude and pleasure, the authors credit the following, who generously gave up their time and/or help.

[1] `https://www.overleaf.com/`

Acknowledgments

Rakesh Verma thanks his family, Rashmi, Sudha and Sneha, his mentors, colleagues, friends, and his students in the Reasoning and Data Analytics for Security (ReDAS) Laboratory: Tanmay Thakur, Nirmala Rai, Vasanthi Vuppuluri, Keith Dyer, An Nguyen, Arthur Dunbar, Boris Chernis, Luis F.T. Moraes, Shahryar Baki, Daniel Lee, Ayman El Aassal, Avisha Das, Vibhu Sharma, Devin Crane, Marc Magnusson, Houtan Faridi, Uroosa Ali, and Nabil Hossain, without whom none of this would have been possible. Thanks to Narasimha Shashidhar Karpoor for early discussions on phishing. Thanks also to the National Science Foundation and the Department of Defense (Army Research Laboratory) for supporting some of the research described in this book.

David Marchette thanks his family, Susan, Steven, Jeffrey and Katy, and his mentors, colleagues and friends, without whom none of this would have been possible.

Both authors would like to thank the following colleagues and students for generously contributing their time and reviewing parts of the manuscript: Ricardo Vilalta for a close reading of the Machine Learning chapter, Wenyaw Chan and Luis Moraes for feedback on the Statistics chapter, Chester Rebeiro for thoughtful comments on the Basics of Security and Security Analytics chapter, Edgar Gabriel for a careful reading and good feedback on the Big Data for Security chapter, Arjun Mukherjee for constructive comments on the Text Mining chapter, Carlos Ordonez and Daniel Lee for reviewing the Data Mining chapter, and Avisha Das for excellent suggestions on the NLP chapter. Of course, we are responsible for all remaining errors and omissions.

Chapter 1

Introduction

This book is an attempt to organize in one place the mathematics, probability, statistics, machine learning, and text processing information that is required for a practitioner of cybersecurity analytics, as well as the basics of cybersecurity needed for a practitioner. Such an undertaking cannot be comprehensive; however, we aim to give some background on the main areas of knowledge necessary for cybersecurity analytics, and provide a basic understanding of the ideas, with references for further study.

There are many books and articles relevant to this topic. To mention just a few, see for example [299, 333, 402, 404, 445]. See also the computer security section of your local bookstore. See [62] for a survey of machine learning methods for security. There are also several books aimed specifically at data analysis and machine learning for cybersecurity. See for example [3, 82, 293]. These are compendiums of papers or chapters written by experts in the field, and provide a number of specialized investigations in areas touched on in this book. For a more introductory text, see [208]. There are also a number of articles that survey machine learning methods for security. See for example [62].

Most of cybersecurity is focused on defense – firewalls, secure coding principles, securing computers through a Security Technical Implementation Guide (STIG), virus scanners, phishing and spam detectors, and training users in best practices. Organizations will often hire a red team, a group of security experts who analyze their network/computers for vulnerabilities and errors, to allow the company to better harden the systems. Others keep security experts on staff, or contract with a security company to provide the information technology infrastructure for them.

Some of this is enhanced through statistical and machine learning algorithms, and it is these algorithms that are the focus of this book. Basic design flaws in software such as buffer overflow vulnerabilities are best addressed at the source, through better coding practices. However, we do not live in a perfect world, mistakes are made, and most software is in binary format without access to the source code. Thus, methods need to be developed to detect vulnerabilities without access to the software, and often this means one must observe the actions of software, network traffic, or user activity to detect "unusual" events, or known attacks. This is where pattern recognition techniques, such as statistical and machine learning methods, come in.

The news is full of the terms Artificial Intelligence (AI) and Machine Learning (ML). Artificial intelligence refers to writing code to mimic intelligent behavior. Code that guides a robot through a maze, chess playing software, and recommendation systems in online commerce may all be examples of artificial intelligence. Note that the first example may be purely algorithmic – there are many path-following algorithms that have been developed – or it may be more sophisticated if for example there are potential obstacles that must be detected and avoided. The early chess-playing machines utilized powerful search methodologies that allowed them to "look ahead" a number of moves, as well as pruning algorithms to attempt to keep the search tree to a manageable size. These were also algorithmic; modern methods have at least some aspect of learning in them – they learn from playing, although even these algorithms will often maintain some core of "look ahead" searching in them.

A key component of artificial intelligence is search – the ability to rapidly find matches to a pattern in a large set of data. In fact, it has been said that *artificial intelligence is search.*[1] An interesting book on the history of artificial intelligence is [306], in which Alan Turing is quoted (page 70) as supporting this sentiment. Clearly AI is more than just better search algorithms, but much of AI can be understood in terms of "searching for patterns" and AI algorithms often contain some aspect of search.

The field of AI encompasses the field of machine learning as a subfield. The distinction is that machine learning utilizes data to "learn" the important patterns or actions, rather than requiring a clever person to invent an algorithm and implement it in code. The machine learns the desired capability through processing *training data* which is representative of the task at hand. Obviously, this requires clever people to come up with training algorithms to allow this to happen, and this book will describe a number of these and illustrate their application for several cybersecurity purposes.

Of course, as noted above, there's more to AI than search, and there's more to ML than learning from data. This book focuses on techniques related to data, and to the types of problems that can be addressed through data analysis and machine learning methodologies. Interested readers should investigate one of the many books written specifically on AI, machine learning, or related subjects, such as [128, 213, 371]. Readers interested in neural networks, which is touched on only briefly in this book, might find [172] of interest.

The idea of a *model* is central to the aspects of AI and machine learning that we are focused on in this book. The model provides a representation of the data that allows one to make predictions about new data. In machine learning these models are statistical in nature, and the parameters of the model (and in some cases the model form itself) are determined from the data. AI methods that utilize rules also have a statistical basis to them, whether the rules are extracted from experts or learned from data. There's always uncertainty involved in any system that interacts with the "real world" and this means that one must incorporate statistical models in the solution.

In essence, one posits a family of models from which the data was drawn or which can adequately approximate the true model and defines an algorithm or set of algorithms to select the appropriate model from the family and fit the parameters of that model. It is often the case in machine learning that the family of models is not made explicit by the algorithm. For example, the random forest algorithm (Section 6.5.1) builds a large number of "decision trees" and provides a vote from these trees. While it is possible to describe the family of models that are encompassed by this algorithm – it is the family of "forests of decision trees" – only theoreticians are likely to care about the details. Similarly, feed forward neural networks (Section 6.8) are a family of models – the hidden layer structure and activation functions determining the model within the family and the weights and biases determining the parameters. In both of these examples, one does not think of the models as *generative* – that is that the data were generated from a model of this type – but rather as approximations that are sufficiently accurate to perform the desired inference.

Several data sets are used throughout the book. Some of these have known problems, and the data are used solely to illustrate the various techniques. One of the data sets, the KDD-Cup data, was constructed from the DARPA 1999 data set, which is an artificial data set originally constructed to act as a testbed for computer intrusion detection algorithms.

[1]We do not know the origin of this phrase, although there are a number of examples of it in literature. See for example `https://clsimplex.com/asset/artificial-intelligence-introduction`.

The original set had known problems (see [291]), and although the KDD-Cup data does not suffer from all of these (see [424]), it is old and not ideal for serious computer security work. With that said, it is a useful data set for illustrating some of the issues and techniques in machine learning, and its wide availability means that it is relatively easy for the interested reader to obtain it and test out the methodologies in this book for herself.

A tool that appears in many places in this book, in some cases prior to it being formally introduced in Chapter 6, is the nearest neighbor classifier. This is one of the most easily understood and implemented algorithms in machine learning, and is a useful tool for illustrating some of the other ideas throughout the book. The algorithm is as follows. Consider the case of classifying an email as spam or not. We are given data $\{x_1, x_2, \ldots, x_n\}$ (emails) for which one has known class labels $\{y_1, y_2, \ldots, y_n\}$ (spam or legitimate – *ham*), and a new observation (email) x which one wishes to classify. One finds the x_i closest to x and assigns class y_i to x. This is trivially generalized to produce a "vote" amongst the k closest x_i. We will use this algorithm in several places to illustrate how one method for analyzing data is/is not better than another for various purposes.

There are a number of important lessons that we wish the reader to take from this book, beyond the basic material and methodologies of analytics for cybersecurity. These are:

1. Without (good) data, there is no knowledge. This is particularly important in cybersecurity. Although a mathematician can start with a set of axioms and develop a theory that follows logically from these axioms, in the real world the axioms are mostly unknown – in spite of centuries of investigation – and are only partly and imperfectly known when we come across them. Thus, one needs data to understand the actual properties and variability of any system one wishes to analyze, and one needs data to determine what types of things one can expect to see. *A corollary to this maxim is that the quality of the data, and of the ground truth, is extremely important.*

2. As Box has said in several places and several different ways ([50, 51]): all models are wrong; some are useful. We will discuss parametric and nonparametric models in this book, and suggest that practitioners investigate different models and examples of both parametric and nonparametric methods when analyzing data or designing algorithms. It is simply not possible to find "the one true model" for anything but the most simplified and constrained situation; however, one can learn a great deal from models that are approximately correct.

3. Simple models are to be preferred to more complex models, except when they aren't. We refer often to David Hand's paper [185] in which he argues this point quite succinctly. This is also a consequence of the bias-variance trade-off: the more complex the model, the more likely it is to overfit the data. While the more complicated model can provide a better fit to the true model, assuming the extra complexity is warranted, it can only do so if there is sufficient data to accurately fit the parameters. Quoting William of Ockham: *"Entia non sunt multiplicanda praeter necessitatem"*.[2]

4. Dr. Hand also notes that the data used to train the system is very often not drawn from the same distribution as the data the system will see once it is deployed. This is particularly important in cybersecurity: the data is nonstationary, meaning that it changes in time. Some of this change is due to growth in usage, some is due to new usage such as novel applications and functionality, and some is due to the adversaries, who continue to invent new ways to compromise systems. Yet, even if the data were stationary, it would still be impossible to collect data that is representative; each

[2] "More things should not be used than are necessary," from Wikipedia, `https://simple.wikipedia.org/wiki/Occam's_razor`, accessed 5/27/2018.

network is different, each computer is used in a slightly different way and has slightly different applications and vulnerabilities, each organization has different policies and concerns. Thus, any algorithm developed for cybersecurity must be robust to the fact that the data that was used to train it is not fully representative of the data it will see when deployed.

5. The above point emphasizes an extremely important issue: evaluating one's solution on new data is critical. In a cybersecurity application, one must be sure that the data used to test the system is representative of the data the system will see. One must be constantly vigilant to changes in the data and changes in the environment that generates the data. This means that adaptive methods that can adjust to new situations may be attractive; care must be taken, though, to ensure that this does not introduce a vulnerability – the ability of an attacker to craft data that causes the system to adapt away from the correct model. See the references on adversarial methods in Chapter 6, and the references therein. We discuss this briefly below as well.

6. There is almost never just one correct answer to any problem. Different methods have different strengths and weaknesses, and knowing these is important to understanding which methods are most appropriate for a given task. On the other hand, utilizing several different methods can provide one with additional information about the problem, and can allow for a hedge against incorrect assumptions. With that said, there has been quite a bit of work done to investigate the performance of various algorithms against various data sets. Although there is no one universal answer, there are often guidelines that can be obtained by reading this literature. For example, in the area of classification (supervised learning), an extensive evaluation of existing methods ([150]) shows that there are a few algorithms that are consistently good: ensemble methods like random forests, and kernel methods such as support vector machines.

7. The more you understand the underlying mechanism that produces the data, the better you can design an algorithm to solve the problem you wish to solve. Similarly, the more you understand the underlying theory and applicability of a given method, the better you can apply that method in a rigorous manner to new problems.

8. The flip side of the above point is that it is important to understand the data you observe, and this requires tools to visualize and plot the data. There are a number of techniques discussed in this book, and there are many other very good books on visualization and graphics applied to data analysis. A good place to start are the books by Edward Tufte, [437, 438, 439]. See also [96, 514, 515]. The grammar of graphics ([488]) lays out the basic structure of graphics in a coherent manner, and provides insight into how the pieces of a good graphic fit together and are related. This has had a revolutionary effect on statistical graphics.

The issue of parsimony of model, or overfitting of the data, is illustrated in Figure 1.1. To illustrate the problem of choosing a model that is too complex, we fit a degree 20 polynomial to the data. Note that the polynomial interpolates through all the data points, which is clearly not desirable here. In effect, the polynomial is "memorizing" the data. Note further that outside of the data the model can be wildly different than the correct value (indicated by the solid line), and even wildly different from the nearby data points. The range of values attained by the polynomial fit in this region is $[-4279576, 6886]$, hence between the nearly vertical lines in the figure the curve is massively wrong.

To illustrate that the problem isn't just the number of parameters, we show as a dashed curve a nonparametric estimate ([161]). This uses a window to compute a locally linear fit

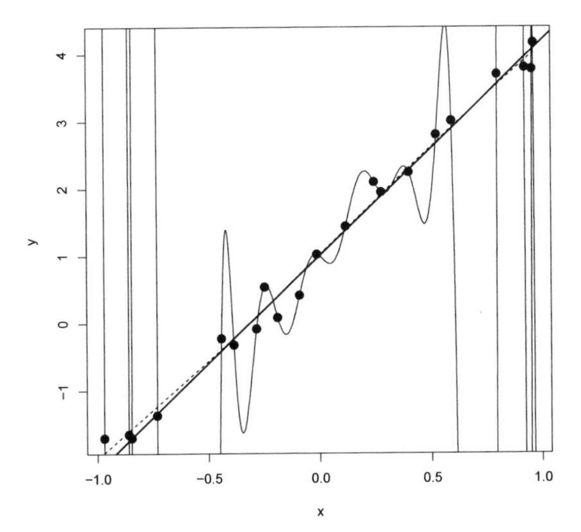

Figure 1.1: A collection of 21 points, drawn from the model denoted by the solid line with added noise. A polynomial fit of degree 20 is shown as a solid curve. A nonparametric smoother is shown as a dashed line. The vertical lines on the ends of the interval are indicative of extremely large values for the polynomial fit in these regions.

to the data, and although in a sense the number of parameters is large – in fact it is equal to the number of observations times 2 – the overall model does a good job of fitting the model for this data set.

The worst overfitting occurs near the "edge" of the data, while there is much less variability in the interior of the support of the data. However, even there, the fit between points is quite far off the correct model.

The polynomial fit in Figure 1.1 is a good one to keep in mind, particularly in light of Figure 1.2.[3] In the polynomial case, the model family is the set of all degree 20 polynomials on one variable with coefficients in $\mathbb{R}$ (or that subset of $\mathbb{R}$ that can be represented in 64-bit arithmetic). For this example, suppose our family requires that the coefficient of x^{20} be nonzero, say at least some $\epsilon > 0$ in absolute value. While the true model – a line – is not in the family, it is possible to get quite close to the model – at least in the range of the data: $[-1, 1]$ – by setting most of the other coefficients to zero, keeping only the dominant and linear terms. So the extrinsic error is quite small in this case, but one would require a very large number of observations to drive the intrinsic error to something manageable.[4]

The error between points, in regions for which there are no observations, is of concern – it corresponds to how well the algorithm "generalizes" outside of the data used to fit it. This is one of the most important aspects of machine learning: to understand the methods and the models they produce, so that one is confident that the system will operate correctly on new data. There is a new field called "adversarial learning" in which one utilizes an adversary to force a machine learning algorithm to create models that are robust to attacks. This is also used to defeat algorithms, and to highlight their inadequacies. See [173, 283, 340, 372, 420] for some examples. The more complex the model, the more difficult it may be to validate

[3]This figure is inspired by Figure 12.1 of [115].

[4]We are assuming that the error is only computed within the range of the data, since the polynomial will go to infinity at a very high rate outside of this range.

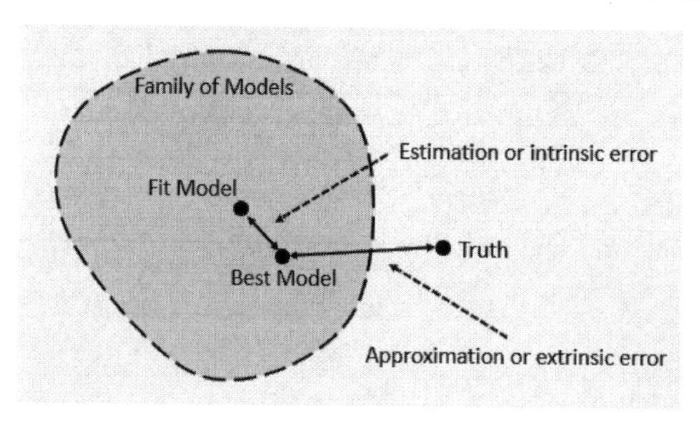

Figure 1.2: The different errors that occur when fitting a model to data. The family of models, whether parametric or nonparametric, is given by the gray region. This contains the model which best corresponds to the true distribution, which (see the Box quote above) is not contained in the region. This results in an error that cannot be reduced, without changing the model family. Additionally, the actual model fit to a data set will not be the best possible model, and so there is error introduced there as well.

it and determine exactly "what it is doing" – and, most importantly, what it will do with new data. Once again, we come back to the lesson that we want the simplest model that performs the desired task.

Algorithms that continue to adapt – for example online learning algorithms discussed in Section (6.12) – also need to be checked to ensure they haven't "gone off the rails". A sufficiently knowledgeable adversary may be able to inject data into the system that causes it to "learn" to ignore the attack that is ultimately mounted by the adversary. At a minimum, the system needs to be checked to ensure that it still correctly handles the elements of the original data on which it was designed and trained that are still relevant. This is necessary, but not sufficient, and so these systems must be constantly monitored to ensure that they are still performing at the desired accuracy.

In [58] two "cultures" of data analytics or data science are described, which can be thought of as modeling versus algorithms. Essentially, the distinction can be thought of as between those who are focused primarily on obtaining good models, for the purpose in part of better understanding the processes that generated the data, and those who are more interested in the bottom line of performance of the algorithm. One is tempted to think of these as the scientists and the engineers, or academics and practitioners, although these analogies are imperfect. See [123] for an updated discussion of this paper and the issues it brings up.

Both modeling and algorithm performance are key to implementing successful data analytics and machine learning to solve problems in cybersecurity. However, in this book we will tend to err more on the side of the bottom line. Although the models are key to developing and understanding the algorithms, in the final analysis it is the performance of the algorithm that is of fundamental importance. Hence we will illustrate many of the algorithms we discuss on real world cybersecurity problems and data and compare and contrast their performance on these data sets, rather than focus more on the validity of the models used in the algorithms. There is a strong bias in this book towards nonparametric methods that relax assumptions about the underlying distribution of the data rather than parametric methods that posit a particular parametric family for the distribution.

As noted above, a single book cannot cover all the topics of mathematics, statistics, artificial intelligence and machine learning that are relevant to cybersecurity. We do not cover game theory, which formalizes the concept of a game and investigates strategies for playing these idealized games. It is possible to model some of cybersecurity as a game with the security analyst "playing" against the attackers. See [310] for an introduction to some of the basic ideas.

We cover only the basics of cryptography; whole books are devoted to this subject. It may be one of the most important aspects of cybersecurity, but it is not a panacea. Ignoring the possibility that quantum computers may make some cryptography obsolete,[5] it is possible to extract information from encrypted traffic, if sufficient care is not taken in the implementation of the cryptographic system. See [491, 492] for some methods of extracting information from encrypted streams through statistical machine learning, and [493] for a potential solution. For another approach to defeating cryptographic protocols without stealing keys or breaking the encryption, the reader should consult, e.g., [273, 274, 284], and the many papers on cryptographic protocol verification.

We also do not discuss the work in artificial immune systems, see for example [106, 194, 422, 432]. This work utilizes analogies with biological systems to develop methodologies to detect "infections" such as malware or malicious users – the *insider threat.*

This is also not a book that covers all of cybersecurity. Chapter 3 provides the basic introduction to cybersecurity and gives an overview of some of the relevant issues. Examples of cybersecurity data and analyses are provided throughout the book to illustrate different methods. For a complete understanding of cybersecurity, a book dedicated to the subject, such as [349, 404], is a place to start.

The best way to learn data analytics is to do it. Several repositories of cybersecurity data are discussed in the text, and the reader is encouraged to obtain any of these data sets and utilize the various methods discussed in the text. The programming languages R [358] and Python have extensive packages that implement the algorithms discussed in the text. Most other languages will also have packages or libraries that implement these methods as well.

[5]Or, they may not. The jury is not yet in on this.

Chapter 2

What Is Data Analytics?

Data analytics usually refers to the analysis of data for business applications, although it is certainly used more generally. Viewed more broadly, it is sometimes referred to as data science, data analysis, or data mining, and in a very real sense one could argue that it is just another name for statistics. However, there are aspects of data analytics that are a superset of statistics, for instance: data storage and retrieval, machine learning, streaming data analysis and frameworks for big data analysis.[1] The term *data science* usually refers to the full spectrum of the data analytic process, including the selection of models, experimental design, and interpretation of the analysis, while *data analytics* refers to the analysis portion.

Leo Breiman [58] split what we would today call data analysts or data scientists into two "cultures": the data modeling culture and the algorithmic modeling culture. David Donoho [123] refers to these as the generative modelers and the predictive modelers. The main distinction of these two cultures is that the data/generative culture is primarily focused on understanding how the data was generated, what models are appropriate to model the data, and how can these models (and parameters fitted to the models) provide useful information about the system that generated the data; the algorithmic/predictive culture is much more interested in the bottom line – how well can my model predict future data? Obviously these are not mutually exclusive, but, like another well-known split – Bayesians versus frequentists – there can be strongly held views on one side or the other. This book leans more toward the algorithmic/predictive end of the scale, although we believe that it is important for users to understand the models underlying their algorithms, and to think carefully about what models (algorithms) are appropriate for a given problem.

Data analytics consists of

1. Ingesting;
2. Cleaning;
3. Visualization and Exploratory Analysis;
4. Feature Extraction and Selection;
5. Modeling;
6. Evaluation;
7. Inference.

[1]One of the authors (DJM) would still argue that this is also mostly statistics, with a smattering of computer science thrown in.

We will consider each of these in turn, although in practice there is a lot of interaction between these. These steps are neither discrete nor independent. For example, there is definitely feedback between modeling and feature extraction, and between both of these and cleaning. How we ingest the data, where and how we store it, and what we collect from the sensor, is driven by the ultimate goal, and also by information learned as we analyze the data. Also, as we will see, visualization and exploratory data analysis (EDA) occurs throughout the process.

The final step, inference, also called exploitation or decision making, is the ultimate goal of the analysis and drives all the previous steps. This is the topic of the rest of the chapters in this book.

It is important to realize that this last step may not be known at the time of collection. In "traditional" statistics[2] one has a scientific or practical question one wants to answer and an experiment is designed to collect and analyze data to answer the question. In cybersecurity, we know the question is something along the lines of "protect the network" – or computer or private data – but from what? One collects very different data to defend against network attacks like denial of service or man-in-the-middle attacks than against phishing, or viruses, or nefarious insiders, or .... What data should we be collecting now, so that we have the data we need to develop protections against the attacks of five years from now? The data volumes and cost of instrumentation to collect data mean that we can't collect everything. Ideally, one always starts with a problem (desired inference) and collects the data to support the inference. In practice, one often relies on existing data, at least initially.

One is often restricted in the data one can collect, either by the vendors that provide the infrastructure of the network, by policy and privacy consideration, or by physical constraints. In reality, we may not even be able to collect the data we would really like to have. For example, if the problem is an insider threat, privacy issues may make it impossible to collect everything the user does. The decisions on what data are collected may be outside the purview of the cybersecurity analyst. The analysis of existing data in the context of the desired inference can be used to argue for changes to allow for the necessary data to be collected, or it can provide information on the fundamental limits on the performance of the inference for the given problem.

This chapter will lay out the basic ideas and provide some illustrations of them for cybersecurity. Each of these ideas will be revisited throughout the book. One aspect of cybersecurity analysis that we would like the reader to keep in mind is that the tools of data analytics provide ways to determine the utility of particular data streams, and can be used to guide decisions of what data should be collected, and how that data should be stored and maintained.

2.1 Data Ingestion

Data comes to the analyst in many ways, not all of them in a nice easily processed form. While in cybersecurity we generally don't have to worry about inputting data scrawled in lab books, we do often have to read system logs (in various formats), extract data from pdfs and web pages, extract information from packet headers and packet payloads (implemented in various protocols and formats), analyze source code and binary files, process unformatted text, analyze network topologies (physical, logical, and defined by various types of communication such as email or social networks); the list of data types and data sources is long and growing.

A very simple example will help set the stage. Many organizations generate reports of cyber intrusions or attacks, and these are often provided as human-readable documents.

[2]There is considerable controversy in the term *traditional statistics*, hence the scare quotes.

These documents do not always follow a strict format, and so writing a parser to extract the data can be challenging. The parser must: convert the pdf to text; locate IP addresses within the document; find and process the section that describes the attack; extract entries from tables, and possibly figures. Even the problem of linking the IP addresses reliably to attacker/victim can be nontrivial.

Ideally, the provider of the data (whether a router, network sniffer, intrusion detection system, or web scraper) has converted the data into a structured and known format, and the problem of ingestion is relatively simple. Even better (possibly) is that the data reside in a database and can be accessed through this database. For online[3] security systems this latter will generally not be available, as the online system will be accessing data directly from a sensor. In this case, it is expected that the sensor will provide the data in a simple and easy to parse format.

For systems that scrape the web, or that monitor email or social network posts or blog posts, ingesting the data can be quite challenging. The data are often in many different formats, with both structured and unstructured components, and in the case of web pages, may be dynamic, with the structure changing in time. In these cases, it may not be simple to determine which data are most relevant.

Relevance depends on the purpose for which the data were collected. In many cases, one collects data which will be used for several different purposes (some of which may not be known at the time of collection), so in addition to the problem of parsing the data is the one of deciding what data to keep and what to discard. For example, in monitoring social networks, does one keep information about the ads that are presented to the user? This will depend on the purpose for which the data are to be used.

2.2 Data Processing and Cleaning

We note at this point that the word "cleaning" carries an unfortunate stigma, implying that the data is "dirty" or in error, and so, as the title of this section indicates, there is more to this step than removing errors. However, some data actually is bad, due to transcription errors or misconfiguration of the sensor, and these need to be removed, or at a minimum identified and labeled as bad. As a rule, we advise against permanently removing "bad" data, although one may temporarily remove it for a given calculation.

If any of the data was input by a human, there is the possibility of error either inadvertent or deliberate. When processing email for spam, phishing, evidence of an insider threat, leaking of proprietary information, one has to cope with the enormous variability in spelling, use of acronyms and short-cuts, slang, typos, dialects, character sets (and emoticons, emojis, and whatever the newest thing is as you read this), the ever growing lexicon, not to mention borrowing from or writing in all the different languages that humans use.

Even data collected automatically can contain errors. Consider the New York Taxi data sets.[4] These contain information about a large number of taxi trips within New York City and the environs, with pickup/dropoff times, latitudes and longitudes for the pickups and dropoffs, and other data. Many of the latitudes and longitudes are reported as $(0, 0)$ which is off the coast of Africa – some GPS units report this when they malfunction or are unable to obtain a GPS signal. This is not the only error in the GPS data, however – there are values in Europe and other U.S. states – and even for those values which are theoretically possible, such as a position in Maine, the reported cost proves that these are indeed errors.

[3] An online, or streaming, security system is one which collects data as it arrives (packets over the network, emails as they are delivered, keystrokes as they are typed) and processes each observation for an immediate decision (real-time or near real-time). Typically, online systems do not store and retrieve older data, although some short-term memory is often available.

[4] Available at `http://www.nyc.gov/html/tlc/html/about/trip_record_data.shtml`.

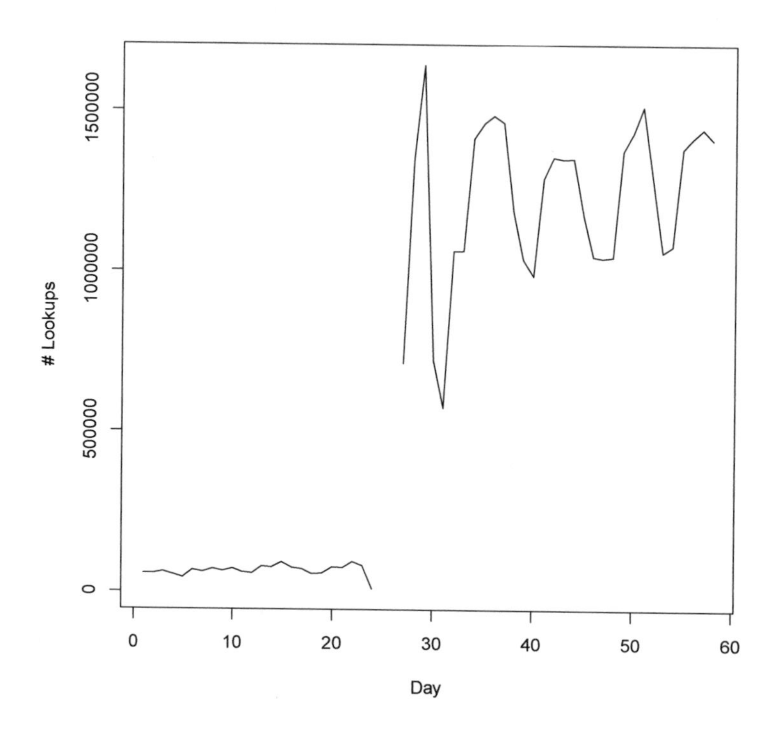

Figure 2.1: Number of DNS lookups by day.

Some of the pickup times occur after the dropoff times, which is likely to be a result of human error, but may also be an error in the device that records the times.

These data also have the problem that different companies provide the data in slightly different formats or with slightly different values. Even with a well-curated data set, there can be a number of problems that must be addressed before any inferences can be made.

It is also not always evident which parts of the data are errors or need to be "cleaned". Consider the Domain Name Service (DNS) data[5] [232, 233]. The data correspond to 40,821,591 DNS lookups between 15,296 computers over a period of 58 days. The first three lines of the data file ("dns.txt") are:

```
2,C4653,C5030
2,C5782,C16712
6,C1191,C419
```

The first number represents time, the second string is the source computer and the third is the label of the computer resolved. Figure 2.1 gives an overview of the data.

Several things are clear from the picture. First, there is a change that occurs after day 24 – there are two missing days, and then the volume of data increases dramatically. Second, in that later half of the data, one can clearly see a periodicity of 7 days – a weekly pattern.

These data are very clean. They have been curated by the researchers at Los Alamos and contain no errors (that are detectable by us). However, the data may still need to be cleaned, depending on the purpose for which they are to be used.

[5] Available at `http://csr.lanl.gov/data/cyber1/`.

One can view the DNS lookups as evidence of communications between the source and destination systems. Consider the problem of determining which computer is the most active. First one needs to define "active". If one defines it as the total number of DNS lookups, the most active computer is C585 with 1,624,999 requests. The next two are C743 with 1,149,807 requests and C1823 with 391,719 requests. Note that these data contain times where a source looks up itself. This happens 865,039 times in these data! It might be reasonable for some purposes to remove these – "clean" them – even though they are not errors. In this example, this kind of cleaning does not make a difference in the results.

Another possible definition of "active" is the computer that connects (looks up) the most distinct computers. There are 425,025 unique records, so on average a computer looks up another specific computer about 96 times. Under this definition, the most active computer is C561 connecting to 12,103 other computers with the next two most active computers being C395 connecting to 11,920 computers, and C754 connecting to 1,067 computers.

This example illustrates two of the most common preprocessing or cleaning methods: removing duplicates and removing observations not relevant or useful for the inference task, either because they contain errors, or in this instance because they correspond to something we choose not to consider.[6]

Another interesting data set available from Los Alamos is the network flows data.[7] These are connections between computers containing the time and duration of the connection, the source and destination machines and ports, the protocol used, and the number of packets and bytes. Source and destination ports are integers between 0 and 2^{16}; however, a perusal of the data shows that many of these start with the letter "N." This is a flag indicating that the port has been anonymized to protect information specific to this network. This points out a problem that cybersecurity analytics must grapple with: it is difficult to provide data that is timely, broadly useful, and which does not provide information specific to the system that can be used to compromise it. There are 129,977,412 flows in these data, each flow corresponding to one direction of a flow of data between two computers.

2.3 Visualization and Exploratory Analysis

Exploratory data analysis (EDA) is the most important aspect to any data analysis. One cannot properly model data without understanding the data, and we are a visual species that obtains most of our information through our eyes. While one can use many machine learning tools as black box algorithms – blindly apply them to a data set and see what comes out – this is generally not a good idea. By the same token, one can cross streets without obeying the traffic laws or looking for oncoming traffic. We strongly advise against either of these decisions.

2.3.1 Scatterplots

The simplest, most often used, and in many ways most powerful EDA tool is the simple scatterplot. Here we plot as dots all the points of two-dimensional data. This idea can be extended to plots of multivariate data, although the more variables one plots, the more difficult the interpretation.

Consider the network flow data from the previous section. Two of the fields are the number of bytes and the number of packets in the session. Logically, these should be

[6] One could view this latter as more of a subsetting problem. This is true, but it is an example of a preprocessing step. For another example: in the taxi data, one might remove data whose pickup and dropoff locations are the same (within some tolerance), or whose locations fall outside a given distance from the center of New York City.

[7] Also available at `http://csr.lanl.gov/data/cyber1/`.

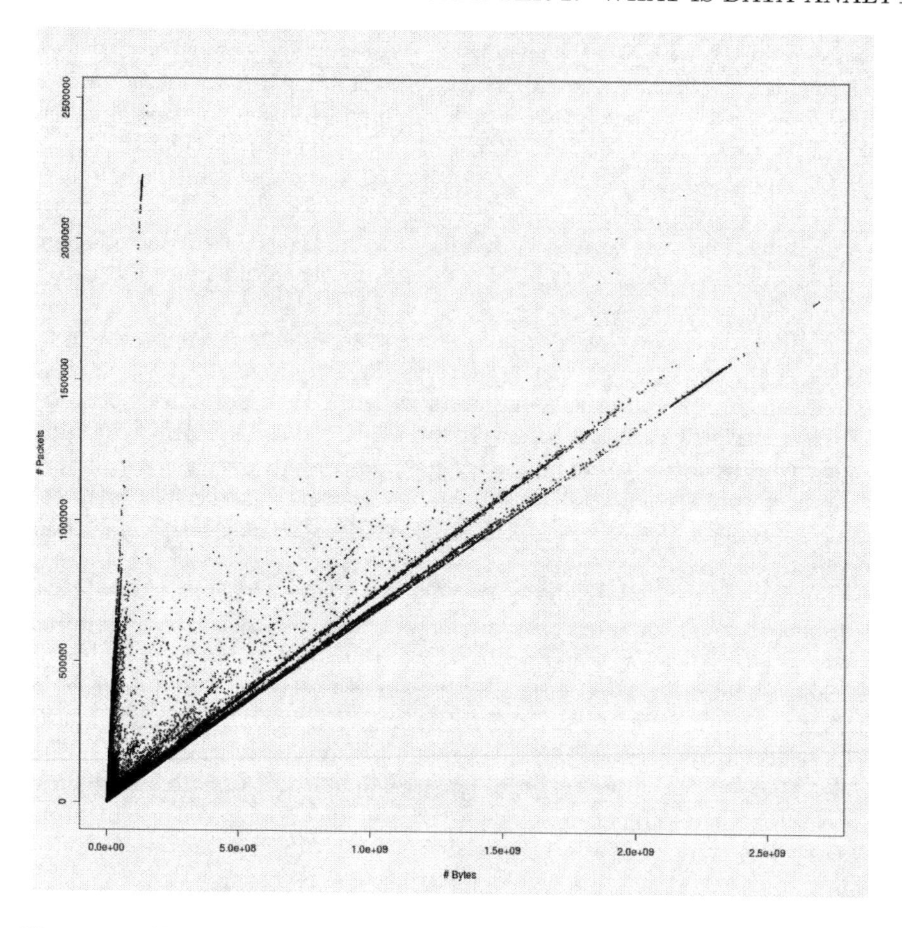

Figure 2.2: Number of bytes versus the number of packets in the flows data.

(approximately) linearly correlated, and we can investigate this hypothesis through a scatterplot. Figure 2.2 plots the number of bytes against the number of packets. We observe distinct lines, with slope corresponding to a relatively constant number of bytes per packet. A hypothesis is that the different lines correspond to different applications.

Next consider the ports used by the flows. Ports can be considered categorical (certain ports are assigned to certain applications and vice versa) and ordinal (port ranges are used freely as source or destination ports for any service that needs a port). Figure 2.3 shows the source and destination ports in time. Three things are immediately obvious. First, these plots look identical (to the eye) – this is because these data correspond to two-way communications, and a source port for one way is a destination port going the other way. This shows up clearly in these scatterplots. Second, there are definite differences in time in the pattern of the ports in each of these plots, as seen by the fact that the gap between the largest observed values of the port numbers and the majority of the observed values increases in time. Finally, the full range of ports are not used. Note that we are not claiming that this is an error in the data – this type of analysis can only provide information about the data; any explanations must come from the data source.

The scatterplot tells us quite a bit about the data. As discussed above, in [233] we are informed that while some ports (ones commonly tied to a set of specific common applications) are reported accurately, many of the ports are "anonymized." This anonymization process is not described, but it is not necessarily the case that the anonymization is the

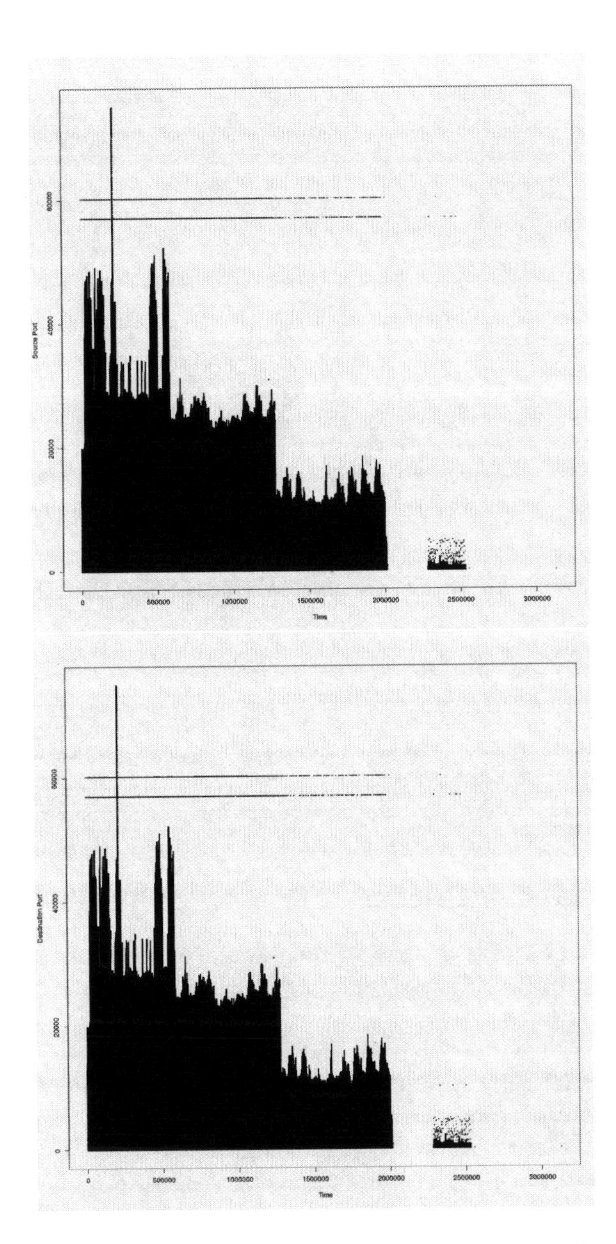

Figure 2.3: Source ports and destination ports in time.

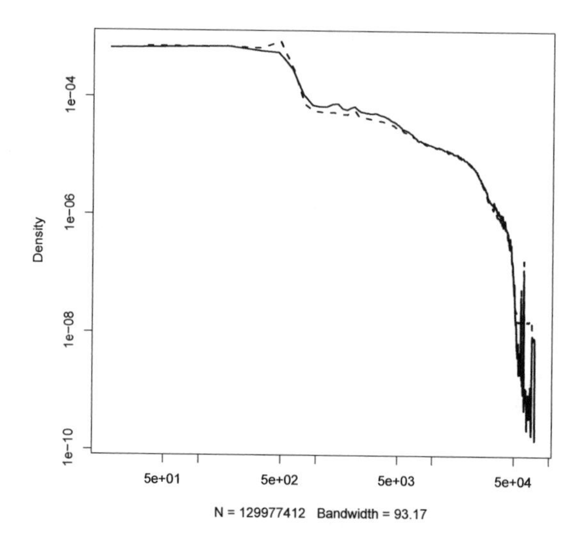

Figure 2.4: Estimate of the probability density function of ports (source ports in the solid curve, destination ports in the dashed curve) on a log-log scale for the network flow data.

reason for the changes in time we observe in Figure 2.3. Whether the differences are the result of changes in the anonymization process or in the flows on the network cannot be known just from this analysis.

In Figure 2.4 we depict the probability density functions (estimated using a kernel estimator, see Chapter 4) for the source and destination ports. There is a dip around 500 and another around 10,000. The ports less than 500 are pretty much all assigned, and thus likely to be left alone by the anonymization process, and also likely to be commonly seen on the network, as is indicated by the flatness of the curves in this region.

2.4 Pattern Recognition

Before considering the rest of the topics in this chapter, we need to introduce the key set of ideas that allow us to design algorithms to detect cyber attacks. These ideas will be revisited in more detail in Chapters 5 and 6.

We are interested in malware detection and classification, phishing, detecting an insider threat, detecting unusual activity, and characterization of activities and network sessions. The types of pattern recognition that we are typically interested in for cybersecurity fall into two general categories: Classification (also called supervised learning) and clustering (a form of unsupervised learning). We are also concerned with outlier detection and with the characterization of "normal" activities, which properly fall into the more general area of statistics.

In classification we have an input (the data) and an output (the class label). There is a version of supervised learning that is distinct from classification: regression. Here the output is a value (which may be continuous or discrete) that we want to match to the input. For example, the reader is probably familiar with linear regression, where the goal is to fit a linear model that predicts the output from the input. For example, consider predicting the number of packets used to transfer a given number of bytes in a flow. See Figure 2.6. Obviously there is more than one line here, and so one must subset the data – by the direction of the flow, for example. The figure indicates that, with a little work, it may well be possible to design a set of algorithms that perform this prediction, which may be useful in network design.

2.4.1 Classification

A classification algorithm is one which takes data (features extracted from some set of data) and assigns one of a set of class labels to each observation. This can be a "hard" assignment – the algorithm returns a single class label – or a "soft" assignment – the algorithm returns a set of class labels with weights corresponding to the algorithm's assessment of the probability that the observation was generated from the given class.

Classification starts with training data. These are observations for which the true class is known. Given these labeled observations, the goal is to produce an algorithm that, to the extent possible, will label new (unlabeled) observations correctly.

Formally, we are given a training set $(x_1, y_1), \ldots, (x_n, y_n)$. The $x_i \in \mathcal{X}$ are observations (features), the $y_i \in \mathcal{Y}$ are class labels. We seek to find a function $g : \mathcal{X} \to \mathcal{Y}$ that is "good". We'll consider some definitions of "good", and how to estimate them, in Section 2.7. For now, we might want to minimize the probability of incorrect classification (the error):

$$P[g(X) \neq Y]. \tag{2.1}$$

To do this, we'd like to solve

$$\arg\max_{g} P[g(X) \neq Y]. \tag{2.2}$$

That is, we want g to be the *Bayes classifier*: the classifier with the lowest error, the best probability of correct classification.

We may at times consider different criteria for performance. For example, the cost of incorrectly flagging an email as phishing may be very different from missing the phishing attempt, and so we may assign different weights to the different types of errors, or we may allow more errors of one type in order to reduce the number of errors of the other type. We will discuss this in more detail below.

2.4.2 Clustering

Clustering, or unsupervised learning, takes unlabeled data and returns a grouping of the data. We are not given any a priori grouping or class labels. Instead we seek to find the "natural" groups, called *clusters*, within the data.

As might be suspected, this is not well defined. First, what does it mean to be a cluster? Consider Figure 2.5. How many clusters are in this figure, and what is a cluster? Clearly there are "round-ish" clusters (4 of them?). There are also "chains", but it is not clear how many of these there are – is the central "plus" a single cluster, or 4 (or more) "chain" clusters? Are the arms of the spiral 3 clusters, or more?

Now consider Figure 2.5 at a higher level: clearly there are three "groups" in this: the plus, the spiral, and the balls. This illustrates the hierarchical nature of clustering – there may be different clusterings at different scales, and thus a hierarchy of clusters.

Clustering thus has two fundamental, and difficult, problems. First, one must define what one means by a cluster, and the definition will drive what types of clusters that one can find. Then, one has to figure out how many clusters there are. Neither of these tasks is trivial, and while there are many tools that aid in this analysis, there is no universally optimal procedure. This is in contrast to classification, in which there are universal classifiers – those which attain the optimal classification performance – with the nontrivial caveat that this is an asymptotic statement, which may not be particularly useful in the real world of finite training sets.

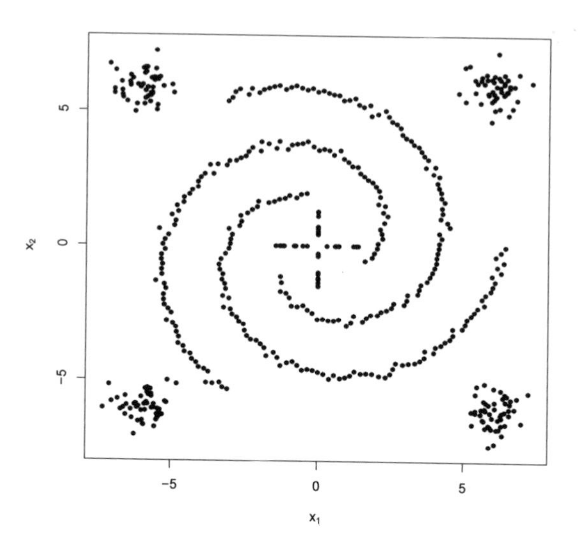

Figure 2.5: Random data illustrating one of the difficulties in defining "clusters" in data.

2.5 Feature Extraction

A measurement is a value extracted from some object that one is investigating. For example, in health care this might be blood pressure or temperature or some aspect of health history. A feature is a function (possibly the identity function) of one or more measurements that is to be used to perform inference. Given a problem domain – such as network traffic – there are potentially many measurements one could take, some more useful for a given inference task than others. Given a set of measurements, there are many (infinitely many) features one could produce. Feature extraction will refer to both choosing the appropriate measurements to extract, and determining the features that are relevant to the problem. Feature selection is choosing the best – under some criterion – of features to use for inference.

Consider the problem of monitoring network traffic to detect attacks on the network. The data we observe, at the most basic level, are packets. There are values that we need to extract from these packets – port numbers, IP addresses, protocol flags, byte counts, packet payload information. Further, we may be interested in aggregates of packets, such as TCP/IP sessions or flows. Each measurement we choose to take is a feature. Further, we can combine features, such as taking the number of bytes and dividing by the number of packets in a flow to produce a new feature – the bytes-per-packet ratio.

Feature extraction is very much driven by expert knowledge of both the problem domain and the specific problem. For example, if one is interested in detecting distributed denial of service attacks, features such as the number of connection attempts that are not completed might be a useful feature, but the number of connections containing source ports above $30,000$ may not be particularly useful.

2.5.1 Feature Selection

Even when the data analyst is herself a domain expert, or has access to one, there will be features that the expert knows are important, and others that are unnecessary for the desired inference, or that are redundant. Also, each feature we measure contains some amount of noise, either a purely random process – such as the time between packets or the actual value of a source port assigned randomly – or best modeled as random – such as the number of packets in a particular web-flow, or the duration of an SSH connection.

It may seem obvious that the more features we collect, the better we can do at inference. This is false, and not just because we are imperfect humans who cannot utilize the features optimally. Trunk [211, 436] showed that even if a new feature that we might wish to collect has valuable information for the inference task that no other variable contains, and we use the optimal model for the inference, it can be the case that utilizing this new variable will reduce the quality of the inference. In other words, sometimes adding new useful information can hurt more than it helps.

This seemingly counter-intuitive statement is an example of the bias-variance trade-off, which we will discuss in more detail in Chapter 4. The intuition comes from the fact mentioned above that features contain noise. The idea is that if the amount of noise is high compared to the benefit of the feature, this noise can add enough "confusion" to our estimates to overcome any benefit of the new feature. Thus, when we add features, we want them to do more than add useful information, we need them to add enough useful information to overcome the inherent variability (noise or error) in the feature.

There are a number of feature selection methods, and they fall into two main types: true selection methods seek to choose a subset of the features among all the features measured, and projection or embedding methods compute linear (or nonlinear) combinations of the features, and then select a small set of these combinations.

It should be noted that feature selection is never performed in a vacuum. It is important to keep the ultimate goal in mind: what inference are you trying to make? Features that are ideal for detecting phishing in email are likely to be useless for detecting an insider emailing proprietary information from the company.

Suppose one observes a session between two computers and wishes to determine whether the session is web traffic. The obvious thing to do is look at the ports – ports 80, 8080 and 443 are the most common web ports. In this example, however, we observe traffic between unknown ports, and suspect a covert web server has been set up. Assume we do not observe the packet payload – all we observe are the time of the session, the duration of the session, number of packets and number of bytes. We'll use the flows data to investigate this, as an illustration of feature selection methods.

The time of the sessions is in seconds from an arbitrary and unknown start time, so for the purposes of this study, we will assume that the first packet in the data occurred at time $0:00$ and convert all the times to hour of the day from this baseline, taking on integer values from 0 to 23. The method we will use for determining what type of session we are observing is to compute the mean of each of the features for web traffic – those in our data to/from web ports – and nonweb traffic – everything else. To do this properly, we would use a subset of each of these data sets to construct the estimates, then use the other subsets to compute our performance under different choices of features. Since we are merely illustrating the ideas, we won't bother to do this. Our estimates are a type of "resubstitution" estimate, which we'll describe below, and give a biased estimate of performance, but because of the large numbers of observations this bias is not of large concern for this example.

To be precise, we will compute the following statistic:

$$c = \arg\min_{m} d(x, m), \tag{2.3}$$

where the minimum is taken over the two means, web and nonweb traffic, and d is the euclidean distance. So we compute the distances to the means shown in Table 2.1, and whichever mean is closest to our data determines whether we call the traffic web or not.

Table 2.1: Means Computed on the Web and Non-web Traffic in the Flows Data

Type	Duration	Packets	Bytes	Hour
Web	8.136343	18.43661	14299.12	11.30068
Non-web	10.07475	281.0832	238138.4	11.40093

The optimal feature selection method is to try all combinations and select the one that performs best. In general, this is impossible due to the fact that there are $2^n - 1$ possible combinations of n features (ignoring the empty set). One needs to be a bit careful here, since as we've indicated above, adding features increases variance, and so it has to improve things "enough" to be of value, but for now we will ignore this point. In the case at hand there are 4 features, and $2^4 - 1$ or 15 different combinations of between 1 and 4 features, and it is well within our computing resources to try all combinations. On a reasonably new laptop, it took a little less than 3 minutes to compute the performance of all the feature combinations.

We'll approach the feature selection iteratively, or step-wise. Starting with all the features, we obtain an error of 0.8969873.[8] We then check the error we obtain by removing a single variable, and discover that removing the number of bytes produces an error of 0.8855153. Now we repeat, removing each feature and computing the error. The next feature to remove is the number of packets, giving an error of 0.6739519. Finally, we find that removing duration results in an error of 0.5055502 using only the hour of the day, and this is optimal.

The above procedure can be reversed: select the single feature which results in the best performance, then add the feature (if any) which best improves the performance, stopping when no new feature improves the performance.

In this experiment, both procedures result in the same answer: the hour of the day is the best predictor of web traffic. A reasonable explanation of this is that web traffic is driven by humans, and humans operate primarily during certain hours of the day (and web surf more during some hours than others). Knowing this, one might change the data to record the hour at the initiating computer. This requires data we have not collected – we need to know which computers are on the local network and which are external, and we need to know where the external computers are. This illustrates a point discussed earlier: the data one should collect is dependent on the inference task, and it is often the case that only through data analysis can one determine what missing data might be of value.

Another point that we didn't consider in the above discussion is that each connection consists of two flows, and with web traffic the client side of the flow tends to contain far fewer bytes/packets than the server side. So what we should have done is to combine the two flows into a single session, and split the bytes and packets into client and server values. Further, we might think that it makes sense to consider the ratios of these to each other rather than (or in addition to) the raw values.

Figure 2.6 depicts the number of bytes against the number of packets for the web traffic. The open circles, which mainly occur next to the vertical axis on the left, correspond to client-to-server flows, and are very close to the $y = x$ line. The '+' symbols correspond to server-to-client flows. The lower diagonal line evident in this plot corresponds to flows averaging about 1497 bytes per packet. This is a clear indication that the flows in the two directions are substantially different for web traffic – a fact well known to a domain expert – which suggests that we would be justified in treating the flows differently, and combining the flows (and the features extracted from them) into sessions.

[8]Note that the value of this error is immaterial – the method we are using to discriminate between the two types of sessions may be very bad – this experiment is for illustrative purposes only.

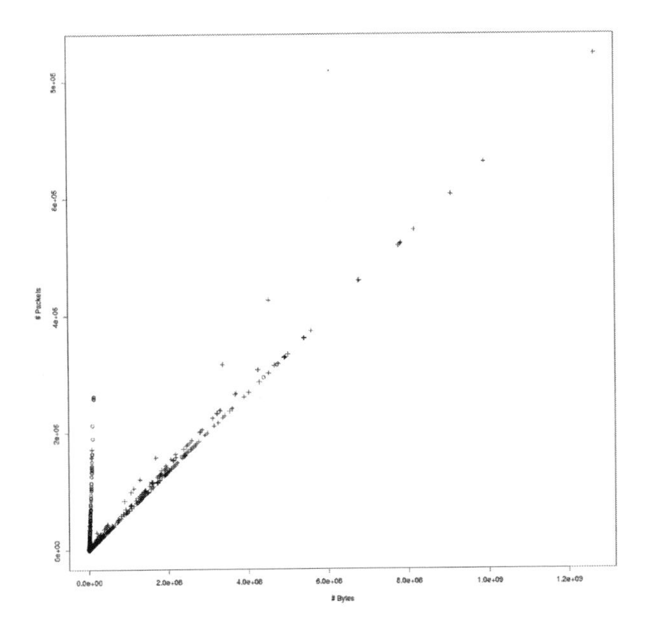

Figure 2.6: The number of bytes against the number of packets for web traffic. The open circles correspond to client-to-server flows, the pluses correspond to server-to-client flows.

It may also be worth looking at the range of source ports – perhaps web traffic (or certain implementations of the web client) has a certain range of source ports it prefers. More importantly, sessions between the two computers that occur close in time to the suspect session and have close source ports may be an indication of a web session – all the connections spawned by a visit to a web site – and so we could imagine computing features to bring this information into our feature vector. Adding such features to the mix would (presumably) improve the performance of the discrimination algorithm, while increasing the computational complexity of computing all possible combinations of features.[9]

The fact that both top-down and bottom-up feature selection algorithms resulted in the optimal choice of features is by no means typical. In general, these iterative methods are suboptimal. However, as the number of features increases, it can quickly become infeasible to compute all possible subsets of features.

A more interesting example is a case of malware discrimination. Kaggle has a contest of malware[10] classified into 9 families, with the data provided as hex-dumps of the bytes (see [366]). Some of the bytes are obfuscated with '??' as a way of ensuring that the malware cannot be re-binarized into executable files. For the purposes of this experiment, we processed the data by enumerating the proportion of times each distinct byte value (ignoring the '??' values) appears in the file. Thus we are extracting 256 features for each file.

It is nontrivial to visualize data of this high dimensionality. One method that has seen some interest in recent years is the t-SNE method ([290]). Some care must be taken to use this appropriately, see for example [477]. However, it can produce interesting pictures, as indicated in Figure 2.7.

[9]This toy problem assumes that we do not have access to the payloads, or that the session is encrypted in a way that makes it useless for determining the session type.

[10]`https://www.kaggle.com/c/malware-classification`

Figure 2.7: The 9 families of the Kaggle malware data, visualized via t-SNE. Each family is depicted using a distinct color/symbol pair. See the book cover for a full-color representation.

There are several points to learn from Figure 2.7 and from further exploration of the t-SNE method. It appears in the figure that the families do separate fairly well, even though this plot is a 2-dimensional representation of the data. The "swirly" nature of the plot, with the long "curves" of observations is quite common in t-SNE plots. Finally, and this is extremely important, always set the seed of your random number generator before calling the code. The algorithm is stochastic in nature, and if you do not set the seed prior to calling the code, you will be unable to reproduce the plot. Unfortunately, we (DJM) did not do this, and so the "duck" in the figure is, sadly, not reproducible.

We repeat the experiment above, using the malware from families 1 and 2 in place of the web/non-web data, and we find that the first procedure, where we iteratively add features, results in Table 2.2 where we show the performance as we select each feature. The resulting classifier would use all three features, and have the best performance. Adding a further feature results in a decrease in performance. Table 2.3 shows the result of going the other way: starting with all the features, then iteratively removing features until the performance stops improving. The final classifier would use all features except those in the table. Removing any further feature increases the error. Thus, starting from one feature and iteratively adding results in 3 features, while starting with all the features and iteratively removing results in 254 features and worse performance. Since $2^{257}-1$ is an unimaginably large number, we have no hope of performing the complete experiment on all possible combinations.

As we have said, this approach, whether starting at one feature and adding, or starting at all features and subtracting, is suboptimal. It can be the case that adding two features improves performance, even though adding any single feature does not. Further, it can be the case that the optimal set of features does not contain the best single feature.

Table 2.2: Sequential Selection of Features on the Malware Classification Problem

Feature	Error
52	0.07638716
55	0.04802190
209	0.04553371

Table 2.3: Removed Features for the Malware Classification Problem

Feature	Error
??	0.06593680
255	0.06494153
16	0.06469271

Instead of trying all combinations, or iterative methods as discussed above, one can use correlations to select features. Given a set of features, compute the correlations of the features with each other and with the dependent variable – for example, the class label. One chooses those variables with a high correlation with class, then considers the between feature correlation to perform further down-selection of features: if two features are highly correlated, one may only need one of them for the inference. This method, like the iterative methods, is suboptimal in general, since it is only looking at pair-wise correlations, but can be useful particularly when there are a large number of features or when some features are of little utility for the inference.

Feature selection can also be performed through regularization, which is basically a method of introducing some information to reduce the complexity of a model, reduce overfitting, or improve estimation. This will be discussed in more detail in Section 4.9. In its simplest form, regularization penalizes adding parameters to the model, in order to control the complexity of the model. More “complex” models are required to “justify” the extra complexity by demonstrating that the extra complexity provides “enough” improvement to be warranted. Different interpretations of the words in quotes result in different regularization algorithms.

2.5.2 Random Projections

One popular method for reducing the dimensionality of a data set is through random projections. The idea is to take random linear combinations of the features and use these in place of the original features. At first blush, this seems nonsensical: how can a random projection be useful? In fact, the colloquial use of the word “random”, in English at least, suggests nonsense.

It turns out, however, that if one cares about the relative distances between points, which is the basis for many of the algorithms that have been developed for data analytics tasks, then a random projection can be shown to retain these distances! The Johnson-Lindenstrauss theorem ([107, 220]) basically states that if one projects points into a lower dimensional space using random (Gaussian) projection vectors, the distortion between the points is controlled provided enough projections are used. The basic idea is that if we don't want the distortion to be worse than ϵ we embed into (on the order of) $\log(n/\epsilon^2)$-dimensional space.

2.6 Modeling

There are five basic aspects of modeling relevant to data analytics:

1. Specification: select the family or families from which to choose a model. For example, one might decide to use Gaussian mixture models to model one's data.

2. Selection: choose from within the set of models. In the mixture model case, this would be selecting the number of components to the mixture, and perhaps the shapes of the mixture covariances, or other constraints.

3. Fitting: fit the parameters of the model to the data. For the mixture case, one needs to determine the means, covariances, and mixture proportions.

4. Assessment: determine whether the model is appropriate for the data.

5. Inference: make the appropriate decision using the results from the above steps.

We will illustrate these steps on cybersecurity problems in the chapters to follow. There is far more to this subject than can be covered in a single book. The reader is encouraged to consider the myriad of references on the subject such as [39, 64, 189, 213, 489], as well as the Bibliography section of this book. In this section we will sketch out some of the issues that arise and give some examples.

2.6.1 Model Specification

Consider the plot in Figure 2.8. The points correspond to web flows of at least 1 MB of data. It seems that there are four or so lines in the data, corresponding to flows of approximately the same number of bytes per packet. We can compute this bytes-per-packet value as the slope of the line.

$$\theta = \arctan\left(\frac{\#\text{ Packets}}{\#\text{ Bytes}}\right). \tag{2.4}$$

Using Equation (2.4) to compute the bytes-per-packet rate, and computing its log, we obtain the data in Figure 2.9. The lower plot depicts the kernel estimator for the probability density of the log of θ – we'll discuss this in more detail in Chapter 4. For now, think of this as a smoothed version of a histogram.

Now we seek to specify a family of models to account for the data. We will focus on the single variable $\log(\theta)$. We could rely on domain expertise if it is available: a domain expert could indicate that it is known that data of this type tend to follow a given distribution. We could rely on a statistical model of how the data were generated: if the data were waiting times in a queue, an exponential model might be appropriate. We can "look at" the data: plots like Figure 2.9 can give indications of potential models, which we can then fit to the data and determine whether the model is a good fit or not. This latter is almost always what we are forced to do for most complex data problems (such as occur in cybersecurity) although input from domain experts and statistical models is always important. We could take a "model-free" approach, and simply observe that there are two main peaks in the data, and select regions that encompass each peak: if an observation is in $[a, b]$ it is in the first peak, if in $[c, d]$ it is in the second, etc. The latter approach quickly becomes infeasible when we consider more than one variable at a time.

It seems clear that from Figure 2.9 that the (log of) bytes-per-packet has two main peaks, possibly two small peaks just past the value of -7.0 and a very long tail. We might posit a model for this as a mixture of 5–10 normals (Gaussian distributions) with one for each of the peaks, and one very large variance component to model the long tail.

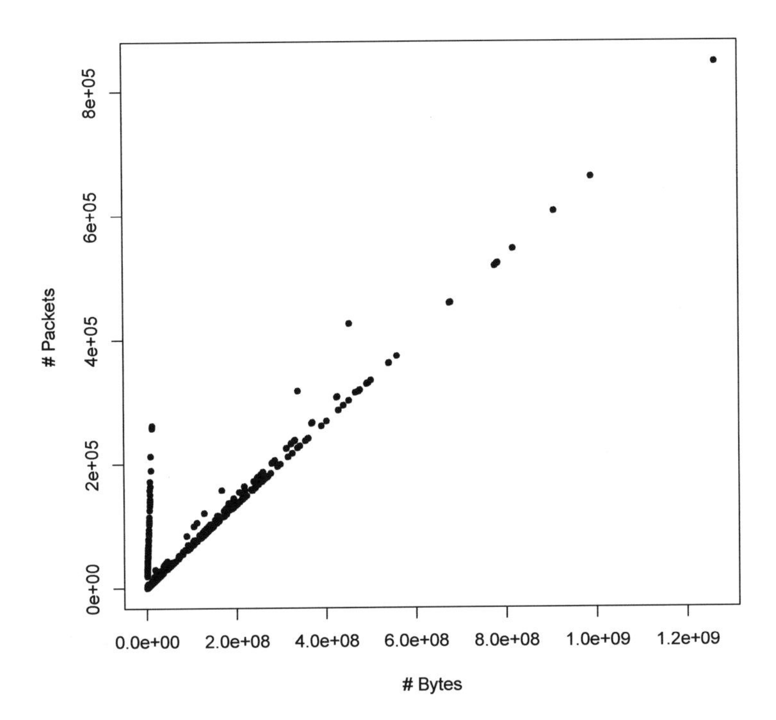

Figure 2.8: Web flows consisting of 1 MB of data or more.

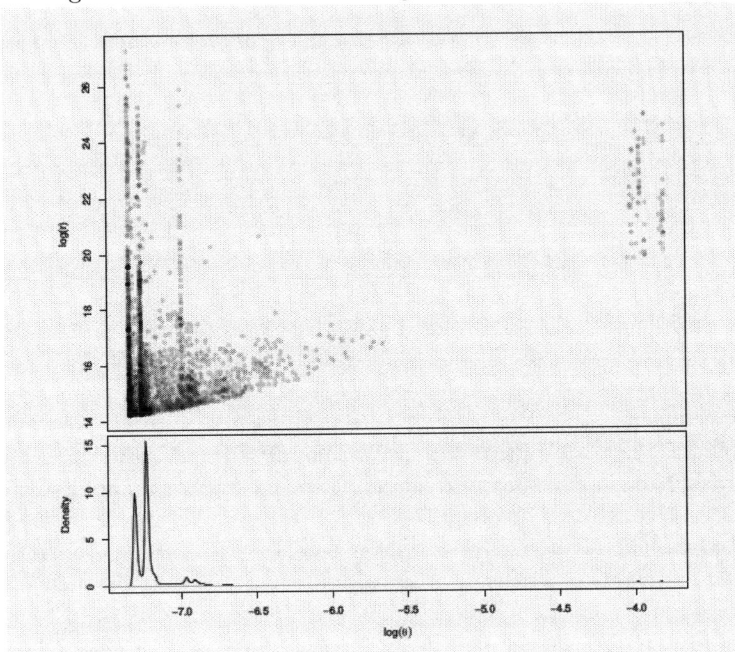

Figure 2.9: Plot of $(\log(r), \log(\theta))$ for the large web flows of Figure 2.8. Alpha blending has been used to reduce the effect of overplotting. The bottom plot is a kernel density estimate of the logged θ values – the bytes-per-packet.

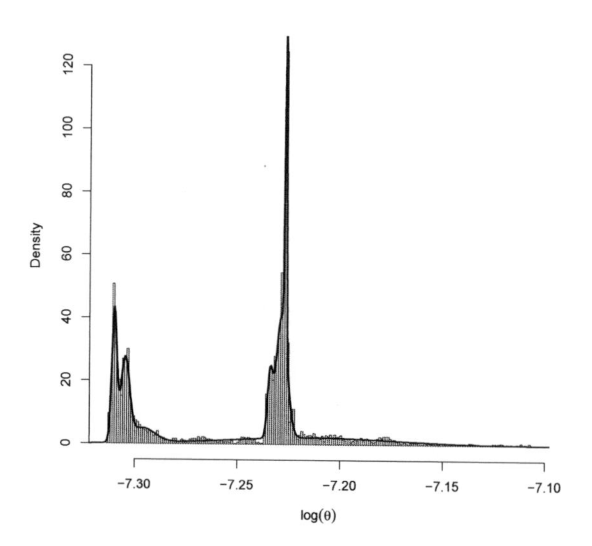

Figure 2.10: Plot of a histogram of $\log(\theta)$ from Figure 2.8 in the region of the two main peaks. A Gaussian mixture model with 7 components fit to the data in this region is shown as the curve.

An m-term univariate Gaussian mixture model is defined as:

$$f_{gm}(x) = \sum_{i=1}^{m} \pi_i \phi(x; \mu, \sigma^2), \tag{2.5}$$

where $\phi(x, \mu, \sigma^2)$ is the standard normal density with mean μ and variance σ^2 (standard deviation σ):

$$\phi(x, \mu, \sigma^2) = \frac{1}{2\pi\sigma} e^{\frac{(x-\mu)^2}{2\sigma}}. \tag{2.6}$$

The mixing proportions $\pi \in [0, 1]$ sum to one.

There are two problems with this model. The first is that it is not at all clear how many of the "bumps" in Figure 2.9 are significant – how many correspond to real populations of web traffic. So we have some doubt as to how many components are appropriate for the data.

The second problem is that the "bumps" are not actually well modeled by normal densities. This might be more a problem of our choice of density estimate than a problem with the model. Figure 2.10 shows a mixture model fit to the data, compared to a histogram of the data.

Recall that the point of fitting the model is to get a good representation of the data, for the purposes of inference. For this (toy) example, the desired inference is to group the web traffic into different types via characteristics of flow. For the data in Figure 2.10, is it sufficient to model the data as the two main "peaks", ignoring the internal structure? Or is this internal structure of interest to the inference task? Are the different components of the mixture fit indicative of different "types" of flows?

Answers to these questions rely fundamentally on domain knowledge. However, from a data analytics perspective, the structure is there – at least a model and the histogram suggest that it is there – and so we require tools to decide if this structure is "significant". Should we believe it is real, or should we select a more simple model? Why did we pick 7 components? Why not 2 or 5 or 12? The next section discusses this, and we will go into this in a bit more detail in the rest of this book.

2.6.2 Model Selection and Fitting

In Figure 2.10 we see a 7-component Gaussian mixture fit to the subset of the data in the region indicated by the x-axis. We chose Gaussian mixtures in the above section on model specification (although we did not provide much discussion on why this model in particular). Once the model family is selected, we must choose amongst all the possible models in the family. This generally comes down to estimating the parameters of the model.

For concreteness, we'll stick with the normal mixtures model. Recall Equation (2.5). We need to select m, the number of components of the mixture, and then we need to fit the other parameters:

- $\pi = (\pi_1, \ldots, \pi_{m-1})$,
- $\mu = (\mu_1, \ldots, \mu_m)$,
- $\sigma = (\sigma_1, \ldots, \sigma_m)$.

Note that there are only $m - 1$ mixing proportions since they are constrained to sum to 1: any $m - 1$ completely specify the last.

Assuming we have selected m, there are a number of algorithms for fitting the parameters, most of which are variations on the EM (expectation maximization) algorithm. See the books [308, 309] for more information. There are also many software packages that implement these algorithms. The R language ([358]) has several packages devoted to mixture models, in particular *mixtools* and *mclust*, and there are libraries that fit mixture models for most languages.

The problem of choosing m is a bit trickier. The idea is as follows. Given a probability density function (such as the mixture model of Equation (2.5)), the likelihood is defined as the product of the density computed at each observation. The maximum likelihood principle is that the parameters that maximize the likelihood are the "best". So, naively we could compute the mixture model for $m = 1$, compute this likelihood, then do this for $m = 2$, $m = 3$, and so on, and select the value of m for which the likelihood is maximal.

Unfortunately, this won't work. The maximum likelihood is monotonic nondecreasing in m. To see this, note that a model with $m+1$ components contains all m component models (set one of the π_i to zero). So we have to penalize for the complexity: adding a component has to increase the likelihood "enough". There are many such penalties, the most common being the Bayesian Information Criterion (BIC). If L is the likelihood computed on the n observations, BIC is:

$$\text{BIC} = \ln(n)m - 2\ln(L). \tag{2.7}$$

So we select the model with the smallest BIC (note the negative on the likelihood term). The R package mclust implements this, as well as several other criteria, and was used in the fit shown in Figure 2.10, to both select $m = 7$ and fit the parameters.

2.7 Evaluation

We saw above that in classification we are interested in the performance – the probability of correct classification. We'd like to break this down a bit more; particularly for cybersecurity, the number of "normal" or benign events far outweigh the number of attacks, and so one can get near perfect performance simply by always classifying an event as benign. Unfortunately, while this classifier is rarely wrong, it is always wrong for the most important cases!

We borrow terminology from the information retrieval literature. "Precision" is the proportion of retrieved documents that are relevant to the query. That is, if we think of the "retrieved documents" as being the observations classed as "attack", this is the probability that an observation that the classifier calls an attack truly is one. "Recall" is the proportion of documents that are relevant that the query finds. So, it is the probability that an attack will be correctly classified.

In the two-class case (attack, nonattack, say), we think in terms of true positives (TP), that is true attacks classified as an attack, false positives (FP), observations classified as attacks that are not truly attacks, and false negatives (FN), attacks that are missed by the classifier.[11] True negatives (TN) are defined analogously. Then:

$$\text{Precision} = \frac{TP}{TP+FP} \tag{2.8}$$

$$\text{Recall} = \frac{TP}{TP+FN}. \tag{2.9}$$

From these, we also define related terms:

$$\text{True negative rate} = \frac{TN}{TN+FP} \tag{2.10}$$

$$\text{Accuracy} = \frac{TP+TN}{TP+TN+FP+FN}. \tag{2.11}$$

Often, in the two-class case as we are considering here, the ROC (receiver operator characteristic curve) is used, which plots FP against TP. Note that you can choose one of these (usually through a parameter that controls the appropriate rate). For example, if your classifier returns a probability (or "confidence") that the observation is an attack, you can adjust the threshold on this probability until you get the desired false alarm rate. The true positive rate is then fixed by the classifier – it is whatever it is based on that threshold.

At this point, it is important to note that the above is stated as if one can know the false alarm rate – one can't. One can only obtain an estimate. This estimate can be computed using the training data, a *validation set*, which is a subset of the training set that is withheld from training and used for various model fitting or error estimation purposes, or by collecting new data.

With that said, there are two cases where one can guarantee the false alarm rate. Clearly we can always have a 0 false alarm rate (never alarm) or a perfect true positive rate (always alarm) but neither of these are useful. There are generally trade-offs, based on both the cost of a false negative, and of a false positive.

For more than two classes, one often sees a confusion matrix, which provides information about how errors are made. The confusion matrix for a problem with m classes is a table that is $m \times m$, with each row corresponding to one of the true classes and each column corresponding to each of the estimated classes.[12] The entries in the ij^{th} cell is the empirical estimate of:

$$P[\widehat{C} == c_j | C == c_i],$$

where $\widehat{C}$ is the classifier output. In practice, it is the proportion of times the classifier assigns class j to an observation from class i. In the two-class case, the reader can easily see how these map to the TP, TN, FP and FN above.

[11] `https://en.wikipedia.org/wiki/Precision_and_recall`

[12] One will sometimes see the rows and columns switched, in which case the rows correspond to the estimated class and the columns to the true class. This should be made clear in the table. Also, one sometimes adds a column or row giving the class-conditional performance of the estimator.

The issue of how one estimates these error rates is an important one. In a perfect world, one would have access to as much ground truth data (both the observations and the true class label) to design (train) our system as one needed, and similarly as much ground truth data as is needed to test the system. In the real world, getting ground truth data is often expensive and difficult, so one is generally forced to take what one can get.

We want an unbiased estimate of the performance; that is, we want a way of computing an estimate so that the expected value of the estimate is the true error. In the worse case, we'd settle for a method that would be asymptotically unbiased – if we have enough test data, it will converge to the right value.

If we are lucky, and have access to plenty of data (with plenty of observations from each class), we can randomly split the data into a training set and a testing set. We build the classifier on the training set, and once we are satisfied that we have the best classifier we can build, we test it on the testing set, and report our results. This is an unbiased estimate of the error, assuming the training and testing data were drawn from the same distribution.[13]

In the real world, we take what we can get. If there isn't enough data to split it into two large-enough sets, we have two choices: resubstitution and crossvalidation.

Resubstitution, as the name implies, is passing the training data back through the newly created classifier, and computing the error from this. It is biased, and how biased it is depends on the classifier we are using. The nearest neighbor classifier is as bad as it can be for this approach. If the observations are unique, then the resubstitution error for the 1-nearest neighbor classifier is always 0, which is (essentially) never a very good estimate of true performance. On the other hand, if the number of observations is large, using resubstitution to estimate the performance of a linear classifier has a relatively small bias, and so will not be too far off from the true value.

Crossvalidation is similar to the training/testing set approach. Given k, the withhold-k crossvalidation[14] procedure is as follows. The data are split into sets of size k. In turn, each such set is treated as a testing set, and all the remaining observations become the training set. The classifier is built – the parameters estimated – using the training set, and the testing set is run through to determine the accuracy on that set. This is repeated for all the sets. A similar approach is referred to as k-fold crossvalidation. In this, the data is split into k-sets – usually these sets are of roughly the same size – and each set is withheld and used as the test.

Crossvalidation trades off the bias for variance: since each time a slightly different classifier is built – slightly different parameters result from the parameter estimation – there is in effect a small amount of "estimation noise" inserted in the error estimate. As a rule, crossvalidation is a much more accurate and reliable method than resubstitution, but it comes at the cost of having to do many different parameter estimation steps. For very complex and computationally expensive classifier algorithms, this may make crossvalidation untenable. For these situations, one may utilize k-fold crossvalidation so that one need only fit k models.

2.8 Strengths and Limitations

One of the authors (DJM) likes to present students with the following task. Consider Figure 2.11. How many groups are there? What patterns do you see? What structures can you find in the data? Do you think these groups/patterns/structures are "real", or just an artifact of randomness? How can you decide?

[13]Note that at this point, when we see that the test results aren't as good as we had hoped, there is a tendency to return to the training data and try again. This is cheating! The results on the test set will no longer be an unbiased estimate of performance.

[14]Crossvalidation (without specifying k) refers to withhold-1 crossvalidation.

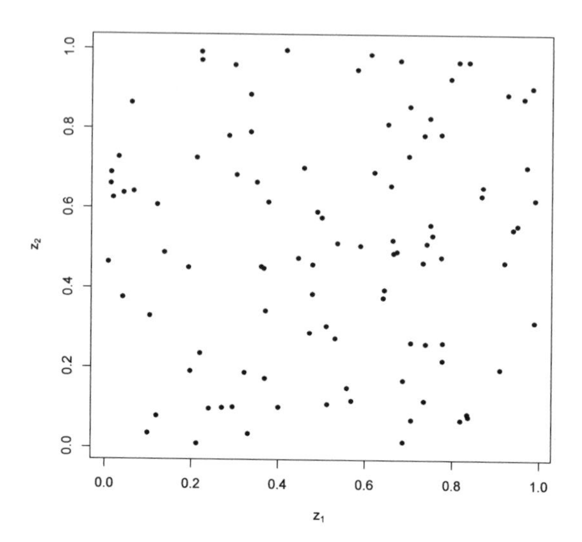

Figure 2.11: Another data set.

Consider the number of triples of points in the figure that form straight (or nearly straight) lines. Is this more than one would expect in random data? Or is there something causing this? Look at all the pairs of points that appear to be approximately the same distance apart. Should we be surprised at this, or is there a process that is generating these pairs "on purpose"?

Stare at the figure long enough, and you'll likely start to see animals, faces, and other structures. This is the same mechanism that finds familiar shapes in clouds, or pictures of famous people in pieces of toast.

Data analytics gives us tools for detecting patterns like these. It also (through statistical analysis) provides tools for testing whether these patterns are "real" – that is, whether we are finding patterns that we would not expect. Unfortunately, without a clear model for the data, this analysis can be quite challenging.

The data of Figure 2.11 was produced by the following R code ([358]):

```
set.seed(3233)
z <- cbind(runif(100),runif(100))
```

There is no structure to the data, but humans have been evolving to detect patterns for millennia, with a very severe penalty[15] for poor recognition. The tools of data analytics are also good at finding patterns like these, particularly when guided by a human who may have an underlying (possibly unconscious) reason for finding such patterns.

With any data that has a random component, whether because of measurement error, or because the data are fundamentally random, there will be outliers and spurious "patterns". Consider Figure 2.5 in Section 2.4.2. There are two points midway between the lower left "clump" and the arm of the spiral. These are outliers from the "clump" (we know this since we generated the data) – but note that there are three points on the spiral that "point at" these points, which could lead one to think that the five points lie near a line, and that this line may be an "interesting" structure in the data. Shifting the two points just a little to

[15]To wit: failing to obtain, or becoming, dinner.

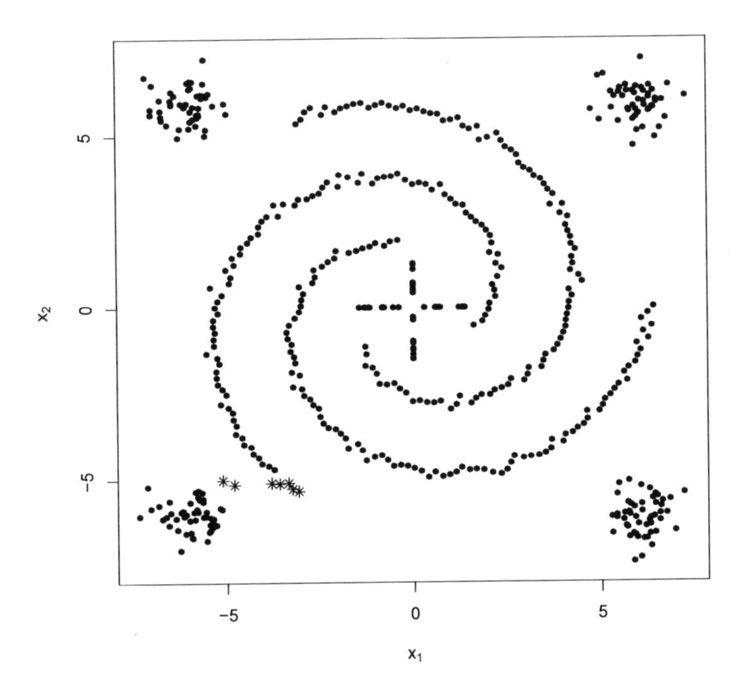

Figure 2.12: Random data illustrating one of the difficulties in defining "clusters" in data. An apparent "pattern" discussed in the text is depicted with the asterisk symbols.

the right would reinforce this "pattern", and the fact that there is a gap in the spiral at just this point might bring the final two points on the spiral arm into the pattern.

Figure 2.12 depicts this "pattern" using a different plotting symbol for these points. The reader would be excused for thinking there may be something interesting there, particularly if patterns like this happened to be the target of the analysis. Data analytics can provide tools for detecting patterns, and it can provide tools to assess the significance of these patterns, but the tools can only provide some indications of reality. Certainty is not attainable, except in very benign cases.

Thus data analytics provides algorithms for finding patterns, both "normal" and "unusual", and tools for determining whether these patterns are likely to be spurious – the result of random variation – or indicative of something different (and possibly important) in the data. It also provides tools for processing massive amounts of data, and very high-dimensional data. Pattern recognition algorithms provide tools for organizing the data (clustering) and for determining whether the data represent something of interest – an attack – or not (classification). The best pattern classifier known to man (the human brain) is easily fooled, and the data analytic techniques provide some hedge against being fooled, and allow us to detect things that even a human has a hard time detecting.

2.8.1 The Curse of Dimensionality

Our intuitions are based on our 3-dimensional world.[16] Once we start investigating higher dimensional data, our ability to view, understand, and make inferences about the data

[16] Some would say 4-dimensional, because of time.

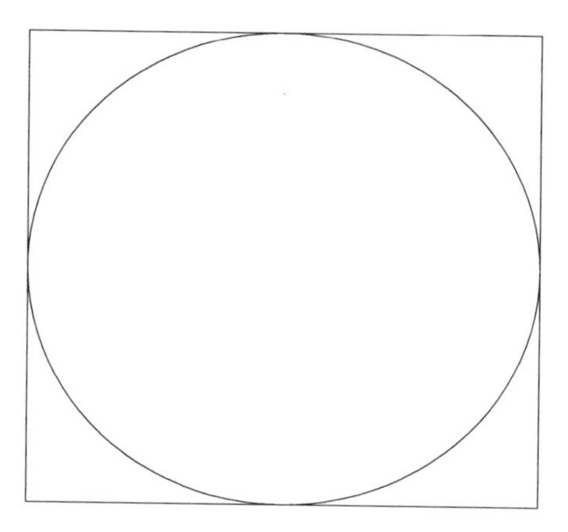

Figure 2.13: A circle inside a box in $\mathbb{R}^2$.

become extremely limited. Our intuitions from low dimensional space simply aren't correct for higher dimensional spaces. This is sometimes referred to as the *curse of dimensionality*, and we'll look at a couple of counter-intuitive examples of this in this section.

Consider a disk inside a square (Figure 2.13). It is apparent that most of the box is filled by the circle, and this is true for a ball inside a cube. However, as the dimension increases, the proportion of the hyper-cube taken up by the hyper-sphere goes to zero! In a sense, everything is in the corners!

Actually, it is (slightly) more accurate to say that everything is along the sides. To get some intuition for this "everything is along the sides" statement, consider the following: given d, draw d values from $[0, 1]$ uniformly. Think of this as a point in $\mathbb{R}^d$. Now, what does it mean for a point to be near a side of the d-dimensional hyper-cube? It means that one of the d values is close to either 0 or 1. However, given any ϵ, the probability that one of the d numbers is within ϵ of either of these points goes to 1 as d goes to infinity. Said another way, for a point to be in the middle of the "box", every value must be in an interval around the center, and the probability of this happening is the length of the interval raised to the power d. Since the length of the interval is less than one, the probability goes to zero as d increases. Our intuition is that data drawn uniformly in the interval "fills out the interval" evenly. As d gets large, this idea of "filling the box evenly" completely breaks down.

An example from kernel estimation[17] is given in [393]. We will describe this estimator in detail in Chapter 4. The set-up is data drawn from a normal distribution, using a normal kernel estimator to estimate the density. With a fixed measure of expected error, the author sets it up so that 4 points are sufficient to have this expected error for univariate data. By the time $d = 10$, 842,000 points are required to attain the same expected error.

A final example, from [227], is the prisoner in a cell. There are spherical guards in the corners of the cell who are as large as they can be (see Figure 2.14). The prisoner is as large as he can be in the space between the guards, provided that he and the guards are all disks. In 2-D, the prisoner cannot see out of the cell (imagine that the cell walls are glass, so the guards can see outside). Now consider what happens as we increase the

[17]See 4.1.

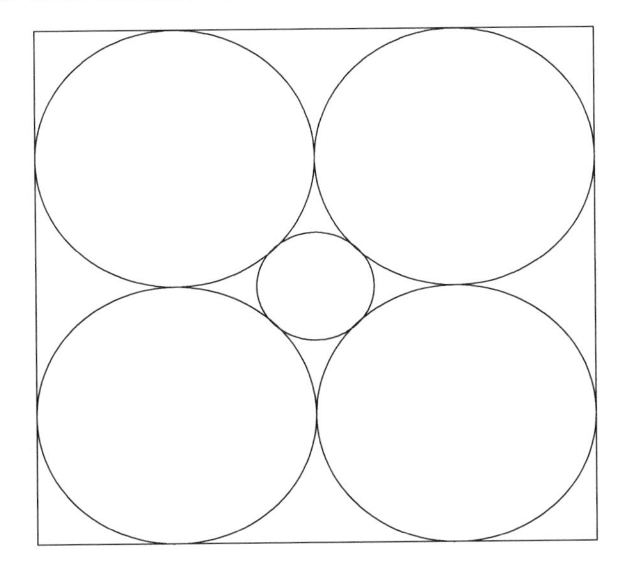

Figure 2.14: A 2-dimensional prisoner (central disk) surrounded by guards.

dimension, keeping everyone involved as hyper-spheres. At $d = 3$ (where real prisoners live), the prisoner actually has space between the guards and can see out of the cell (there are now 8 guards instead of 4). As d increases, the ratio of the volume of the prisoner to that of the cell increases and by $d = 10$ the prisoner is actually bigger than the cell (hence some of him is outside). As d increases, more and more of him is outside the cell!

This fact is so counter-intuitive that it deserves more discussion. Let's let the "jail cell" have wall-length 2, and center it at 0; thus the guards all have diameter 1, hence radius 1/2, and the radius of the prisoner is the distance from the origin to the nearest guard. The distance from the center of any guard to a corner is $\sqrt{d}/2$, and so the distance out from the corner that is covered by the guard is the guard's radius plus this: $1/2 + \sqrt{d}/2$. The distance from the center of the prisoner to the corner is $\sqrt{d}$ and so the radius of the prisoner is $(\sqrt{d} - 1)/2$. When $d = 9$, this is 1, and so the prisoner touches the walls of the cell. For $d > 9$, the radius of the prisoner is larger than 1, and so some of the prisoner is outside the cell. It is worth calculating the ratio of the volume of the cell to that of the prisoner, and observe that as $d \to \infty$ the ratio goes to zero, and hence the proportion of the prisoner inside the cell shrinks to zero.

The take-away message is that while we must be cautious to rely too heavily on our intuitions, especially for high dimensional or complex data, the tools described in this book can be used with great success on these data, provided that we take care to apply them properly and understand how they work, when they fail, and how to properly evaluate them. For more information about the curse of dimensionality, quirks of high dimensional data, and related issues, the reader is encouraged to investigate [124]; more mathematically sophisticated readers may find [461] and [467] of interest. In the latter, the terminology used is *concentration of measure*.

Chapter 3

Security: Basics and Security Analytics

In this chapter, we give a tour of computer security through the data science lens. It covers the basic security background needed to understand the later chapters, and goes a little beyond this necessary background. We also briefly mention some interesting, nontrivial "applications" of data science techniques from the previous chapter to security. We then define security analytics and discuss the unique challenges imposed by the security domain on data analytics. Besides exposing the reader to the security background needed to understand this book and giving notable applications of data science to security challenges, the chapter also covers the basic knowledge needed by a reader who wants to secure his/her own computer.

Before the advent of the Internet, securing a computer was not such a big challenge. However, all that changed with the Internet age and the use of communication tools such as electronic mail. Things are getting progressively harder with mobile phones and the Internet of Things.

All the above technological advances have led to an explosion of attacks and the defenders have had to scale up their responses quickly. As a consequence, computer security is a vast topic now, so there is no way we can do justice to it in a single chapter. Hence, we refer the interested reader to good books on security for more details about the following topics [108, 403]. A good book on networks and protocols, e.g., [347], would also be helpful for understanding the networking concepts and protocols, e.g., TCP, SSL/TLS, etc., referred to in this chapter.

3.1 Basics of Security

The main characteristic of security is that it is holistic. For example, your laptop may have all the latest tools and protection mechanisms, but if you left it in the coffee shop, you have lost it. Now, unless you encrypted all the data on it with a really good encryption algorithm and key, you may as well assume that someone else has your data.[1] Similarly, a house may be secure, but if the financial statements are thrown in the trash, then a dumpster-diving attack can retrieve this information, which can then be exploited by thieves. We call such aspects of security, *physical security*, and do not cover them in this book. What we present most of in this book is usually called *technological security*, by which we mean the security of the operating system, the software, the data, the network, and the software environments

[1] The password on your laptop is usually easily cracked.

of hosts on the network as a whole. We will also touch upon human aspects of security occasionally in this book.

3.1.1 Know Thy Enemy – Attackers and Their Motivations

As Sun-Tzu is supposed to have written in "The Art of War," ([442]) thou should "know thy enemy."[2] It helps in security if we keep in mind the different kinds of attackers that are out there and their motivations. How many different types of attackers are there? We have what are called *script kiddies*, which are attackers who use scripts and code written by other attackers. We have the *cyber terrorist*, whose motivation is political or religious usually, and whose goal is to cripple the "enemy." There are also *hacktivists*, whose main motivation is to embarrass. We have *nation states* and we have the *sophisticated hackers*. The last, but not the least, group of attackers is that of *disgruntled insiders*, disaffected employees, who turn rogue on their organizations for various real or imagined grievances.

What are the motivations of the attackers? The motivations range from revenge, for the disgruntled insider and nation state, to money and thrill for sophisticated hackers and script kiddies. There is also the pursuit of information for various goals that include: domination, tarnishing reputation, free entertainment, or monetization. Crippling the enemy is a goal of the cyber terrorist and the nation state.

How determined are the attackers? Attackers, such as sophisticated hackers, disgruntled insiders and nation states, can be extremely determined. Some interesting examples of their determination can be found in [117]. On the other hand, script kiddies may be easily deterred and look for people or organizations that are easier to attack.

Besides knowing the enemy, one must know oneself. Hence, the defender should take an inventory of the assets that are worth protecting and the kinds of protections needed for each asset. This is usually called Threat Modeling or Analysis. It brings us to our first security principle, viz., *Security is a game of economics.* This means that one should not spend $10 defending something whose value is only one dollar. It also means that by raising the bar, one can discourage all except the most determined or foolhardy attackers. We will look at the different types of protections first.

3.1.2 Security Goals

For the technological aspects of security, there are three fundamental security goals, with the neat acronym CIA:

1. Confidentiality: Preventing unauthorized access to data, messages, or programs.

2. Integrity: Making sure that data, messages, or programs, are not modified or corrupted.

3. Availability: The resources, which may be machine(s), web site(s) or application(s), are up, functioning correctly, and providing the services at the capacity and rate for which they were designed and configured.

In addition to these three fundamental goals, there are also a number of other important security goals, resulting in the acronym: CIAAAAN, or CIA^4N for short:

1. Authentication: Proving one's identity to another person or machine.

2. Authorization/Access Control: Different people and programs are authorized to do different functions in an organization or computer network. This is called an access control policy. Defining a correct policy and making sure that the policy is not being violated are part of this goal.

[2] "If you know the enemy and know yourself, you need not fear the result of a hundred battles."

3. Accountability: If something goes wrong, then who or what is at fault.
4. Nonrepudiation/Nondeniability: If an agent takes some action, then it should not be able to deny taking that action later.

Having goals is not enough. One must develop or implement procedures or mechanisms to achieve them. While developing these mechanisms, we must make sure that we keep our adversary in mind, as stated before. In security, this is called adversarial modeling. We refer the interested reader to Lockheed Martin's Cyber Kill Chain framework and Mitre's ATT&CK[3] Matrix for more on adversarial modeling.

We now briefly consider the mechanisms that have been devised for achieving these goals. While reading this section, one should carefully consider the attacker and the possible attacks, and then evaluate the relevance and the effectiveness of these mechanisms.

3.2 Mechanisms for Ensuring Security Goals

The goal of confidentiality is a very important one, and many of the solutions that have been developed for this goal are useful in other security goals as well. Hence, we will cover this in more depth than some of the other goals.

3.2.1 Confidentiality

The security goal of keeping things secret, or confidentiality, has been around ever since humans have been around. As far as recorded history is concerned, the Roman general and emperor Julius Caesar used what is now called the "Caesar Cipher" for secrecy on the battlefield. The idea of this cipher is to encode each letter by the third letter that comes after it and to treat the alphabet as written on a circle, so that the letter 'A' is encoded by the letter 'D,' 'B' by 'E' and 'Z' by 'C,' since the letters wraparound after Z and C is the third letter after Z. Thus, the plaintext message *zebra* would be encoded as *cheud.*

The basic definitions of cryptography are as follows. Data or messages to be kept secret are called *plaintext.* The usual mechanism to ensure this goal is through *cryptography*, which is the study of ciphers or codes. Keeping a message secret involves *encrypting* it and the reverse process of going from an encrypted message to the plaintext is called *decryption*, and the encrypted message is called *ciphertext.* Both the encryption and decryption algorithm take as input a key, usually a random sequence of bits or digits, besides the plaintext input for encryption and the ciphertext input for decryption. The encryption and decryption algorithms are usually published for the benefit of getting them "debugged" through the wisdom of the crowd. The theory is that the more the scrutiny of the algorithms, the more robust they will be at resisting attacks. Thus, the secrecy of the ciphertext rests on the key. This is also known as Kerckhoff's principle. A cryptosystem should be secure even if everything about it is public, except the key. This also brings us to another security principle, viz., security cannot be achieved through obscurity.

In *symmetric key/private key* cryptography, there is just one key, which is shared by all participants involved in exchanging messages, and it is kept secret. In *asymmetric key/public key* cryptography, each participant has two keys: one for encryption, which is public, and the other for decryption, which must be kept private.

Attacks on cryptographic schemes can be categorized based on the attacker's knowledge or power. The least amount of attacker knowledge is required in the case of *ciphertext-only attacks.* In these attacks, the attacker has access to only ciphertexts. Next in the hierarchy are *known plaintext attacks*, where the attacker has pairs of plaintexts and corresponding ciphertexts. If the attacker can choose plaintexts and get them encrypted using the cryptographic algorithm as a black-box, we get what are called *chosen plaintext attacks.* The

[3]Adversarial Tactics, Techniques, and Common Knowledge.

most basic requirement for a cryptographic scheme is that it should resist at least known plaintext attacks.

The two basic techniques for encryption are *substitution* and *permutation*. For example, the so-called Caesar Cipher is a substitution cipher in which every letter is encoded by the third letter after it, treating the English alphabet circularly, so that the letters, A, B, C, "follow" the letter Z. If we encode the 26 letters using numbers 0 to 25 with 0 representing letter A and 25 representing Z, then we can write that the ciphertext number $N_c = N_p + 3 \ mod \ 26$ when the plaintext number is N_p. Similarly the equation for decryption is straightforward. The Caesar Cipher is called a *monoalphabetic* cipher since each letter is encoded by a single letter and the encoding is fixed. *Polyalphabetic* cipher schemes may encode several letters together or use different letters as encodings of the same letter or both. Polyalphabetic ciphers use a set of related monoalphabetic ciphers. Examples of polyalphabetic ciphers are the Vignere cipher and the Vernam cipher in which the user selects a key, which is then repeated if it is shorter than the message. Each letter of the key is an index into a monoalphabetic cipher and the substitution for the plaintext letter is done using the substitution rule of the chosen monoalphabetic cipher.

The one-time pad is a special kind of Vernam cipher in which the key is a random sequence as long as the message. It is best explained using a bit encoding of messages. Imagine that the plaintext message is the bit sequence: 0110101 and the key is the random sequence: 1101001. Then, the ciphertext sequence is obtained by bit-wise exclusive-or of the key and the plaintext, 1011100. As long as the key is a truly random sequence and as long as the message, the one-time pad provides what is called unconditional security. However, the one-time pad is not a practical cipher scheme since generating truly random long sequences is very difficult,[4] the key must be different for different messages (hence the name one-time pad), and one also needs a secure mechanism for sharing the key.

A *block cipher* is one in which the plaintext is divided into blocks and each block is used to produce a block of ciphertext. For example, the Data Encryption Standard is a block cipher and works on blocks of 64 bits. A *stream cipher* on the other hand encrypts a stream of plaintext one bit or byte at a time.

Practical symmetric-key cryptographic schemes are the Data Encryption Standard (DES), the triple-DES and the Advanced Encryption Standard (AES) or Rijndael after the names of its developers (Vincent Rijmen and Joan Daemen). The DES is based on what is known as a Feistel structure and the AES consists of a substitution-permutation network.

Unfortunately, the monoalphabetic substitution ciphers and the permutation ciphers are easily broken since they preserve the underlying statistics of the plaintext language such as English. Therefore, the DES combines both these operations and applies them over several rounds to the plaintext. The two important objectives behind this elaborate construction are called confusion and diffusion, and they were introduced by Claude Shannon in an influential paper [384], which was classified when it was written.

Definition 3.2.1. *1. Confusion refers to the relationship between the ciphertext and the key. It says that an attacker should not be able to discover the key given only the ciphertext.*

2. Diffusion refers to the relationship between the ciphertext and the plaintext. It says that an attacker should not be able to discover the plaintext from the ciphertext. In other words, the underlying plaintext language statistics must be "diffused" in the ciphertext.

We give the basic ideas involved in designing block ciphers, following the presentation of [237]. A block cipher is defined by a function, $E_k : \mathcal{P} \rightarrow \mathcal{C}$, that maps a plaintext p chosen from a plaintext space $\mathcal{P}$ to a ciphertext c in a ciphertext space $\mathcal{C}$ (i.e., $c = E_k(p)$). The

[4]Recently, there has been some exciting progress on this front.

Table 3.1: Block Cipher Parameters

N	Number of rounds
l	Size of round input/output
m	Number of overlapped parts in round input/output
w	Number of bits in each part ($l = mw$)
M	Number of functions

function E_k is typically a permutation (hence $\mathcal{P} = \mathcal{C}$), which is specified by an encryption algorithm and a secret key k chosen from a key space $\mathcal{K}$.

The encryption algorithm in a block cipher works in N rounds. Each round has an l bit input and creates an l bit output. The plaintext is the input for the first round, while all subsequent rounds take their input from the previous round's output. The output from the final round is the ciphertext. Additionally, each round also takes a second input called the round key, which is derived from the secret key using a key expansion function. Each l bit input of a round in the block cipher consists of m nonoverlapped parts of w bits each. These parts are subjected to a key addition, confusion, and diffusion in each round.

Confusion, which is typically implemented by w bit substitution boxes (S-boxes), is the only nonlinear operation. It helps to obscure the S-box outputs with respect to its inputs. The S-boxes are designed in such a way that the correlation between the input and output is minimized. Diffusion and key addition are linear functions. Diffusion causes information to spread to multiple bits in the round output while key addition ensures that the round output depends on the secret key.

One can formalize the block cipher as a composition of M functions, which are applied to the plaintext and produce the corresponding ciphertext, as follows.

$$c = E_k(p) = f_M \circ f_{M-l} \circ \ldots \circ f_2 \circ f_1(p) \tag{3.1}$$

The functions are Boolean and could be either linear or nonlinear. Every *linear* function in Equation 3.1 is of the form:

$$f_i(x_1, \ldots, x_l) = \bigoplus_{j=1}^{l} a_j x_j \tag{3.2}$$

where the addition is the exclusive-or function or addition modulo 2. Every nonlinear function in Equation 3.1 is of the form:

$$f_i(x_1, \ldots, x_l) = \bigoplus_{j=1}^{l} a_j \prod_{z \in Z} x_z \tag{3.3}$$

Here the a_j's are binary constants and $Z \subseteq \{1, 2, \ldots, l\}$.

The l function inputs are grouped into m components of w bits each. We call each of these w bit components a part. Functions act separately on each w bit part. The result of having multiple such rounds in the cipher is that the ciphertext is statistically indistinguishable from a pseudo-random number. Table 3.1 summarizes the block cipher parameters, while Table 3.2 provides specific values of parameters for some block ciphers.

AES has a specification that can accommodate different key sizes. We give a brief specification of AES with 128 bit key, AES128. AES128 has 10 round functions and a secret key k, plaintext p and ciphertext c each of 128 bits such that $k, p, c \in \{0, 1\}^{128}$ and hence

Table 3.2: Parameter Values for Some Block Ciphers (Material based on [237])

	N	l	m	w	M
AES	10	128	16	8	40
CLEFIA	18	128	16	8	74
SMS4	32	128	16	8	128

Table 3.3: The Rail-fence Cipher with Two Rails

W		a		e		o		t
	e		r		l		s	

$l = 128$. After an initial key whitening (key is Xor'ed with the plaintext, goal is to increase security), each round splits its input into 16 parts of one byte each ($w = 8$). The first nine rounds have four operations: AddRoundKey (ARK), SubBytes (SB), ShiftRows (SR), and MixColumns (MC). The final round is similar except that it does not have the MixColumns operation. The ARK, SR, and MC functions are linear, while the SB is a nonlinear function. A composition of these 4 functions, each repeated several times, is applied to the plaintext to generate the ciphertext. In all, AES128 applies 40 functions ($M = 40$ in Equation 3.1) to the plaintext one after the other before the ciphertext is obtained. We refer the reader to books such as [402] for the details.

For a basic permutation cipher, we turn to the Rail-fence Cipher in which the message is written diagonally down and up on a number of horizontal lines or rows depicting rail-fences. Then the ciphertext is obtained by reading row by row. For example, we write the plaintext message "We are lost" on two fences as shown in Table 3.3.

Reading row by row, we get from the top row: *Waeoterls* as the ciphertext. Note that such a scheme does not diffuse any of the frequencies of the individual characters in the plaintext and so it is vulnerable to statistical analysis.

Popular public-key cryptography schemes are RSA and Elliptic-curve Cryptography. In public-key cryptography, each principal has two keys, one public and the other private. The public key of Alice is used by Bob (or whoever wants to send a message to Alice confidentially) to encrypt messages intended for Alice, who uses her private key to decrypt them. RSA is named after its inventors, Ronald Rivest, Adi Shamir and Leonard Adleman, and based on number theory, specifically modular exponentiation and Euler/Fermat theorem. Following are the steps in RSA.

1. Both Alice and Bob independently choose two large prime numbers (over 2048 bits each) and compute their product. Let p_A, q_A be Alice's primes and p_B, q_B the primes chosen by Bob. Let $n_A = p_A q_A$ and $n_B = p_B q_B$.

2. Next, both compute the Euler totient function of the products n_A and n_B. These are easily computed as $\phi(n_A) = (p_A - 1)(q_A - 1)$ and similarly for $\phi(n_B)$.

3. Alice chooses exponent e_A coprime to $\phi(n_A)$ and Bob chooses e_B coprime to $\phi(n_B)$. Alice publishes her public key, which is (e_A, n_A) and Bob publishes his public key, which is (e_B, n_B).

Table 3.4: Pros and Cons of Symmetric Key and Public-key Cryptography

	Symmetric	Public-key
Complexity	Linear	Quadratic
Key Set-Up	Harder	Easier

Table 3.5: Three Key Number Theory Problems

Cryptosystem	**Problem**	**Caveat**
RSA	Factoring N = pq	Quantum Computing (QC)
DSA	Discrete log $y = g^x$ mod n	QC, "low" complexity for certain fields
EC-DSA	EC Discrete log $Q = x.P$	QC

4. Both of them also compute secret exponent d_A and d_B as the multiplicative inverses of e_A and e_B, respectively, i.e., $e_A d_A \equiv 1 \ mod \ \phi(n_A)$ and $e_B d_B \equiv 1 \ mod \ \phi(n_B)$. These computations can be done using the Extended Greatest Common Divisor algorithm.

Now suppose that Alice wants to send a secret message M to Bob. Assume that the message is a number (e.g., ASCII representation of text) and divided into blocks $M = \langle M_1, M_2, \ldots, M_l \rangle$ so that each block $M_i < n_B$. She encrypts each block by computing $E(M_i) = M_i^{e_B} \ mod \ n_B$ and sends the encrypted blocks to Bob. Notice that she uses *Bob's public key* to encrypt the message, since only Bob has the corresponding secret key for his public key. To decrypt, Bob computes $D(E(M_i)) = E(M_i)^{d_B} \ mod \ n_B$, and retrieves M_i. This follows from the properties of modular exponentiation and the choice of the exponents. Of course, things are not so simple in practice, since there are a number of mathematical and side-channel attacks on RSA,[5] such as tracking the time taken for modular exponentiation. The difficulty of a brute-force attack on RSA is considered to be of similar difficulty as the problem of factoring a number. However, so far, no one has been able to show that factoring is a difficult problem.

RSA is much slower than DES and AES; because of its much larger key size modular exponentiation cannot be done completely in hardware. This has led researchers to seek more efficient public-key cryptography algorithms. Recently, the elliptic curve cryptographic (ECC) algorithm has emerged as a strong contender. ECC can achieve similar security as RSA with much smaller key size. For example, 256 bit ECC provides the same level of security as 3,072 bit RSA. We refer the reader to [403] for the details of ECC.

We close the confidentiality section with a comparison of symmetric and public-key cryptography in Table 3.4. In practice, e.g., the SSL/TLS protocol, the public-key cryptographic algorithm is used to set up a shared secret session key, after which symmetric key cryptography is used.

Popular public-key cryptosystems rely on the difficulty of three problems from number theory: factoring, discrete-log and the elliptic curve discrete-log problem. However, factoring and discrete-log are both easy ("polynomial-time solvable") for quantum computers as shown in [389]. In 2013, significant speed-ups were achieved for the discrete log problem in certain kinds of fields on ordinary computers [224]. In Table 3.5, DSA stands for the Digital Signature Algorithm.

The mechanisms that have been developed for confidentiality can also help in authentication of the sender or the receiver and in integrity. For example, RSA has the symmetry

[5]Computations on physical devices can leak information through what are called side-channels, e.g., power consumption, acoustic noise, electromagnetic radiation, etc.

property, which means that a message that is encrypted by the private key can be decrypted by the public key. Now if the private key is known only to the person, say Alice, who generated it, then this gives a way to prove that a message was sent by Alice if it is encrypted with her private key. This is called a "digital signature". Of course, such a message is not confidential since anyone can get hold of the public key and decrypt it. If both goals are required, then Alice can sign the message with her private key and then encrypt it again using Bob's public key. Usually, the whole message is not signed, but only a cryptographic hash of it.

3.2.2 Integrity

The goal of message integrity can be achieved by a message authentication code (MAC) or a cryptographic hash function. The basic idea is simple: we add redundancy to the message. The following example will clarify the basic idea. Imagine that a message block of fixed length is transmitted over a noisy channel and suppose that noise on the channel is intermittent and rare, so that only one bit of the message block is changed during transmission (sometimes a 0 gets changed to a 1 or a 1 gets changed to a 0, but not both). At the receiving end, we need to detect whether the noise has affected the message or not. For this purpose, we can compute a parity bit, bitwise exclusive-or of all the bits in the message, for the message and transmit the message along with the parity bit. Assuming that the parity bit is not affected by the noise, we can detect single-bit errors at the receiving end, since the parity of the message garbled by noise will not be the same as the parity bit of the original message. The idea of a MAC or a hash is similar, but more sophisticated, since a number of practical considerations are involved.

Older MACs were based on symmetric cipher schemes such as DES, but modern MACs are based on cryptographic hash functions. RFC 2104 defines the hash-based MAC called HMAC as follows. Let H be a cryptographic hash function which hashes data by iterating a basic compression function on data blocks of byte-length B and K a secret key. Then,

$$HMAC(K, M) = H(K \oplus \ opad \ || \ H(K \oplus \ ipad \ || \ M)),$$

where *ipad* is the byte 0x36 repeated B times, *opad* is the byte 0x5C repeated B times, usually $B = 64$. We append zeros to the left end of K as necessary so that its length is B. If K is longer than B, then it can be hashed using H and then the resulting string can be used as the key for the HMAC. The symbol || denotes concatenation.

3.2.3 Availability

Availability is concerned with ensuring that the computing infrastructure is up and performing as it should. Power failures, disk crashes, and network outages are some of the disruptive faults that can bring a system to its knees. Plus there are Denial of Service (DoS) attacks, which are conducted by commandeering a group of computers, called zombies, by a wily attacker. Typically, they make a very large volume of requests in a very short time, so that a server is unable to handle the load and becomes unresponsive to legitimate requests for service. The most virulent form of this is called a Distributed Denial of Service (DDoS) attack, in which scores of machines are used by an attacker. Recently, using IoT (Internet of Things) botnets, DDoS attacks have become commonplace and truly worrisome in scale. "In each attack, hundreds of thousands of devices connected to the Internet are being leveraged," according to a securelist.com article.[6] Attacks are also becoming more complex, e.g., the Moscow stock exchange was attacked with SYN, TCP Connect, HTTP-flood and UDP flood [234].

[6] https://securelist.com/ddos-attacks-in-q1-2017/78285/ - Accessed on July 18, 2017.

What are the mechanisms for ensuring availability? The most basic protection mechanisms are captchas, since they prevent one source of DOS attacks. For a DOS or DDOS attack, identification of suspicious traffic is critical. For this purpose, it is crucial to build a profile of normal traffic and then catch deviations from this profile. This is tricky to do, since indications of an attack can be subtle and one needs to devise new features based on existing information.

For example, in the case of a SYN flood attack, the attacker sends a lot of connection requests quickly, but does not complete the handshakes. The idea is to soak up a lot of resources at the target machine, so that it will not be able to service the legitimate requests. The usual features of connection records, e.g., timestamp, destination host, service, etc., cannot distinguish SYN flood connections from normal ones [265]. An S0 flag signals a connection request, where the handshake is incomplete. However, this can happen in a normal connection also, e.g., due to traffic congestion. But, if there is a large percentage of S0 flags within a short time period for connections to the same service on the same host, it is likely a SYN flood attack.

It is a good practice to overprovision bandwidth, since traffic during advertising campaigns can rise suddenly and overwhelm the web server. Other mechanisms are:[7]

1. Having a rate limit at the router to prevent the web server from getting overwhelmed.

2. Adding filters at the router to drop packets stemming from obvious sources of attacks.

3. Timing out half-open connections aggressively.

4. Dropping spoofed or malformed packets.

5. Setting lower SYN, ICMP and UDP flood thresholds.

The next step is to call the ISP or hosting provider and report the attack. There are also companies that specialize in DDoS attack mitigation.

3.2.4 Authentication

Authentication is the goal of proving one's identity to a machine or another human individual or organization. We are all familiar with this task since all of us have many computer accounts and the login process for these accounts requires us to use our passwords to prove our identities. A password is something that we know. When two agents, Alice and Bob, want to exchange confidential information remotely, then one possibility is for one of them to call the other using the other's phone number. Again, this is based on something that they know. If they want to communicate over the Internet instead, then they typically use what is called a cryptographic, challenge-response authentication protocol. Again these protocols are based on something that is known to only one of them, so that it uniquely identifies the person who uses that knowledge. The design of such protocols is error-prone and they may contain subtle errors so that even if the cryptographic algorithms are never broken, an attacker can convince the honest agents that they are communicating with each other, when they are only talking to the attacker, e.g., see the seminal paper by [284].

Other possibilities for achieving authentication are *something that you have* and *something that you are*. Examples of the first are identity cards, smart cards and bank ATM cards. For the latter, we have biometric authentication schemes such as iris scans, retina scans, fingerprints, palm prints, etc. All of them have their relative advantages and disadvantages. For instance, iris and retina scans are more intrusive, but harder to replicate.

[7] `https://www.esecurityplanet.com/network-security/5-tips-for-fighting-ddos-attacks.html`

The latest authentication mechanisms use multiple factors for authentication. For example, two-factor authentication could combine what you know with what you have. Two-factor authentication schemes are used by banks, for example, in their ATM cards. The customer has to use the card (something that you have) along with a Personal Identification Number (PIN, something that you know). Two-factor authentication is now recommended for reading email and other Internet tasks as well. However, regardless of how many factors are used for authentication, one is still vulnerable if one downloads attachments in emails or clicks on links.

3.2.5 Access Control

The first step in access control is devising a policy that meets the security requirements. Many organizations use what is known as the Bell La Padula model. In this model, each resource (e.g., a file) has a certain level of protection assigned to it and every employee has a certain level of access as well. Typically, four levels are used: public, classified, secret and top secret. In addition, there are three simple rules:

1. No read up. A person at level L cannot access resources at levels higher than L.
2. No write down. A person at level L cannot "write" information into resources at levels lower than L.
3. Tranquility property. A resource's level cannot be reclassified, if it is being accessed by someone.

After a suitable policy is devised, the next step is enforcing the policy, or access control. The basic data structure for access control, or who is authorized to do what on a computer system, is the access control list (ACL), which one can think as a sequence of pairs of the form (user, permission on a specific resource). Resources may be files, printers, etc. However, this is unwieldy to manage. Lookups and updates are slow and since everything is excruciatingly explicit, there is a huge amount of tedious work, depending on the system resources, when a new user joins the system or an existing user leaves the system. So we introduce abstractions such as role-based access control and attribute-based access control. In role-based access control, users have what are called roles on the system and the roles have associated permissions. In essence, the ACL is now split into two lists: the user-role list and the role-permission list. This makes things more manageable and less tedious.

In attribute-based access control, which gives finer granularity of control, each attribute may have permissions associated with it. For example, each column in a table may require a different level of access.

3.2.6 Accountability

When something goes wrong on a computer system or network, we need to determine who or what was at fault, the extent and details of loss or damage, and then learn appropriate lessons for the future. This is the goal of accountability. For this purpose, the basic mechanism is an "audit log," which is a record of all the activities on the system. Conceptually, we think of a single audit log, but, in practice there is more than one log and there may be an archiving mechanism as well for the log. For example, there may be web server logs, Lightweight Directory Access Protocol logs, command logs, access logs, error logs and audit logs.

The integrity of the logs on the system is very important, otherwise any forensic investigation would be jeopardized; of course, even system administrators should not be able to turn off the logging mechanisms on the system. The latter requirement is hard to ensure in practice. The basic unit of storage in the logs is an audit record. It consists of a command and its parameters along with any exceptions as a result of executing it and the resources that were required for its execution.

3.2.7 Nonrepudiation

Suppose a customer calls his stock broker and places an order to buy a large number, say 10,000, of shares of a company at a stock price of $100 per share and subsequently the stock drops by say $10 at the end of the day. The customer has lost $100,000 in a single day. The customer could then decide to sue the stockbroker claiming in court that he had not placed the order. To counter this claim, the stockbroker can produce a recording of the call. The principle that an agent who commits to a transaction should not be able to deny it later is the goal of nonrepudiation. Specialized protocols, generally involving trusted third parties, are designed for this purpose. Another practical example is the use of a notary public to witness signatures. This ensures both identity checking of the signatory, authentication, and nondeniability. This is one of the most challenging goals to ensure, especially when there are no trusted third parties.

3.3 Threats, Attacks and Impacts

Now that we have a bird's-eye view of security goals and mechanisms, let us survey the threat landscape. There are many definitions of the word "threat." According to ISO27K FAQ: information risk management[8] *threats* "are the actors (insider and outsiders) and natural events that might cause *incidents* if they acted on *vulnerabilities* causing impacts." Vulnerabilities "are the inherent weaknesses within our facilities, technologies, processes (including information risk management itself!), people and relationships, some of which are not even recognized as such." Thus vulnerabilities include bad programming practices (e.g., using unsafe C string libraries), bad design (e.g., default option or configuration of a system is insecure), short and easy to break passwords, unsecured machines and networks (open ports, unpatched services), etc. Incident and impacts are easily understood terms in the above definitions.

Since threats are defined as actors, we could use terms such as intruders, who could be misfeasors, masqueraders, or clandestine users [402]. The broad, preferred term in the literature that includes all these classes of actors is attackers, who could be insiders or outsiders. Attackers employ various techniques to attack systems, networks or organizations. We can categorize these techniques into the following groups: human-focused, device/system (includes hardware and software dimensions) focused, network architecture focused, and service focused (e.g., web server, SSH, telnet, database, etc.) attacks, and the rest. Examples of each category are in Table 3.6. In this table, DNS denotes domain name service, XSRF denotes cross-site request forgery and XSS denotes cross-site scripting attack.

In addition to the familiar impacts of attacks such as theft or loss of a system, economic impacts, loss of human time and/or productivity, and loss of reputation, we have the following technological impacts of attacks: theft of data (e.g., credentials of customers or intellectual property), loss of use of a system or network, and the use of systems or networks for attack multiplication.

According to the process diagram for information risk management in the FAQ: The design of a system should begin by identifying the risks to it, evaluating the risks, treating the risks and handling changes. In what follows, we use the phrase "security challenges" to include vulnerabilities, attacks, or techniques employed by attackers.

[8] `http://www.iso27001security.com/html/risk_mgmt.html` - Accessed on 12 July, 2017.

Table 3.6: Examples of Techniques/Attacks

Dimension	Examples
Human focused	Hacking accounts, masquerade, phishing and its variants
Device/System focused	S/w vulnerability exploits, malware, intrusions
Network focused	Port scanning, tunneling, DNS cache poisoning
Service focused	Command/script injection, XSRF, XSS
Miscellaneous	Botnets, Denial-of-service

We cover some of the biggest challenges, again from the viewpoint of a person who wants to secure his or her machine.

3.3.1 Passwords

One of the biggest security challenges for computer users is choosing and remembering strong passwords. The dilemma is that a short and easy-to-remember password is usually weak, especially if it is a single word in the native language of the user. An attacker can just use a dictionary attack on such passwords. Strong passwords are typically more random and contain a diverse mix of characters, which makes them harder to remember. The use of one password for every account is risky, but if it is chosen with care, it may not be such a bad idea from the memorization point of view. However, we do not recommend this approach. We recommend that people use multifactor authentication whenever possible, and if they must rely on passwords for some accounts, then they design an algorithm that generates passwords for them based on the characteristics of the account. This way they only need to memorize an algorithm, not the actual password, which can be generated based on certain input features of the account.

Good password crackers such as John the Ripper[9] and Cain and Abel[10] are now available and should be used periodically by computer users and system administrators.

3.3.2 Malware

Nowadays, malware is one of the biggest security challenges for anyone who is connected to the Internet. Malware refers to bad software and includes: viruses, worms, trojans, logic bombs, keyloggers, rootkits, backdoors, ransomware, etc. Of these, viruses have caught the popular imagination so much so that anti-malware systems are referred to as anti-virus systems.

A virus is a piece of software that needs a host and contains: a method for checking whether a file is already infected, a method for infecting a file, a payload, which is the damage that the virus does, and a mechanism to find other objects, typically files, to infect. The name "virus" is evocative of this replicating behavior. The more sophisticated types of viruses also change their form, so-called polymorphic and metamorphic viruses, and/or use encryption/decryption and other obfuscation mechanisms. Worms are pieces of malware that are similar to viruses, in that they also replicate, but in contrast to viruses, they are independent, i.e., they do not need a host, and they also have mechanisms to propagate themselves to other hosts on the network and through the Internet. Typical propagation mechanisms include: electronic mail, remote execution and remote login facilities.

[9] http://www.openwall.com/john/

[10] http://www.oxid.it/cain.html

Table 3.7: Some Famous Malware Examples

Type of Malware	Examples
Virus	Melissa, Boot Sector, Macro
Worms	Morris, Code Red, Nimda, Conficker
Trojans	Attack on CBS executive [431]
Keyloggers	Actual Spy, Family Keylogger
Rootkits	Sony BMG copy protection, HackerDefender
Backdoors	Guest and Admin accounts on routers and IoT devices
Logic Bombs	Tim Lloyd [166]
Ransomware	WannaCry, Petya, NotPetya[a]

[a] Although we put NotPetya in the ransomware category, its objective was not money, but destruction and chaos. An interesting article on NotPetya is: `https://www.wired.com/story/notpetya-cyberattack-ukraine-russia-code-crashed-the-world/`

Nonreplicating malware includes trojans, logic bombs, keyloggers, rootkits, and backdoors. A *logic bomb* refers to a bad action that is triggered when a condition is satisfied. For example, on a certain date, the file system may be wiped out if a logic bomb is embedded in a program that runs on the system with administrator level access. A *backdoor*, as the name implies, is a control path in a piece of software that may give unauthorized access to a resource. These are usually created by programmers to circumvent a series of security checks during debugging and code development phase. They are left by mistake in software, which is shipped to customers. Recently, researchers have pointed out that backdoors can exist in hardware as well. Trojans, as the name implies, are malware designed to look as useful pieces of software. For example, a program that is ostensibly for playing games may have capability to open a backdoor into a system that could then be exploited by a wily attacker. Trojans can exist in hardware as well. A *keylogger* is spyware that records every keystroke typed by users of a system. In this way, login names, passwords, credit card numbers, etc., can be recorded without the user's knowledge and then periodically shipped to the attacker. *Rootkits* are sets of tools used after an attacker has gained administrator level access to a system. Several malware attacks have been reported in the popular press and media. Some examples of these are given in Table 3.7. Most recently, a lot of people and companies have reported ransomware attacks. The local government of the city of Atlanta is the latest high-profile victim. In *ransomware*, the data is encrypted by the malware, and the victim must pay the attackers, usually in cryptocurrencies such as Bitcoin, to get the decryption key.

Malware defenses include constructing signatures, which are patterns found in the malware code. Ideally, such pattern should be discriminative, i.e., the signatures should flag malware, but not benign software, sometimes also called hamware (borrowed from the use of the term ham for nonspam email). Once defenders devised signatures, attackers started to use obfuscation techniques to defeat them. Such methods include inserting no-ops, reordering independent instructions, e.g., the two instruction $\{x := 1; y := 2\}$ can be executed in any order without changing the semantics or pragmatics of the program; using equivalent operations, e.g., x := 2 * x and x := x + x, etc. This arms-race between the attacker and the defender is a hallmark of all security challenges. To defeat obfuscation techniques, defenders devised so-called behavior detection, i.e., monitoring the behavior of software rather than pattern matching using regular expressions. Monitoring the behavior requires processes that observe network connections, memory consumption, CPU usage, and I/O requests, for example. Most modern malware connects to command and control servers for various purposes such as code update or instructions on where to attack next. To avoid using the same domain/IP address, the malware uses domain generation algorithms.

3.3.3 Spam, Phishing and its Variants

Spam usually refers to emails containing advertisements and is usually an irritant and a time/productivity sink more than anything else. It is estimated that almost 80% of emails nowadays is spam. Authors of spam emails usually do not disguise their emails, hence they are easier to detect. The best defense against spam is a good anti-spam filter and care in selecting services when signing up for something. These filters usually work by learning keywords in spam that occur frequently [312]. Some of them also use other techniques such as analyzing email headers.[11] An example of a spam email is shown in the box below. Observe how it has been marked as spam by the spam filter employed. Note also that the email header is abbreviated.

Date: Tue, 14 Jan 2014 05:52:18 -0500
From: Lupix shopping < *example@gmail.com* >
To: undisclosed-recipients: ;
Subject: *****SPAM***** INQUIRE FROM LUPIX SHOPPING,

Dear Sir/Madam,
We hope this finds you well. We are interested in purchasing your company's products, we do hope to establish a long term relationship. However, we would like to see your company's latest catalogs with the ,
<...> - *omitted*
Our e-mail contact; example@gmail.com
We hope to hear from you soon.
...

Email that is not spam is sometimes referred to as *ham* in some data repositories of spam, and some machine learning papers that use these data. We will use this term as well throughout to refer to nonspam (presumably legitimate) email.

Phishing, in its most popular attack vector, is a form of email masquerade attack in which the attacker tries to gain sensitive information such as digital identity, credit card numbers, etc., by convincing recipients that the email comes from a genuine entity. Phishing is much more difficult to identify automatically than spam because of the use of deception. Phishing attacks are designed to look like legitimate requests for information/action or a legitimate web site that steals the information entered by visitors or downloads malware on the visitors' systems. Phishing detectors are still under development and a study of commercial detectors shows that much more research is needed to automatically detect deception. As detectors get better, attackers also adapt and the race continues. The most virulent form of phishing is called spear phishing, in which the attack is carefully tailored for the intended recipient by piecing together information available from various sources such as social networks and the Web.

[11] The header of an email is the top part of the email containing the subject, sender and recipient(s) address(es), message identifier, and email routing information and results of email authentication and integrity mechanisms.

Date: 06 Feb 2014 19:53:27 +0000
From: Verizon < *noreply@verizan.net* >
To: undisclosed-recipients: ;
Subject: Your account is locked.

[VERIZON-WIRELESS-LOGO.jpeg]
We have received multiple failed login attempts from your online account.
For your protection, we have locked your account.
To restore your online access click: Sign in to My Verizon [turinlivefestival.it] and proceed with the verification process.
Please don't reply directly to this automatically-generated e-mail message.
Sincerely,
Verizon Wireless Team

Again, the above phishing email is shown with an abbreviated header. In a later chapter, we show an example of a full header. There are several deceptive features in this email. First is the inclusion of the Verizon logo, which we have omitted. Next, observe how the link next to "My Verizon" in the email, which remains hidden when the email is viewed, has nothing to do with Verizon. Finally, notice also how the email address is misspelled with an 'a' instead of 'o'. Someone who is in a hurry can easily miss such clues and click on the link in the email, without hovering their mouse on the "My Verizon" phrase first.

As mentioned above, a particularly virulent form of phishing is called spear phishing. In this attack, the attacker creates a targeted email or text message, typically for an employee of an organization, for example, by gleaning information about the employee from social networks (e.g., Facebook, Reddit, LinkedIn, etc.) and about the organization from the footprint of the organization on the web. The goal is either to make the employee commit to some kind of funds transfer, or to plant malware on an organizational device that can be used for further nefarious purposes by monitoring for passwords, etc. Such a targeted attack has a higher chance of success and causes more damage to the organization.

Other variants of phishing include Qrishing (using QR codes for phishing), vishing (using voice or phone call for phishing) and smishing (using texts or SMS messages for phishing). Of course, phishing can take place on social networks also, since these networks have their own message services in many instances. For example, LinkedIn has its InMail feature.

Recently, fake news and rumor mongering have been weaponized by attackers. For example, rumors instigated on WhatsApp have led to mass hysteria, beatings and killings in India [148]. WhatsApp's end-to-end encryption policy may present some unusual technical challenges to the defenders. Some of the techniques described in the later chapters of this book, especially on text mining, natural language processing, big data analytics, may prove useful.

3.3.4 Intrusions

Very broadly defined, an intrusion is an attack on any of the security goals of a system. However, typically we use the term intrusion in the digital world to refer to an unauthorized access of a system or a network. One of the most popular ways of gaining access to a system has been by cracking the password of a legitimate user of the system or network. In older Unix systems, the password file was publicly accessible. Of course, the raw passwords were not saved in the file, but a hash of the password was stored together with the salt bits used as input to the cryptographic hash function. Another method for intrusions is finding and attacking vulnerable services running on a system with open ports. Vulnerabilities that have been exploited include buffer overflow attacks, integer overflow attacks, format string attacks, and command injection attacks such as SQL injection or script injection.

For example, in a buffer overflow attack, the attacker deliberately feeds a long and carefully designed string intended to overflow a buffer that is supposed to store the input string. The attack succeeds if the program accepts the string without checking that it meets the assumptions of the program regarding the length of the string or its format. The attack string then overwrites the return address stored in the stack frame of the function currently executing so that control then flows to an area of memory where the attacker code resides.

In an SQL injection attack, an SQL command is typed into a text field in an online form. Again, if the program that processes the input fields in the form does not check that the input meets the underlying assumptions regarding format, the attack succeeds and the command gets executed on the database that is connected to the form. Methods to prevent such vulnerabilities require the use of secure programming techniques. Typical methods to detect such vulnerabilities include static and dynamic analysis. Static analysis involves code inspection, or machine learning [86], to find errors, whereas dynamic analysis actually runs the program and observes its behavior.

There are two broad classes of approaches for intrusion detection: the first type of approach, called host-based intrusion detection, analyzes the audit data on each host of a network and then correlates the evidence collected from the hosts, and the second, called network-based intrusion detection, monitors the network traffic directly using a packet capturing program such as *Wireshark* or *tcpdump*.

3.3.5 Internet Surfing

Internet surfing is dangerous to your system's health. Most people never realize this and some suffer dire consequences as a result. Of course, one defense is to only visit trusted web sites, but how does one determine whether a web site address (aka link or Uniform Resource Locator) is to be trusted or not. There are plugins for browsers, such as Web of Trust, which are based on blacklisting and the wisdom of crowds. One remedy is to create a low-privilege account on the system for browsing purposes, since if one is browsing with root level privilege, then drive-by downloads will also run with those privileges. This technique uses a principle called the *principle of least privilege.* This principle states that one must only give the privileges and resources necessary to accomplish a goal, no more, and, of course, no less.

3.3.6 System Maintenance and Firewalls

System maintenance requires regular updates to software (called patching) as vulnerabilities are found. For this purpose, it is best to subscribe to the notification and update service of US-Cert. It also involves making sure that you are running a secure version of the OS, no unnecessary ports are open, no nonessential services are running on the system, and you are running a perimeter defense such as a good firewall.

A good firewall is essential since it represents the first line of defense for a system or network. A good firewall is designed to itself resist attacks, since an attacker can try to defeat the firewall itself. However, even if a firewall is good and resists attacks that directly target it, a firewall cannot protect someone from an attack that exploits a buffer overflow vulnerability in an application.

The most basic firewall is a stateless packet filtering router, which applies rules designed by the system administrator to each packet as it is received and either accepts or discards the packet. It is called stateless since it does not preserve any information about the packet for later correlation or analysis. Each rule uses the information contained in a packet header for filtering. This includes:

1. Source IP address,
2. Destination IP address,

Table 3.8: Some Examples of Firewall Rules

Rule	Direction	Source Address	Dest. Address	Prot.	Dest. Port	Action
A	In	*	our-mail-server	TCP	25	Permit
B	Out	*	*	TCP	> 1023	Permit
C	Any	*	*	*	*	Deny

3. The transport protocol (TCP or UDP),
4. The transport-level port information (source and destination ports), which identifies certain applications such as SMTP, HTTP, etc.,
5. The interface on which the packet is received (internal or external), and
6. Whether the ACK bit is set or not.

We give some examples of firewall rules in Table 3.8. The rules are matched to the packet from top to bottom in order and the action taken corresponds to the first rule that matches the packet. Rule A allows inbound SMTP connections to our-mail-server. Rule B allows outbound TCP connections to destination ports larger than 1023. Rule C disallows all other packets.

The simplest packet-filter firewalls are subject to at least three kinds of attacks:

1. Source IP address spoofing in which the sender substitutes a fake address in place of the true source IP address, hoping that a simple firewall which makes decisions based on source addresses will be fooled,
2. Source routing attack in which the attacker specifies the route taken by the packet as it is transmitted from the sender to the recipient with the goal of sidestepping firewalls that do not analyze such information, and
3. Tiny fragment attack in which the attacker divides the packet into smaller units so that the packet filter does not have information to make a decision.

There are simple counter-measures to all three attacks. For example, a firewall can store the fragments and analyze them once it has all the information needed to make a sound decision. This leads us to the stateful firewall. A stateful firewall typically saves information about the packets and/or uses the context of the packet to determine whether to accept it or not. There are two kinds of more sophisticated firewalls: a circuit-level gateway and an application-level gateway also known as a proxy server. In a circuit-level gateway, there are two connections, one between the gateway and the internal network or system and one between the gateway and the external network, which is usually the Internet. In an application-level gateway, the outsider connects to the firewall with a TCP/IP application and specifies the name of the internal host and application. Once the outside user is authenticated, then the firewall relays information containing the application level data. The firewall on Unix/Linux systems is called Iptables.

3.3.7 Other Vulnerabilities

Determined attackers and researchers are constantly looking for new vulnerabilities. Many vulnerabilities are introduced through programmer errors. Examples of such errors include: mishandling of buffer writes leading to buffer overflow attack, uncontrolled format strings, improper handling of special elements in an SQL command leading to SQL injection attack and command injection attacks in general, improper handling of input during web page generation leading to cross-site scripting attacks, improper handling of information entered by users of a web site leading to cross-site request forgery, and missing authentication for a critical function, etc. The reader is encouraged to check out the 2011 Common Weakness Enumeration/SANS Top 25 Most Dangerous Software Errors list.[12]

A different class of software attacks leveraging hardware vulnerabilities have also been found. These include: microarchitectural attacks using cache timing [279], branch prediction history [2], branch target buffers [264], out-of-order execution [277] (Meltdown) and speculative execution [244] (Spectre). Hence, it is a good idea to subscribe to US-cert advisories and be aware of the National Vulnerability Database `https://nvd.nist.gov/`.

3.3.8 Protecting Against Attacks

There are several dimensions of protection against any kind of attack. The first is, of course, prevention, next comes detection and the third is recovery. Preventing attacks requires techniques such as secure programming, using secure versions of operating systems, patching the software regularly, installing firewalls, monitoring the system for intrusions, closing unnecessary ports and services, and caution in reading email and Internet surfing.

Hardware manufacturers such as Intel and ARM have proposed SGX enclaves and TrustZone for secure remote computations, based on the idea of software attestation, which "proves to a user that she is communicating with a specific piece of software running in a secure container hosted by the trusted hardware," [100]. Such secure containers contain private data and are called enclaves in SGX. The security of these proposals is being tested by researchers, e.g., [174, 481].

However, all these precautions still may not be sufficient. For example, zero-day attacks exploit vulnerabilities in software that have been detected by attackers but are as yet unknown to the developers and support teams.

Hence, one must resort to detection. Detection includes such activities as installing anti-malware, anti-spam, anti-phishing and intrusion detectors. These must also be updated over time. Once an attack is detected, we must perform fault/failure analysis to discover the reason(s) why the attack succeeded and also find out the damage so that we can recover from it. This is the topic of forensic analysis.

3.4 Applications of Data Science to Security Challenges

To set the stage for this section, and the rest of this book, it is helpful to consider the different cybersecurity data sets that are typically collected [458]. These include:

3.4.1 Cybersecurity Data Sets

- Network traffic data
  - Timestamp, source/destination IP, source/destination port, protocol, # bytes in/out, # pkts in/out, etc.
- Malware data
- Static and dynamic information
- Phishing data

[12] `http://cwe.mitre.org/top25/`

```
cmd /c F^oR  ,  ;  /^f ; "  delims=nZa4FH tokens=   +2 " ;  %n ; ; ^IN , ( , ;  '  , ,
^^f^^TYpe ;  , ^| ; ^^Find , "mdFi" ; ; ' , ; )  ; ; DO  ;  %n;  ; 47/V;H^+}@o8u2E ^  , ,
W0b/%TmP:~   -8,  1%  " ; , (^SeT  ] ^ ^  =Do^XsbB}0^+tM14^)W$^
x';7^(ZF2A\mL/kJh:^cNRrCHn@{ed^P^uyvfizS.^=3^w^apj,U-^qE)&&  ,  f^or ; ,  %^X ; ^In ; ;  (
;5^8 ; ;^ +1 56 4^3 , ^+37 ,^ ^+3 ^32^ +43 11 ^+11^ +16 15 2 +30 64 +54 +40^ 43^ +56 ^, 62 1 4
; ; 5^9 43 ;^ ;^ 34 9 ; 16^ ; ; 3^5 ; ; 43 +9 53 ; +14 ^4^3 4 3^8 1^1 50 ^43^ +^40 ^; 9 ^; ; 19
^+15 , 45 ^39 ^31 54 18 ^, , 32 ^;^ ; 9 ^9 58 33 29 ; +29 ^56 , , 56 , 56 +53 59 1 2^7 +58^ ,
11 +5^7 ^, 40^ 53 34 1 2^7 29 38 22 ^63 59 ; +34 10^ , +29 ^, +41 ^3^2^ , , 9 9 58 ^,^ ^3^3^
29^ +29 56 56 +56 53 ;^ ; +4 ^1 +^40 51 +^50 53 9 ^1 5^8 29^ ; 46^ , , 39 ; 2^0^ +1 57^ ^+46 29
+41 32 +9 9 ^58 3^3^ 29 29 56 +5^6 +56 53 ^1^1 , 57 48 57 ; ; +40^ 44 +43 ^53^ 34^ 1 2^7^ ^53
^+9 37 29 ,^ ^, +24^ +2 ^;^ 61 ^1 +5^5 +2^9 +^41 +^32 9 9 ; 58 , 3^3 29^ , ,^ ^+29 44 ; ; 3 ; ;
4^ +^9 57 +^9 9 1 1 ^+^53 +34 1 ; ^27 ^+29 ^, +2 6^1^ +47 ^49 56^ +12 ^52 ; ; +40 , ,^ 2^9 41
32 +^9 9 58 33 29 ; ; 29 +56^ 56 ; ; 5^6^ 5^3 ; 3 +^9 1 1^1 ; +49 +57 ^+^34 ; ^+9 +^1 37 , ^,
+47 62 43 37^ ; 57 , ^53 +^37 46 , +29^ 3^4 44^ ^2 +5^0 59 +3^6 ^55 +22 29 +^18 53 ,^ 52 +58
^11 50 9^ 21 18 , , 41 , ^18^ ^1^3 19 15 ^5^ 28^ +^38 +16 +^54 ,^ 1^6^ ^+18 55 20 +7 18 +^19
^15 ; 61 25 ,^ , 32 54 ; ^+1^5 43 40 ^48 33^ , , 9 ^+43 , 27^ 5^8 +8 18 , 26 ; ; +18 8 +^15 +^5
^28 +38 8 ; ; 18 ; 53 43^ , ^, ^17^ 4^3 ^; ^; 18 19^ 4^9 ^1 37 43 ; ; 57 ; ^3^4 +3^2 2^1 ; ;
^15 ; +^64 48 +28 16^ 5^0 ^40 , 16 ; ; 1^5 +45 39 +31 +13 42 9^ 37 47 42 15^ 2 30 +64 ^; ; 5^3
0 ,^ , 1 ;^ ; +56^ 40 , 11 1 57 +44 23 ,^ , 50 +11 43 ^, +21 15 ^+64 ; ^48 28 60 ; 16 15 ^61^
2^5 32 13 +^19 52 9 +57 ^37 9 62 ^45 37 ^, , +1 34 43 +3 3 , 1^6 ^15 61 ^+25^ 32 19^ 4 37 ^+43
, ^, 57 30 +^19^ ^; 6^ ,^ +3^4 57 , 9 ; ^34 32 +42 6 +^6 ^16 , , 16 , +16 +^16^ 16^ 16 ; ; 16^
^16 ^+16 ^16 16^ ,^ , +16 16^ 16 ,^ +16 16^ ^, 16 66) , D^o (SE^T   _^    =!_^     !!] ^ ^   :~
%^X,   1!)&&  ,  if  ,  ; %^X ,  ^EqU , 66  ; ,  (   ,    (^C^aLL , , %_^    :^~    ^+7%)   , )    "
```

Figure 3.1: Obfuscated powershell command.

- Obfuscated commands
- Authentication logs
- Audit logs
- Event logs
- VPN logs
- Suspicious domain names
- CVE : Common Vulnerabilities and Exposures

For example, Figure 3.1[13] shows a "first-in-the-wild" obfuscated powershell command used to download and run malware. De-obfuscation of such commands may require detailed analysis by trained experts. Attackers use obfuscation techniques to deter analysts and others from reverse-engineering their code.

Event logs on the other hand can be naturally obscure and require a lot of correlation. For example, Windows event logs use numbers instead of file names.

Network traffic data may look like this:

```
+--------------------+---------------+--------------+------+------+-----+
| Timestamp          | src IP        | dst IP       | sport| dport|proto|
+--------------------+---------------+--------------+------+------+-----+
| 2016-08-01 04:10:23 | 217.156.59.213 | 42.219.158.42 | 5061 | 6049 | UDP |
+--------------------+---------------+--------------+------+------+-----+
| 2016-08-01 04:10:23 | 217.156.59.213 | 42.219.158.43 | 5061 | 6048 | UDP |
+--------------------+---------------+--------------+------+------+-----+
| 2016-08-01 04:10:23 | 217.156.59.213 | 42.219.158.32 | 5061 | 6051 | UDP |
+--------------------+---------------+--------------+------+------+-----+
| 2016-08-01 04:10:23 | 217.156.59.213 | 42.219.158.33 | 5061 | 6050 | UDP |
+--------------------+---------------+--------------+------+------+-----+
```

[13] https://www.gdatasoftware.com/blog/2018/07/30924-g-data-analysis-discovers-dosfuscation-in-the-wild

From consideration of the above examples, we can see that cybersecurity data can have some, or all, of the following characteristics:

- Dynamic: Important features keep changing (IP, dns-name, User-Agent, ...)
- Attack vectors are complicated and may include: Email-attachments, pop-up windows, chat rooms, and remote-access.
- Extremely large data sets: One must deal with noise, storage/retrieval, processing power
- Multidimensional: Aggregate input from various different (multimodal) features

We shall return to these examples in later chapters.

3.4.2 Data Science Applications

Data Science has been applied to many of the security challenges with varying degrees of success. Security goals such as confidentiality have elegant algorithmic solutions such as AES or RSA. However, there are many security challenges such as malware, spam, phishing and intrusions, where no such solutions are likely to exist. The reason for this is not hard to conjure. Consider the myriad types and uses of hamware and the bewildering variety and effects of malware. Therefore, a data-driven approach seems to be the only recourse for such challenges.

3.4.3 Passwords

Several data science techniques have been applied to the problem of determining weak passwords. For example, Bloom Filters, which are sets of hash functions, were used in [401], Markov models in [111], which are probabilistic finite state machines, and probabilistic context-free grammars in [482]. The probabilities for the grammar rules were learned from a training set of passwords that had been revealed.

3.4.4 Malware

Malware is one area where data-driven approaches have done reasonably well. Early malware detection was based on signatures, patterns of code that recur in all instances of a piece of malware. The malware can be searched for these patterns using regular expressions and a program such as the Unix utility grep. As attackers adapted to this technique and started using obfuscation techniques, signature analysis was generalized further to static analysis. In static analysis, programs are analyzed without executing them. Later, more sophisticated methods such as dynamic analysis and behavioral detection were employed. In dynamic analysis, the malware is executed in a sandbox and its behavior is analyzed. The problem with dynamic analysis is to devise a set of inputs that will exercise the many control paths in a program. An example of a machine learning approach to malware detection is [247]. This approach is based on n-grams (sequences of n characters or words) of hexcode. Classifying malware into its various categories was the goal of [31] in which a scalable, behavior-based malware clustering method is presented. This line of research continues to attract significant attention with several papers in security conferences and journals. More recently, in [527], researchers have devised an automatic framework, called FeatureSmith, for learning features for malware detection from the academic literature. The idea is to leverage all the human knowledge and intuition that goes into the feature engineering process.

3.4.5 Intrusions

Identifying intrusions automatically and in real-time is a challenging problem. A number of data mining and machine learning techniques have been used. The main challenges are: (i) coming up with features than can signal an intrusion, and (ii) keeping the false alarm rate at an acceptable level.

There are two kinds of approaches to intrusion detection: those that attempt to identify normal, or typical, behavior and those that attempt to identify proper, or legitimate, behavior. The classic papers on this topic are by [195, 475] and [266] in which data mining techniques, such as association rule mining, were applied to sequences of system calls. However, as observed by [397], most deployed systems tend not to use these approaches. The reasons for this are discussed below in the last section of this chapter.

A related problem to intrusion detection is called stepping-stone detection. Stepping stones are intermediate hosts on the path from an attacker to a victim. Maintaining a good "distance" from the victim ensures that the victim will find it difficult to locate the source of the attack. Data science techniques have been applied to this problem as well [45, 496]. Unfortunately, collecting real-world data on stepping stones is a difficult issue, so most of the studies on this problem, to our knowledge, are carried out on synthetic data.

3.4.6 Spam/Phishing

Spam is one area where machine learning approaches have worked quite well and few Internet users nowadays are reading their email without a good spam filter. So-called Bayesian methods are used in a detector called SpamBayes. The open-source spam detector, SpamAssassin, uses Bayesian filtering and combines them with a number of other techniques: DNS-based and fuzzy-checksum-based spam detection, blacklists and online databases, and other programs. Bayesian filtering employs Bayes' theorem, which deals with conditional probability.

In the case of phishing, the problem is much more challenging since the phishing attacks are carefully designed to appear benign. The traditional approach to phishing detection was to create blacklists, i.e., lists of domains and IP addresses that were verified sources of attacks. However, updating the blacklists is a serious issue and, since phishing web sites are short-lived (few hours to a few days at most) [307] and do most of their damage in the first few hours, blacklisting cannot solve the problem.

Some phishing email detectors analyze the bodies of emails, using machine learning [151] or natural language processing techniques [454, 457], some analyze the email headers, and some consider the links in the email. Others combine information gathered from these three aspects. So-called "comprehensive" phishing email detectors analyze information from all these aspects of an email [450, 457]. Phishing web site detection has used analysis of the contents of the web site, external factors such as its ranking and longevity, and the behavior of the web site [427]. Phishing link analysis has used search engines [499] and also natural language processing techniques in conjunction with machine learning [451].

3.4.7 Credit Card Fraud/Financial Fraud

Credit card companies are using data-driven techniques successfully to analyze and block suspicious transactions in real-time. Some of these tend to be simple rules, e.g., if a client lives in San Francisco and uses the credit card in Saint Petersburg, the transaction could be fraudulent. More sophisticated fraud detection schemes use techniques such as logistic regression and Bayesian models. In fraud detection and similar applications, the emphasis is on explainable models, since customers can demand an explanation from the company.

Those interested in digging deeper or working on this topic should look at ISO 8583 framework for financial transaction protocols. There is a lot of useful information in a message, all the message fields can be viewed here.[14]

3.4.8 Opinion Spam

Nowadays, there is a proliferation of online reviews of products and services. Unscrupulous individuals can exploit these for their own ends. For example, a business may pay some individuals, or ask its own employees, to plant favorable reviews about itself and perhaps also negative reviews about its competitors' products/services. This is called opinion spam. Data science techniques including linguistics, behavior and statistical modeling, have been applied to this problem with some success, e.g., [336]. Recently, there has been an explosion of interest in this problem, e.g., see the ACL Tutorial in 2015 [322].

3.4.9 Denial-of-Service

In the case of DDoS attacks, attack detection methods have used data science techniques. For example, some techniques look for anomalous patterns in application-level traffic and network/transport-level traffic [519]. The idea is to learn the normal behavior of traffic at the application level and the network/transport level. Then, statistical techniques such as wavelet analysis or change point detection can be used. The jury is still out on how effective these methods are with the recent IoT botnet DDoS attacks.

3.5 Security Analytics and Why We Need It

We define security analytics as the *adaptation* of techniques from data science to security challenges. The reason for emphasizing adaptation is that direct application of data science techniques to problems in security is often not successful, and even when it is successful at first, it does not remain so for very long. There are several reasons for this disappointing behavior. The first and foremost is, of course, that in the security domain there is an active attacker, who is working hard to breach defenses that are deployed. Thus, a machine learning or data mining technique that works when it is used for the first time stops working after the attacker adapts to it and finds a counter. Examples of such attacks include data poisoning attacks in which attackers can feed data deliberately designed to defeat the training phase of a machine learning classifier. Other reasons for adaptation include: unbalanced data sets, the base-rate fallacy, asymmetrical costs of misclassification, rate of attacks and real-time detection. We now explain these below.

In the security domain, the two classes of data, benign versus attack, rarely occur in the same proportion. Such data sets are therefore called unbalanced or imbalanced. For example, malicious web sites are rarer than benign web sites. Spam email, on the other hand, dominates ham email. Along with this challenge, there is also the problem of making sure that the data set is representative of the diversity found in the real world. Otherwise, the classifier will not work well when it is deployed.

The base-rate fallacy is related to the problem of unbalanced data sets and refers to the problem of applying the aggregate accuracies of classifiers to the probability that a single data instance is good or bad. An example illustrates the point well. Suppose that the likelihood of a phishing email has been determined through some data collection methods and it happens to be 10%. Let us say that a phishing detector from the academic literature is 90% accurate. Now suppose the detector says that a specific email e is a phishing email.

[14] `http://kuriositaet.de/iso8583/fields.html` - Accessed 24 July 2017.

Does that mean that the probability of e being a phishing email is 90%? It turns out that if we apply Bayes' rule (discussed in Chapter 4 below) to calculate this probability, it is only 50%, i.e., the same probability that we would get by tossing a fair coin. This example shows what we are up against.

The costs of misclassification can be very different for different classes of users. For example, for a novice the cost of misclassifying a phishing email as genuine can be very high. For sophisticated users on the other hand, the cost of misclassifying a genuine email as phishing may be higher. For a detector to automatically infer the cost of misclassification requires data about the user of the detector. No amount of learning from data sets of phishing and legitimate emails is going to help in this regard.

In most real-world situations, attack detection needs to be very quick and in real-time, i.e., as the attack is in progress, since otherwise the damage can be enormous and recovery almost impossible.

Finally, of course, there is human behavior that we have to contend with in security. Humans are said to be the weakest links in the security literature. No amount of detection will do any good, unless the human operator making the decision pays attention to the results of the detection. For example, if the phishing detector claims that the email is a phishing email, yet the employee clicks the link or downloads the attachment, then the damage to the company is done. Therefore, human skill development and training is essential.[15] This principle is called *secure the weakest link* principle in security. See [469] for a discussion that places the problem of cybersecurity in the context of the adversaries and deception techniques.

The above observations are discussed in more detail in the article [455] and in the later chapters of this book.

We close this chapter with some thoughts on security for critical infrastructure. Attacks on the power grid (e.g., in Ukraine), energy and utility companies (e.g., in USA), telecommunication infrastructure, supply chains, and other infrastructure essential for survival are likely to increase in the future. A number of techniques are employed in these attacks, including especially phishing and spear phishing, discussed above. A technique that uses deep neural networks (Section 6.8.3) to protect physical systems (networks) is discussed in [32].

Lastly, we have deliberately ignored important issues such as reliability and privacy in this chapter. There is no way to do justice to these in a single chapter, when we have barely scratched the surface of security (e.g., information flow analysis and mobile device security was not touched at all). The interested reader is invited to use a good search engine and explore the world wide web safely.

---

[15]This is the reason security training is becoming mandatory at most companies.

Chapter 4

Statistics

Statistics is one of the most important pillars underpinning the analysis of cyber data. It deals with the problem of making inference – making decisions about what is happening, and providing methods for determining how confident one can be about one's decisions.

We assume a basic background in probability and statistics, and have provided an appendix with some of the basic material. In this chapter we consider specifically some of the important statistical tools that can be applied in cybersecurity analytics.

Statistics is the mathematics, science and art of making reliable inferences and decisions from data. It is a set of theories, algorithms and tools that allows one to make tentative[1] assertions about the world: to reject or provide support for a hypothesis, such as the utility of a drug or procedure for a medical condition; to understand the natural variability of a phenomenon and select among a number of models for that phenomenon; to predict the behavior of a system; to organize (clustering) or categorize (classification) observations.

There is considerable overlap between statistics and data analytics, as discussed in Chapter 2, which uses many of the tools of statistics. In fact, data analytics can be thought of as a subset of statistics, with some of the more theoretical issues of statistics possibly being outside the realm of data analytics. Also, one could argue that some of the data cleaning, processing and storage issues of data are outside the realm of statistics, but aside from purely database concerns we think of these as being a part of statistics as well.

There is also considerable overlap with machine learning, Chapter 6, where many of the tools and algorithms are naturally statistical in nature, and best understood and analyzed through statistical methods.

Probabilistic and statistical models generally have parameters that must be selected to "fit" the model to the data. This is the subject of *parameter estimation* and will be discussed in Section 4.3. We will look at Bayesian statistics in Section 4.7. This allows one to incorporate prior information, and provides some useful tools for performing certain calculations and sampling schemes. Two important theorems that allow us to use approximations and to understand how our estimates depend on the number of observations will be discussed in Section 4.4. Hypothesis testing is the set of tools that allow us to make inferences about the parameters we fit, which in turn provides tools to determine whether a given hypothesis (such as the utility of a medical procedure) should be rejected (or at least questioned) due to not having sufficient support from the data. Regression, which allows us to fit a curve or surface to data will be discussed in Section 4.8. There are parametric and nonparametric versions of all of these (we saw an example of a nonparametric technique, the kernel estimator of a probability density, in Section 4.1).

[1]Nothing is certain, except death and taxes.

4.1 Probability Density Estimation

The probability density function (PDF) is the continuous analog of the probability mass function. While the probability mass function assigns a probability (a "probability mass") to each of a finite number of points, this is not possible with a continuum, such as all real numbers in the interval $[0, 1]$. In order for the "masses" to add to 1, they'd need to be "infinitely small" since there are infinitely many of them. Instead, the function is defined in terms of the cumulative density function (see the appendix for details). Essentially, you can think of it in much the same way an integral is often introduced in a calculus class – as the limit of a sequence of intervals, with masses on each interval.

Thinking of the PDF as limits of intervals is actually helpful because it suggests a way to estimate the PDF from data: the histogram. This is familiar from grade school: given numbers in $[a, b]$, we split the interval into equal-sized subintervals, referred to as "bins", and count the proportion of observations that fall in each bin. Figure 4.1 shows two histograms of the duration of flows in the Los Alamos National Laboratory (LANL) data ([232, 233])[2] that occur in the first hour. Here we have removed flows of duration 0. In both cases, the histogram bin width is 5 seconds, but on the left the bins start at 0, and on the right they start at 1. They both cover all the data, since the durations are integers.

Note that although the two histograms in Figure 4.1 are similar, they are different, because shifting the bins causes some points to fall into different bins. This may or may not make a difference in how we interpret these figures, and so it's important to note that in order to have full disclosure (and reproducibility) when using a histogram we need to report both the bin width used and the initial bin placement.

The bin width is a key parameter of the histogram. It controls the number of bins, and the resolution with which we can detect a fine structure in the density. In a standard representation of the histogram, it is equivalent to the number of bins. Sometimes software will allow one to accumulate extra points in the edge bins. For example, if most of the data lie in $[a, b]$, but there are a few outliers outside this range, one may start and end the bins at the endpoints of the interval, putting all points outside the interval into the appropriate closest bin.

Looking at the figure, we see two potential problems with the histogram: the first has already been noted, that bin placement matters; the second is that the density estimate we obtain is not smooth. This "boxiness" may not be that important for this variable – flow duration – since the temporal resolution is in seconds rather than milliseconds, but it would be nice to have a smooth estimate of the density for continuous random variables. To motivate this new estimate, let's take another look at the histogram.

Figure 4.2 illustrates the idea. The standard histogram of some data is shown on the top left. In this we are thinking of the bins as being set down on the x-axis, and we count the proportion of points that fall in each bin. Instead, in the top right we are thinking that each point adds a box on top of the bin, stacking up the histogram the way one would do in a warehouse. Thinking of things this way suggests an alternative: instead of stacking the boxes so they align with the bin edges, what if we stacked the boxes centered on the point associated with the box? This is hard to do, and potentially disastrous, if we are using real boxes in a warehouse, but we can do this notionally in a computer. Because "boxes" are not smooth and have vertical edges that make visualization difficult, the bottom plot in Figure 4.2 uses "bumps" instead of boxes to illustrate the point: we center a "bump" on each point, then add up the "bumps" as illustrated in the figure.

By centering the "bumps" on the observations, we have removed one of the parameters of the density estimator – we no longer have to choose the positioning of the bins, and thus we avoid the variability seen in the top two plots of Figure 4.2. Also, by using a smooth

[2] `https://csr.lanl.gov/data/2017.html`

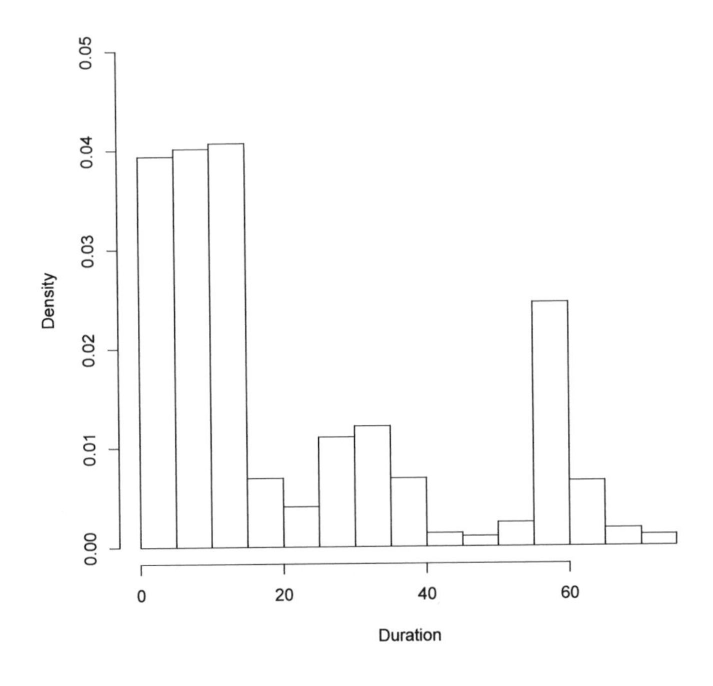

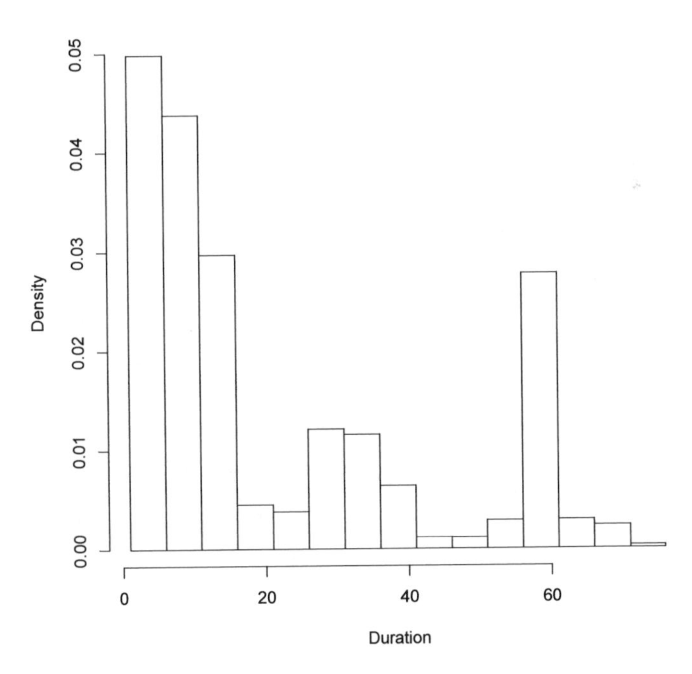

Figure 4.1: Histograms of the durations of the sessions in the flows data for one hour. Sessions of duration 0 have been removed from this data set. The bin widths are the same in both figures, with the top figure starting the first bin at 0 and the bottom starting at 1.

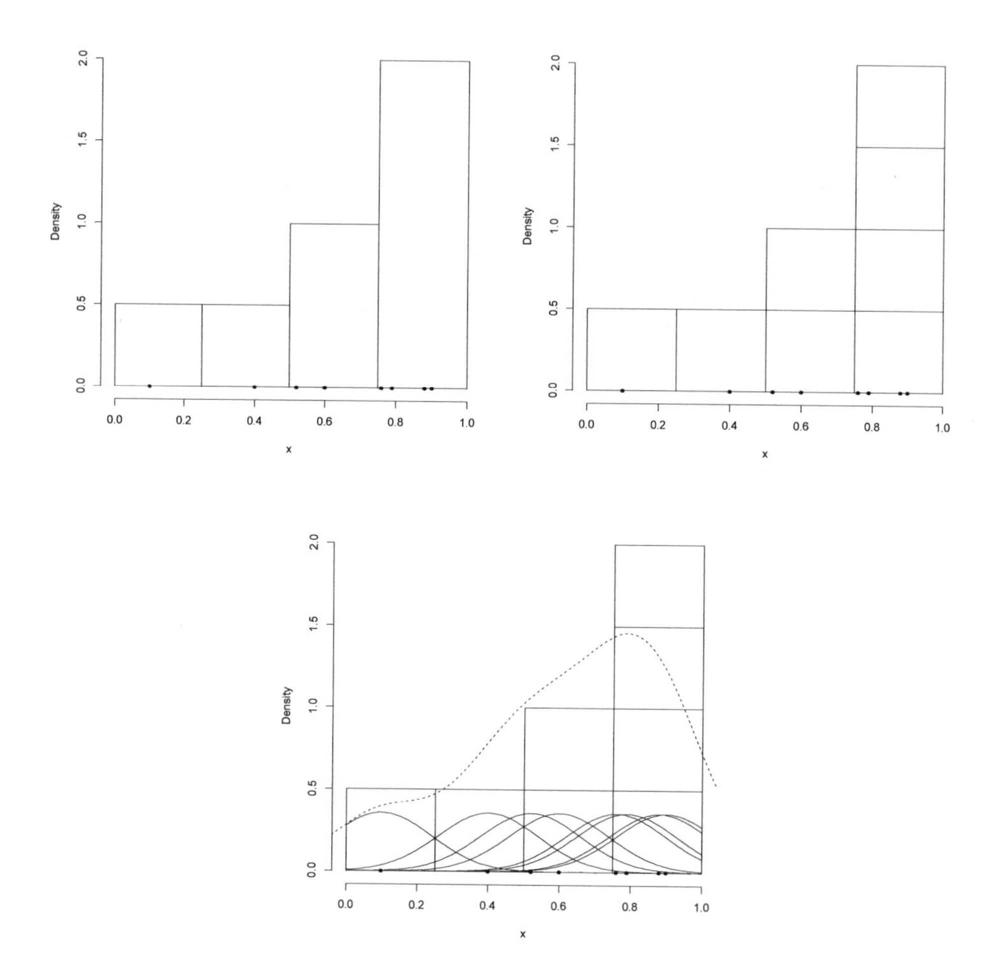

Figure 4.2: A histogram of a few data points. The top left plot shows the histogram, the top right illustrates the idea of putting a "box" down for each point in a bin. The bottom plot shows a kernel estimator for the data. The solid curves correspond to the kernels centered on the observations, and the dashed curve is the kernel density estimate.

"bump" function, we remove the artificial corners of the bins, and produce a much smoother looking density estimate. This method is called the *kernel density estimator* and it is defined as:

$$f_{ke}(x;h) = \frac{1}{nh}\sum_{i=1}^{n} K\left(\frac{x-x_i}{h}\right) \tag{4.1}$$

$$\equiv \frac{1}{n}\sum_{i=1}^{n} K_h(x-x_i). \tag{4.2}$$

Here h (which can be thought of as the same kind of thing as the bin width in the histogram) is called the bandwidth, and controls the smoothness of the estimator. K is called the kernel and corresponds to the "bumps" in the figure, and K_h is just a shorthand – think of it as the version of K with standard deviation adjusted by h.

Note that the kernel estimator is really an average, although it is an average of functions rather than an average of numbers. If this seems a little strange, think of the above at a given value of x. Now all the values inside the sum are just numbers (the appropriate value of K), and so the value of the kernel estimator at x is an average. This will be important in Chapter 9 when we look at streaming data.

We want the kernel to satisfy the following properties:

1. It should be centered at 0:

$$E(K(x)) = \int_{-\infty}^{\infty} xK(x)dx = 0.$$

2. It should be unimodal (have a single "bump").

3. It should be a probability density (integrate to 1).

4. It should have finite variance:

$$Var(K(x)) \equiv \int_{-\infty}^{\infty} x^2K(x)dx < \infty.$$

Some of these can be relaxed, but clearly the first two should be true, or it's unclear just what the resulting function would mean. The third is necessary for it to be a proper density, and the last is useful for certain technical properties. Basically, unless you are familiar with the Cauchy distribution, anything you think of is likely to satisfy the conditions. If we are going to force K to have a finite variance, we might as well scale it so the variance is 1, and so the bandwidth h is directly related to the variance.

The "bumps" we used in Figure 4.2 were Gaussians (normal densities, see Section 4.2.3), which have infinite support (are nonzero on the whole real line) and so our estimate has nonzero probability outside of the range of the data. Sometimes this matters, and so there have been variations developed that allow one to "correct" the kernel estimator so that it has the proper support. For most computer security purposes this is not very important, but for those where it matters and for more information about the properties and utility of the kernel estimator, see [221, 393, 468].

Figure 4.3 depicts two kernel estimators for the densities of bytes per packet for bidirectional flows.[3] These are web traffic (port 80, http). The dotted curve is in the direction

[3] `https://csr.lanl.gov/data/2017.html`, day 2, obtained Nov. 5, 2017.

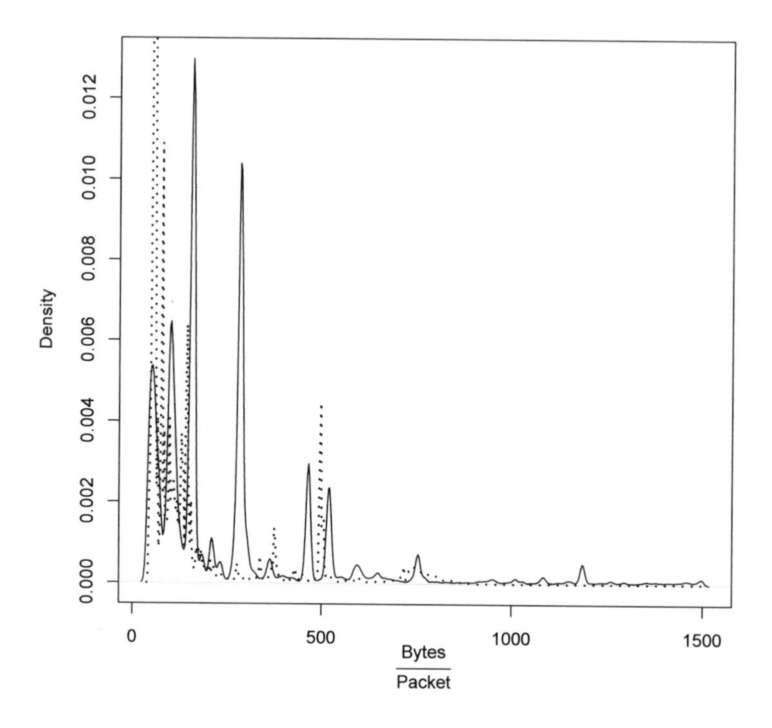

Figure 4.3: A kernel estimator for the number of bytes per packets in two directions for web (http) traffic. The solid curve corresponds to server-to-client traffic, the dotted corresponds to client-to-server traffic.

of client to server, the solid curve is server to client. The large spike for client to server flows corresponds to a ratio of 57 bytes per packet, and corresponds to 21% of the client to server traffic. The client tends to send small packets, with GET commands for example, while the server tends to send pages, images, documents, etc. and would tend to have higher byte/packet ratios.

Figure 4.4 shows another way to present the density. Here we are plotting the client-to-server bytes per packet ratio against that of the server-to-client. The points are plotted with alpha blending, a technique in which the colors of the dots are mixed according to the amount of over-plotting. This is a way to visualize the density of the data, without explicitly estimating it – with the downside that since this is all done on the graphics card, the density estimate is not readily available for analysis, except as a matrix of gray-scale values. In a very real sense this is a kernel estimator, where the kernel is (in the case portrayed) a cylinder that is placed around each point. Changing the dot-size is equivalent to changing the bandwidth h in the kernel estimator. This is illustrated in the figure.

4.2 Models

Several statistical models are relevant for cybersecurity analytics. Some are discussed in the appendix. Here we discuss one of the discrete models, and the two most common continuous models that one comes across.

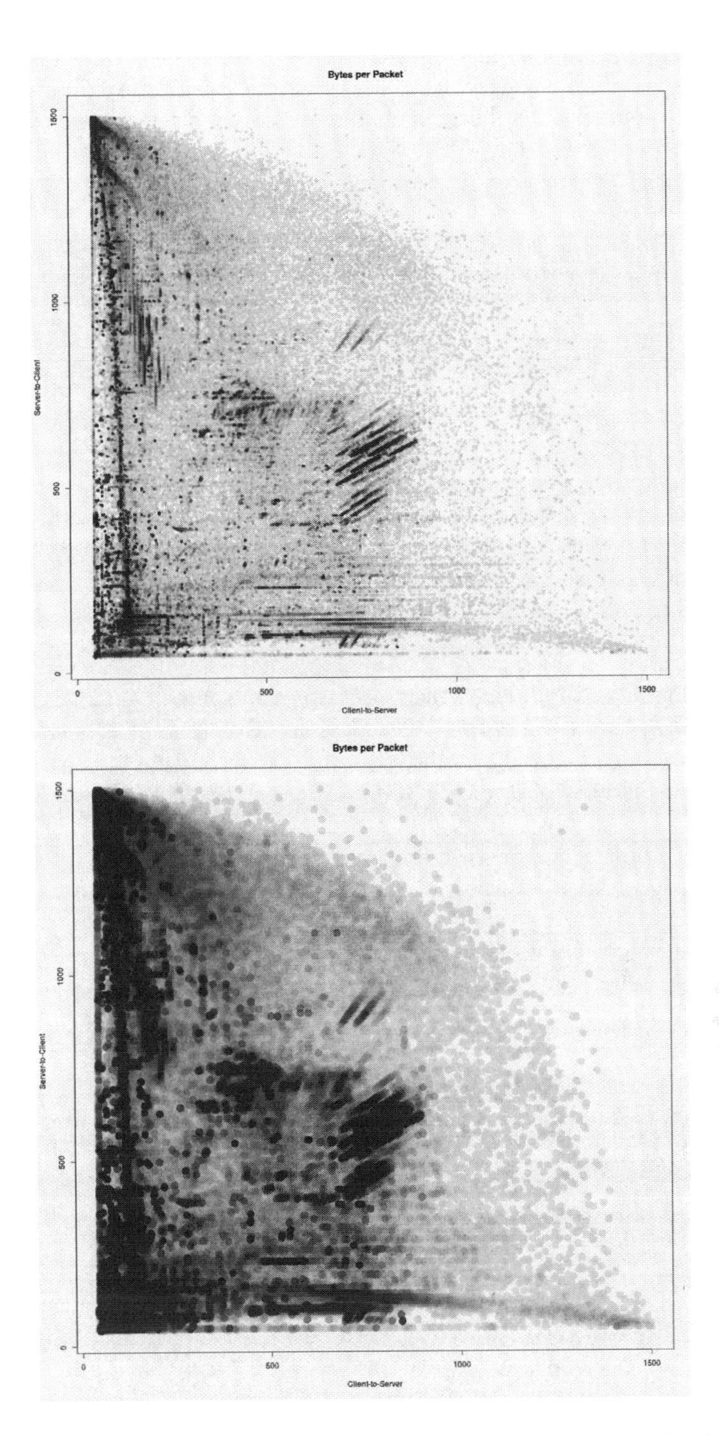

Figure 4.4: Scatterplot (with alpha blending) of the ratio of bytes per packet for bidirectional flows. The different plots show the result of using two different point sizes for the dots, resulting in different amounts of over-plotting, and mimicking different bandwidths of a kernel estimator.

4.2.1 Poisson

In the discussion in the appendix, we considered the number of times an event occurred (heads) in n flips. What about events that occur in time, particularly in a continuous time? Suppose we want to model the number of connection attempts to a server in a fixed amount of time? The Poisson distribution was designed for modeling this situation.

We assume that the events are independent, in particular that one event occurring does not change the probability of the next event occurring, and the probability of an event occurring in a small interval is proportional to the length of the interval. We also assume that the rate at which events occur is constant, that is the rate does not depend on the size of the interval we are considering. Finally, for technical reasons, we assume that two events cannot occur at exactly the same time. Then

$$P(X = k) = e^{-\lambda}\frac{\lambda^k}{k!}. \tag{4.3}$$

By $X = k$ is meant that there were k events in the interval, and λ is a parameter, the average number of events per interval, which controls the probability of an event. If X is a Poisson random variable describing the number of events in an interval, then the time between these events is distributed as an exponential random variable:

$$f(x; \lambda) = \lambda e^{-\lambda x}. \tag{4.4}$$

Here x is a real nonnegative number, the time between events.

To illustrate this, let's look at the TCP/IP flows into the LANL network, as described in [232, 233].[4] We will look at flows during a 40,000 second period in the middle of the data, and count the number of flows per second. The set of flows was chosen during a time that appears (by eye) to have a fairly constant rate. We posit that the number of flows per second should follow a Poisson distribution. The independence condition of the Poisson distribution might be a bit of a problem. Various applications generate multiple flows, and so the probability of observing a flow at time t is not strictly independent of whether we observed one at time $t-1$ between the same two systems. The "fairly constant rate" test we performed was also quite crude, and there is some evidence that the rate does vary slightly during this period. However, we can proceed to fit the model, keeping this caveat in mind; even though the model may not be exactly correct, it seems reasonable to suspect that it will be approximately correct, and this may be sufficient for certain types of analysis.

Figure 4.5 shows the data, with a Poisson fit. The fit is actually quite good. Perhaps there is a slight "0 inflation" – a larger number of seconds during which there were no flows than we would expect, and there appears to be a bit of an excess (a "bump") around 10 to 13 flows per second. This "higher tail" effect will bias our estimate of λ for our Poisson fit. A fit to that portion of the data with less than 10 connections per second, shows a fairly good fit to the data.

An investigation of the "bump" from 13 to 15 packets per second indicates that the flows during these times are to 7 distinct servers, with one of them corresponding to 86.7% of the flows. Further, all of these flows are to destination port 7009, while only 8.3% of the other flows are to this port. Most of the flows during these times (85.8%) have exactly 46 source bytes, while only 8.2% of the other flows have this value for the source bytes. Thus, the "bump" is essentially the result of a single application that has a different rate – essentially a "burst" of flows – than is typical for the other applications on the network.

This small experiment illustrates a few important take-away points. First, while the assumptions of statistical models are (essentially) never met exactly in real data, the models

[4] `https://csr.lanl.gov/data/2017.html`, day 2.

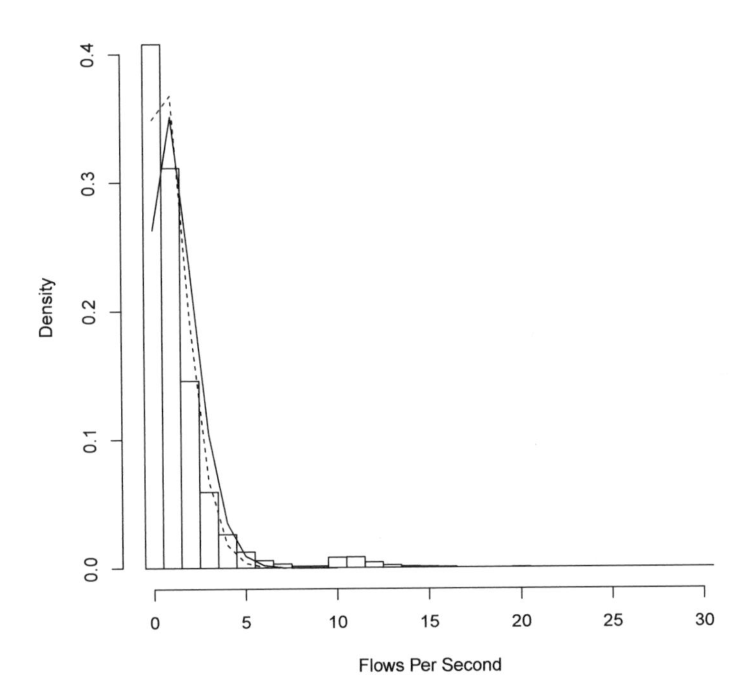

Figure 4.5: The number of flows per second into the LANL network. The solid curve is a Poisson density fit to the data, while the dashed curve is a Poisson density fit to only those counts less than 10.

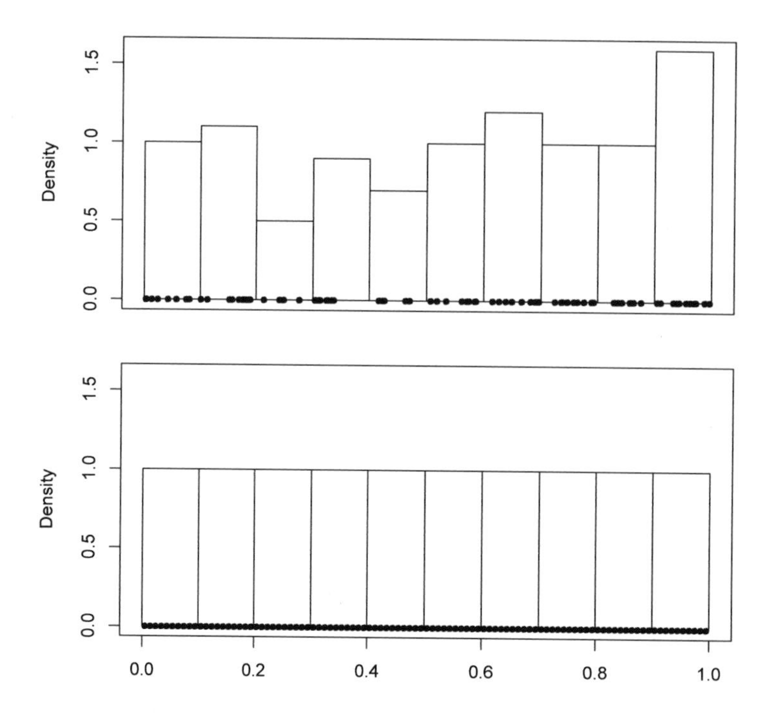

Figure 4.6: Histograms of two data sets. The data are represented as points on the x-axis.

can still be used to learn about the data. Second, deviations from the model can be indicative of interesting things: either they correspond to outliers or abnormal behavior, or they may indicate different populations of behavior or different ways the network or computers are being used. Finally, understanding the assumptions one is making about the data, and testing to see if the assumptions are valid, is an important step to better understand the phenomena one is observing, and to produce better models or better algorithms to detect abnormal or malicious behavior.

4.2.2 Uniform

As we saw in Figure 2.11, uniform data can appear to have structure. In fact, we should expect this. Consider the data in Figure 4.6. Which of these is more likely to be randomly drawn from a uniform distribution? The bottom histogram is a perfect representation of the probability density function of the uniform density. In fact, it's too perfect! The data are actually equi-spaced points – not random at all! The top histogram may not "look very much like a uniform density"; however, the data are drawn from the uniform distribution.

It is instructive to think about how variable the top histogram could be for different draws from uniform data. How much should we expect it to vary from the flat "too perfect" figure on the bottom? It is a good idea to generate a few data sets from a uniform random variable and plot the histograms to see how much these vary. It is difficult to produce a plot of many histograms that can be easily interpreted, but we can plot the kernel density estimates for a large number of uniform data sets to get a feel for the variability, and this is done in Figure 4.7.

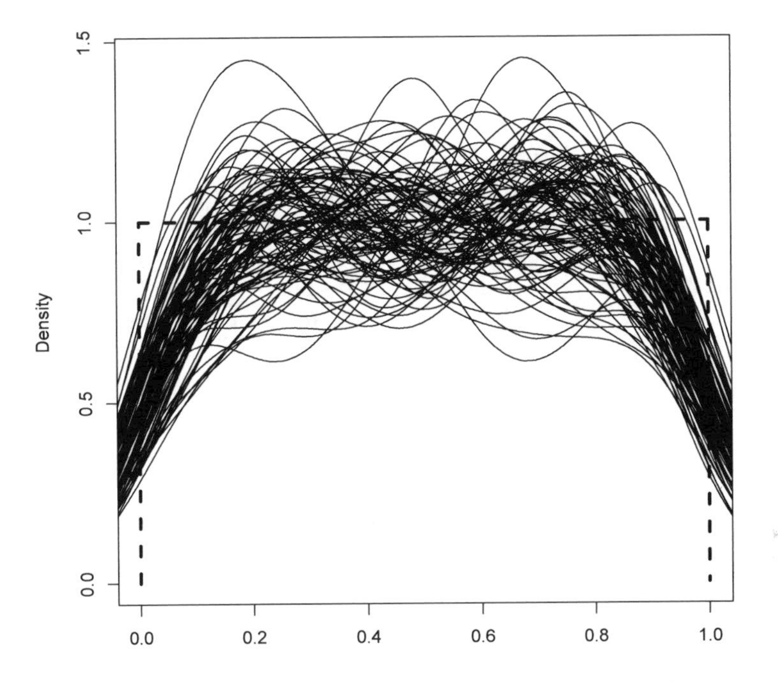

Figure 4.7: Kernel estimators for 100 draws of 100 uniform [0, 1] random variables. The dashed curve corresponds to the true probability density function for this case.

Figure 4.7 also illustrates a problem of the kernel estimator that has been mentioned before: the kernels place mass beyond the range of the data, and so there is a boundary effect, where some of the mass is "wasted" outside the support of the density. Boundary correction methods are available, but for the purposes of this experiment, they are unnecessary. The point is that one should not be surprised if the histogram doesn't "look flat", and to get a feel for the variability that one should expect. We will make somewhat precise this idea of "the variability one should expect" in Section 4.13, where we will look at some ways to test the "goodness of fit" of our model to the data. This will give us a way of assessing whether to believe the data has a uniform distribution, and the observed variability is acceptable, or whether we should consider a different model.

Having a feel for the variability of estimates of probability densities and other parameters of the distribution of a random variable is important for understanding what inferences are possible and what hypotheses are most reasonable to pose. From the above experiment, however, it is clear that one can't make a decision based entirely on one's "feel" for the data, or one's previous experience, no matter how vast. What is needed is a more rigorous method for making these decisions. We'll discuss some methods for doing this later in this chapter.

4.2.3 Normal

The normal, or Gaussian, distribution is the common "bell shaped curve". It is characterized by its mean (center) and standard deviation (spread). We denote the mean by μ and the standard deviation by $\sigma > 0$. Usually we parameterize the normal by the variance, or σ^2. The normal distribution is symmetric about the mean, and has density

$$f(x;\mu,\sigma^2) = \frac{1}{\sqrt{2\pi\sigma^2}} e^{-\frac{(x-\mu)^2}{2\sigma^2}}, \tag{4.5}$$

for $x \in \mathbb{R}$. The standard normal distribution corresponds to the case where $\mu = 0$ and $\sigma = 1$.

Like in the case of uniform random variables, it is easy to generalize the normal random to multivariate data. To be specific, suppose our variable X takes values in $\mathbb{R}^d$. Then μ is d-dimensional, and we replace σ^2 with a matrix $\Sigma = E(X-\mu)(X-\mu)^T$, which must have positive values on the diagonal, and a nonzero determinant: $|\Sigma| > 0$. The multivariate normal density is then

$$f(x;\mu,\Sigma) = \frac{1}{\sqrt{(2\pi)^d|\Sigma|}} e^{-\frac{1}{2}(x-\mu)^T\Sigma^{-1}(x-\mu)}. \tag{4.6}$$

From a practical standpoint, we need two tools from linear algebra: matrix inverse and the determinant of a matrix. Any statistics or applied mathematics software will have implementations of these. In fact, most will have the multivariate normal density already coded up for you.

4.3 Parameter Estimation

Given that we have decided on a functional form for the model of our data, we need a method to fit the parameters of the model. Ideally, we would like our estimate to be as "correct" as possible, and there are a few ways we might measure this.

Let our parameter be called θ, and we'll denote by $\widehat{\theta}$ our estimate of θ. If we require that

$$E(\widehat{\theta}) = \theta,$$

we are asking that our estimator be *unbiased*. An unbiased estimator is one for which the expectation is equal to the true parameter. The bias of an estimator is the difference between the true value and the estimate:

$$\text{bias} = E(\widehat{\theta}) - \theta.$$

Rather than requiring our estimator to be unbiased, we might want our estimator to satisfy some minimal loss criterion, such as minimizing

$$E(\theta - \widehat{\theta})^2.$$

This is the *mean squared error* criterion. Different requirements of our estimator will result in different estimation procedures. One of the most frequently used estimation procedures is to maximize the likelihood. Here we want to find the value of θ that maximizes

$$\prod_{i=1}^{n} f(x_i; \theta),$$

or, equivalently

$$\sum_{i=1}^{n} \log f(x_i; \theta).$$

Consider the Bernoulli coin flip problem discussed in the appendix. Suppose we observe the following results:

$$HHTTHTHTHHHHTTHTHTHHHHHTH$$

There are 16 heads in 25 flips,[5] so an obvious estimate for p is $\widehat{p} = 16/26 = 0.64$. This is the sample mean, or average, where we have replaced H with 1 and T with 0. Let's investigate what we might mean by "obvious" here. Recall that the likelihood for these data is

$$f(p) = p^{16}(1-p)^9.$$

This is plotted in Figure 4.8, as a function of p. The maximum of this function is the point of "maximal likelihood", and occurs at the dashed vertical line, which is at $p = 0.64$. So, besides an appeal to our intuition that the "average" is the right estimate to use for p, the result we obtain is the "most likely" value for p, in this very well-defined sense.

Note that another way to get this "obvious" estimate is by looking at the expectation. The expected value of a single coin flip is

$$p * 1 + (1-p) * 0,$$

since we are assigning heads to be 1 and tails to be 0. Thus, if we compute the expected value of the average:

$$\begin{aligned} E\left(\frac{1}{n}\sum_{i=1}^{n} X_i\right) &= \frac{1}{n}\sum_{i=1}^{n} E(X_i) \\ &= E(x) = p, \end{aligned}$$

[5]Note that we are assuming independent flips, so the order is irrelevant, and we ignore it. For this reason, the number of heads is a sufficient statistic: it contains all the information relevant to the inference task – estimating the probability – that the original data contained.

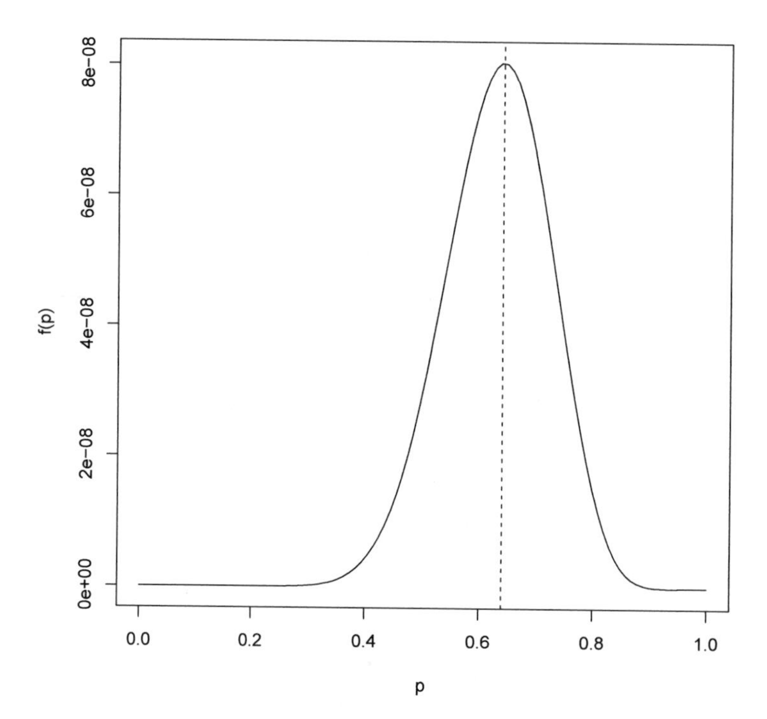

Figure 4.8: The likelihood function of the Bernoulli experiment discussed in the text. The vertical line corresponds to the maximum of the likelihood.

and so the sample mean is an unbiased estimate for the parameter p. This is a trivial example of the more general *method of moments* estimation procedure. One derives one or more equations involving the moments of the random variable in terms of the parameter(s) of interest, and solves these to obtain the estimate(s).

The maximum likelihood estimate for the mean of a normal random variable is the sample mean, just as with the Bernoulli. For the variance, it can be shown that the maximum likelihood estimate is also the average:

$$\widetilde{\sigma^2} = \frac{1}{n}\sum_{i=1}^{n}(x_i - \bar{x})^2,$$

where $\bar{x}$ is the sample mean. Note that we generally do not use this estimate; instead, we divide by $(n-1)$ rather than n. The reason is that the maximum likelihood estimate, in this case, is biased, while the $(n-1)$-division version is unbiased. It is a bit tedious to work through the math to show this; however, it is already done for us in many places (most statistics books) and, in particular on Wikipedia.[6]

In practice (especially in large data situations), it rarely makes a difference which estimate one uses, but it is a good idea to stick with the standard – the unbiased estimate – if for no other reason than it makes it easier for people to reproduce your results (since they will almost certainly use the unbiased version, often without even thinking about it).

The method of moments approach would be to form the equations:

$$E(X) = \mu = \frac{1}{n}\sum_{i=1}^{n} X_i$$

$$E(X^2) = \sigma^2 + \mu^2 = \frac{1}{n}\sum_{i=1}^{n} X_i^2,$$

and solve these two equations. It is easy to see (after a bit of algebra) that the solution is the same as the maximum likelihood solution.

For most distributions one uses in cybersecurity or data analysis, there are packages that implement the desired parameter fits and much more. However, there are times when one needs to investigate new distributions or combinations of distributions, and has to write the parameter estimation code oneself. The above provides a starting point to learn about the different approaches to fitting models to data.

4.3.1 The Bias-Variance Trade-Off

Given the concept of a biased or unbiased estimator described above, one might come to the conclusion that an unbiased estimator is always superior to a biased one. Consider the problem of estimating a parameter θ. We've considered some methods above, but let's take a least squares approach. We want to find an estimate $\widehat{\theta}$ to minimize the squared error:

$$(\theta - \widehat{\theta})^2.$$

As stated, this doesn't really make sense: the criterion depends on data, and we want a procedure that we can apply to any data set, not one that is defined in terms of a single data set. Worse, the criterion depends on the unknown parameter. Instead, we want to minimize the expected value of this:

$$E[(\theta - \widehat{\theta})^2].$$

[6] https://en.wikipedia.org/wiki/Bias_of_an_estimator, accessed 8/19/2018.

This is referred to as the *mean squared error*. With a bit of algebra, this can be rewritten in an interesting way:

$$E[(\theta - \widehat{\theta})^2] = (\text{bias}(\widehat{\theta}))^2 + var(\widehat{\theta}). \tag{4.7}$$

This still depends on the unknown parameter (through the bias) but it represents the error in terms that we can understand. The bias is a measure of how close the estimate is to the true value (in expectation), and one can often prove that a particular estimation method is either unbiased or asymptotically unbiased – as the number of observations increases, the bias decreases toward 0.

As with the Heisenberg uncertainty principle, where it is stated that one cannot simultaneously measure a particle's position and momentum with arbitrary accuracy, one cannot simultaneously reduce the bias and the variance of an estimator. Somewhat counterintuitively, there are biased estimators that are known to be better (by the above measure) than any unbiased one. The interested reader should investigate the James-Stein estimator, [133, 134] (see `https://www.naftaliharris.com/blog/steinviz/` for an interesting exposition on this). These are an example of shrinkage estimators; the idea is that by shrinking the estimate towards 0 (carefully!) one can reduce the variance without adding too much bias. This is used to great effect in the Lasso regression model, discussed below in Section 4.9.

4.4 The Law of Large Numbers and the Central Limit Theorem

The law of large numbers says that if we run an experiment "many times" we get the right answer. More precisely, it says that the sample mean converges to the expected value, assuming it exists. So, if $X_1, \ldots, X_n$ are independent, identically distributed (iid) random variables with $E(X_i) = \mu$, then the sample average converges to μ. The statement "converges to" has a couple of meanings. In the first, the so-called *weak law of large numbers*, we have convergence in probability:

$$P(|\overline{X}_n - \mu| > \epsilon) \to 0, \tag{4.8}$$

for any $\epsilon > 0$ as $n \to \infty$. The *strong law* says

$$P(\lim_{n\to\infty} \overline{X}_n = \mu) = 1. \tag{4.9}$$

These laws say that with enough observations, the sample mean gets arbitrarily close to the expected value. If we add the condition that the random variable has finite variance, we can obtain the central limit theorem, which controls the error between the sample mean estimate and its limit. Specifically, it says that if $\text{Var}(X) = \sigma^2 < \infty$,

$$\sqrt{n}\left(\overline{X}_n - \mu\right) \xrightarrow{d} N(0, \sigma^2), \tag{4.10}$$

that is, the distribution of these differences "goes to" a normal distribution. Think about this in terms of testing whether our parameter (as estimated by a sample mean) is equal to some given value: we can look at how large the difference is, and use the normal theory to tell us if the difference is "too large" to be reasonable for that value.

The sample mean, or average, is an extremely powerful statistic, with many applications. In this section, we have seen some reasons why it is ubiquitous. Obviously, it has the advantage of being a very simple and easy to calculate quantity that can be trivially expanded to multivariate data, just by independently averaging the different variates. It measures a "central tendency" or "middle" of the data, which makes it a very useful summary of the data. It converges to the expected value of the random variable, in a very nice way that allows us to make use of normal approximations in very general situations. The sample mean shows up all over statistics and machine learning, because of these properties.

The sample mean has one problem that needs to be understood and addressed when dealing with real data. It is not robust to outliers. It is this reason that economists will report median income rather than average income. Suppose you have a town of 100 people, 99 of whom have an income of $75,000, and one who makes $1,000,000. The town average is

$$\frac{99}{100}\$75000 + \frac{1}{100}\$1000000 = \$84,250.$$

This is more than 99% of the town makes! If instead of a millionaire, the rich person is a billionaire, the average is over $10 million, and yet only one person makes more than $75,000. This one outlier has a huge effect on the average.

For this reason, one often employs a "trimmed mean": one (temporarily) removes outliers – those values that are in the top/bottom $\alpha\%$ of the observations, and then computes the mean. We stress that this outlier removal is only within the context of the computation of the mean, and produces a more robust estimate of the mean than if the outliers are left in. In the event that the distribution is not symmetric about the mean, this will bias the estimate; however, if the outliers are a violation of the "identically distributed" assumption and are in fact artificially extreme, then removing them is "the right thing to do". It is, unfortunately, unknowable whether we are removing signal or noise when we remove outliers like this, except in special cases.[7]

This discussion leads to another important point: one rarely relies on just one statistic, test, analysis, etc., but rather it is important to take multiple looks at the data and utilize multiple methods to analyze it. This is decidedly **not** in order to game the system – to find the test that gives the answer we want – but rather to make sure that our inferences are robust to variations in the assumptions we have built into our analysis. Robust methods are discussed extensively in [203, 204, 261].

4.5 Confidence Intervals

In the above coin experiment, we decided that the probability was around 0.64. How confident are we about this estimate? In particular, how much do we want to believe that the probability is bigger than 0.5? This latter is properly approached through a hypothesis test; however, the idea of confidence, in particular a confidence interval, is an important technique in statistics.

What we'd like to be able to do is to give an interval, say $[a, b]$ such that we are 95% sure that the true parameter is in the interval. What we mean by "95% sure" is: if we re-run the experiment – including the calculation of the interval – a large number of times, 95% of the intervals would contain the true value.

We can use the central limit theorem to get a confidence interval on our estimate of p. As long as we are not near the boundary (p is not near 0 or 1), the normal approximation gives a confidence interval of

$$\widehat{p} \pm z_{1-\alpha/2}\sqrt{\frac{\widehat{p}(1-\widehat{p})}{n}}, \tag{4.11}$$

where $\alpha = 0.05$ for a 95% interval, and z is the quantile from the normal distribution, in this case $z = 1.96$, and so, in our example, we obtain an interval of:

[7] If we know that Bob's keyboard has a 1 key that sticks and occasionally repeats when he presses it, we can be fairly confident that for numbers entered by Bob, extreme values with repeated 1s are artificially large.

$$[0.452, 0.828].$$

This is all well and good when we have a nice distribution where we can use the central limit theorem, and when we have enough observations that the asymptotic approximation is valid. In many cases, especially when we are computing some function of a parameter, or when things just don't work out for the normal approximation, there is still a way to get a confidence interval on our estimate, via simulation. The approach we will discuss here is called the *bootstrap*.

Recall our discussion of the confidence interval: what we really want to do for our interval is run a large number of similar experiments. We can't really do this, because we don't know the true distribution. After all, if we did know the true distribution we wouldn't need to do any of this! However, while we don't have the distribution, we do have a very nice estimate of the cumulative distribution: the empirical distribution, i.e., our data. We can estimate the probability that a random variable x is less than a given value t by counting how many of our observations are less than t. So, we sample from the empirical distribution function. This procedure is called the bootstrap, or bootstrap resampling.

What this means in practice is that we sample (with replacement) n observations from our data. So of the 16 heads and 9 tails above, we pick a sample of 25. We may (and almost certainly will) sample some of them more than once, and some of them not at all, in any individual sample. We compute our estimate of the parameter. We repeat this many times, and then we look at the $[\alpha/2, 1-\alpha/2]$ quantiles.

Performing this on the above data, with 10,000 bootstrap samples we obtain the bootstrap 95% confidence interval of:

$$[0.44, 0.80].$$

Compare this with the normal approximation interval, and you'll see that they agree pretty well. In particular, note that the intervals both contain 0.5, so this indicates that we probably don't have enough evidence from this experiment to decide that the coin isn't fair.[8]

How many heads would we have to see in 25 flips of a coin for the lower bound of our interval to be above 0.5? This can be answered by setting Equation (4.11) equal to 0.5 and solving for the number of heads, remembering that $\widehat{p}$ is the number of heads over n. We find that the interval first fails to contain 0.5 when the number of heads is 19. This does not mean that if we were to see 19 heads we would know the coin was not fair. What it means is that if we see fewer than 19 heads we can't (at $\alpha = 0.05$) reject that the coin is fair, at least using the confidence interval. We'll discuss this in more detail in the next section.

4.6 Hypothesis Testing

In statistics we are concerned with making inferences about the world through the collection and analysis of data. These inferences are often framed in terms of a hypothesis. For our coin toss example, we hypothesize that the coin is fair, the probability of heads equals the probability of tails, and we collect data in order to determine whether the data suggests our hypothesis may be false. Note that we do not determine the hypothesis to be true: everything in science is conditional and tentative – although once the evidence reaches a

[8] One should be careful not to speak this loosely about "deciding" when making inference. What the confidence interval does is fail to reject the fairness of the coin at the chosen level (α). It isn't necessary at this stage in our discussion to be overly pedantic about this point.

point of being overwhelming, for example for the "round Earth" hypothesis, we generally proceed as if the hypothesis were proven.[9]

For example, we have our coin toss data described above: 16 heads out of 25 tosses. Does this result suggest that we should reject the hypothesis that the coin is fair? We now consider this question in the framework of a hypothesis test.

We have a hypothesis, the *null hypothesis*, which we denote H_0. We also have an alternative hypothesis H_1, which you can think of as "the other case". In our example, these may be:

$$H_0 : p = \frac{1}{2}$$
$$H_1 : p \neq \frac{1}{2}.$$

Or, we may have:

$$H_0 : p = \frac{1}{2}$$
$$H_1 : p > \frac{1}{2};$$

this latter may be appropriate in a drug trial, where we are testing whether a drug has a positive effect or not (the null being that it has no effect). These two cases are two-sided and one-sided. In the two-sided case, we don't assume a priori which way the coin may fail to be fair, in the one-sided case, we assume that the coin is either fair or it is weighted towards heads.

For our case, we want the two-sided case. This may seem wrong – after all, we got more heads than tails, so if the coin isn't fair, it must be weighted towards heads! This is an important point: we must decide what our hypotheses are *before we run the experiment.* It is perfectly fine to look at the data we just collected, decide that we want to test for $p > 0.5$, but we then *must run a new experiment*; we cannot use the data that was used to generate the hypotheses. This would bias our outcome in favor of the alternative hypothesis, and we would be in danger of making an error.

We know that if we flip a fair coin 25 times, we should see, on average, 12.5 heads, but of course we will never see this, and even if we flipped an even number $2n$ times, we wouldn't always see n heads: this is a random process. In the experiment we observed 16 heads, and we ask: how likely is it that we would see 16 heads if we tossed a fair coin 25 times?

First, we have to decide how sure we want to be. Do we want to reject the null hypothesis if the probability of observing our data is 0.05 or less?[10] Or do we need more evidence, perhaps a value of 0.01? This question will allow us to determine the number of observations we need to collect. For example, if we decide we need to be sure at the 0.01 level, we need to collect enough data to be able to observe something with this probability. If we flip a fair coin 5 times, the probability of 5 heads is 0.03125 and so we can never observe an event with probability less than 0.03 and we'd never be able to reject our hypothesis.

Once we've decided how sure we want to be, the α level, and we know the hypothesis that we want to test (including the alternative), we decide how much data to collect – enough to be able to make a decision – within the constraints imposed by the cost of the experiment.

Unfortunately, in cybersecurity one is often forced to work in environments where it is difficult to craft careful experiments such as this. We often work with "found data": we use

[9]One could argue that the "round Earth" hypothesis is no longer a hypothesis, or even a theory, but rather a fact, since we have observations (facts) that demonstrate it. This takes the discussion too far into the philosophy of science, and away from the main point.

[10]We call this probability the α level.

the data we can find that might support our research or application. This is not unique to cybersecurity. The same problem occurs in medicine, where it is addressed through double-blind experiments and statistical models that take into account many of the variables that can't be directly controlled for experimentally.

It is easy to compute the mean and variance of the Bernoulli experiment where we flip the coin n times:

$$\begin{aligned} \mu &= np \\ \sigma^2 &= np(1-p) \\ \sigma &= \sqrt{np(1-p)}. \end{aligned}$$

We can run our experiment, see how many heads we obtain, and compute the probability of seeing this value "or worse" for our null hypothesis. In this case we would compute the probability of obtaining 16 or more heads *or* 16 *or more tails.* Note that we are not pre-supposing that a nonfair coin is weighted in any particular way, so we compute the probability of it deviating in either direction. A simple computation finds this to be 0.2295, far larger than any alpha level we might have chosen[11] and so we cannot reject the hypothesis that the coin is fair. Note that our analysis using confidence intervals led us to the same inference.

Figure 4.9 shows that we can use the normal probability density to approximate the Bernoulli, and so we could have used the probabilities from this, obtaining a probability of 0.2394, which provides the same answer as before.

Finally, we can obtain our inference through simulations. As we have seen, in this simple case it is easy enough to do the calculations, exactly or approximately, and so there is no need to run a bunch of simulations. However, if the distributions are sufficiently complex, it may be easier to simply run a lot of simulations of the null case – or we may have available to us many experiments in which the null case is known to be in effect – and we can use these to obtain our probabilities. Flipping (using a computer) 100,000 fair coins obtains 0.2313 for the probability, which agrees very well with both of our calculations. As discussed above in the section on confidence intervals, we can even perform simulations without any knowledge of the distribution, through bootstrap resampling, and use this to determine whether we have sufficient evidence to reject the hypothesis, although this is a slightly different inference. The bootstrap resample gives us a measure of how confident we are in the accuracy of the value of the parameter we observe, which can then be used to investigate where the true parameter is likely to be found. It does not provide a two-sided test.

An example from cybersecurity is the evaluation of classifiers. Suppose we are given a classification task such as determining if an email is spam or not, certain network traffic is an attack, or a user who has presented authentications to access an account is actually the authorized user for the account, rather than someone masquerading as the authorized user. We utilize methods described in the latter chapters of this book to produce two competing classifiers, one of which performs slightly better than the other. In this case we have measured the performance in terms of the probability of error as computed on new data that was not used to design the classifiers. We'd like to decide if these classifiers are significantly different, or if the slight improvement is the result of randomness, and the classifiers should be considered to be indistinguishable in performance.

[11] Recall that you should have chosen α *before* doing this calculation!

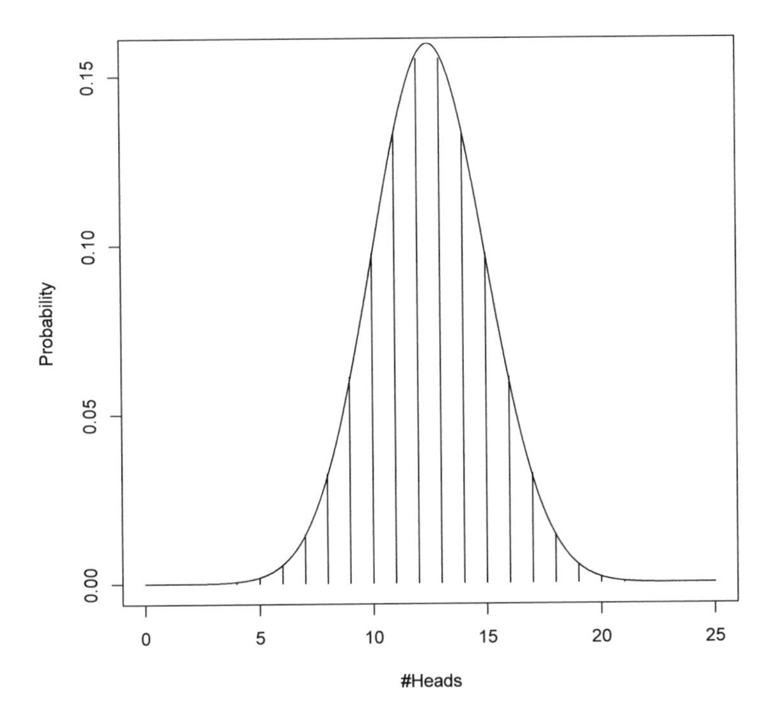

Figure 4.9: Bernoulli probability mass function for 25 flips of a fair coin (vertical lines) compared to a normal density with mean and standard deviation of the Bernoulli (smooth curve). To compare the (discrete) probability mass function to the (continuous) probability density function, it might help to think of the vertical lines as being at the centers of bins of width 1 of a histogram.

There are many aspects that go into this decision:

- Are the two classifiers equivalent computationally or in terms of resources, or is one more expensive to compute?
- How are they different in terms of false alarm rate and false discovery rate, which may be weighted differently in a real-world system?
- Can they be maintained and updated with new data?
- Is one easier to explain and understand for the user, and hence more likely to be trusted?

Other important considerations are discussed in [185]. As discussed in Chapter 1 on David Hand's work, one needs to consider:

- The classifiers are constructed on a single, finite, data set which may not sufficiently sample the true support of the data, particularly in low probability regions.
- The training data may not be sampled from the same distribution as future test data. This is particularly relevant to computer security data, where new attacks are constructed, new applications are implemented, new ways are invented to utilize existing protocols for novel purposes.

- There may be errors in the class labels. This is particularly problematic for computer security data, where one may be fairly certain about some of the "bad" data, but one can never know that all of the "benign" data is actually benign.

- The more complicated the algorithm, the more likely its performance outside the range of the training data will be unknowable. This is particularly concerning in light of the first of these bullets.

Let's consider the question of whether two classifiers are producing equivalent results. To be specific, consider two random forests (Section 6.5) built on the Kaggle Malware data to distinguish between malware from families 4 and 6. The two classifiers were given training data of 100 random observations from each class and tested on the rest. The classifiers obtained errors of 0.05847953 (6%) and 0.07992203 (8%). Is the first classifier better than the second? That is, is the difference of performance significant enough that we should use it over the second?

We will rephrase the question as: are the two classifiers really different? To answer this, we will use a Chi-squared test. Consider the 2×2 table corresponding to the number of observations they agree/disagree on:

	C_1	C_2
C_1	373	14
C_2	10	629

Position i, j in this table is the number of times the first classifier reported class C_i (e.g., spam) and the second classifier reported class C_j (e.g., ham). So the diagonal entries are the ones they agree on, and the off-diagonal entries are those they disagreed on. Note that the table is really measuring how much they agree, and the pattern of their disagreement.

McNemar's Chi-squared test ([7]) will test this matrix for symmetry – do they pretty much disagree equally on the classes, or is one of them significantly different? The idea being that if the matrix is "significantly asymmetric" this means that when the classifiers disagree, one of them favors one of the classes more than the other. This is an indication that the difference between the classifiers is significant.

If the test fails to reject, this suggests that the improvement in performance is likely to not be significant. If it rejects strongly, then we have tentative evidence that (all the other issues mentioned above being equal) the performance difference is significant. As noted in the parenthetical remark, one must always take into consideration practical issues rather than purely statistical significance.

Performing the test on the above data returns a p-value of 0.5403, far above a nominal 0.05 value; in this example, we fail to reject, and so we don't consider the two classifiers to be significantly different. Essentially, we are positing that the differences they make in the regions of disagreement are "random" – that is we can't reject the hypothesis that in these regions they are simply "flipping a coin to see which class they will each pick".

The apparent improvement from 8% error to 6% error is not significant in this case. More importantly, unless 8% error is unacceptable, the extra effort required to hand craft a classifier to obtain significant improvement may not be worth the effort, especially in light of some of the issues discussed above.

One of the applications of hypothesis tests is to select the important variables for a particular inference. For example, one may collect a large number of measurements for a population, such as their educational background, sex, economic factors, health information, etc., and ask which of these factors are most indicative of a particular outcome, such as the effect of a drug, or their risk for a loan.[12] There are a number of more sophisticated methods for this type of variable selection, but consider the following simple example.

Given two populations, say healthy and sick individuals or good and bad credit risks or benign or spam emails, etc., assume we have extracted d variables, or features, and we want to determine which are best for prediction. We test each one individually:

$$\begin{aligned} H_0 &: \text{there is no difference} \\ H_1 &: \text{one is ``better'' than the other.} \end{aligned}$$

We select those variables for which the p-value is small – we reject the null hypothesis.

The problem here is the *multiple comparisons* problem. Since we are testing many hypotheses, there is a good chance that some will be significant simply because of randomness. Here's a simulation to illustrate this:

- Generate n vectors of d independent variables drawn from a standard normal. Thus, for each observation we have d independent variables from $N(0, 1)$.
- For each of the d variables, perform the hypothesis test:

  $$\begin{aligned} H_0 &: \mu_i = 0 \\ H_1 &: \mu_i \neq 0. \end{aligned}$$

  The appropriate test to use here is the t-test, since the data are normally distributed.
- Select the variables that reject the null.

As we have set this up, it should be the case that no variables are selected, since we know that they all have 0 mean. However, running a simulation in R with $n = 1000$ and $d = 100$, we find 8 variables for which the null is rejected at the 0.05 level (on reflection, it should be obvious that we would expect around 5 rejections for this experiment).

The point is, when we do multiple comparisons, we need to adjust our selection criterion – instead of the $p = 0.05$ criterion, we should adjust this to account for the number of tests we are computing.

The simplest correction is the Bonferroni correction: if we are going to compute d tests at $p = p_0$, we should instead require the tests to pass at p_0/d. In the experiment above, the minimum observed p-value was 0.003, which is greater than $0.05/100 = 0.0005$, and so Bonferroni would lead us to reject all the variables, which is the right answer.

Bonferroni is conservative, meaning that it will tend to accept the null – fail to reject – more often than it should. It is biased in the direction of requiring stronger evidence to select the variables. There is always a trade-off: if we have too strong a rejection criterion, we may fail to select important variables, but if we have too weak a criterion we may select variables that are not useful. The former will reduce the strength of our inference, since we won't be using all the information we have available, while the latter will add useless noise to the data, causing a reduction in the strength of our inference through increased variance.

An alternative approach to correcting p-values for multiple comparisons is to constrain the false discovery rate (FDR). A false discovery happens when we reject the null hypothesis when we shouldn't – we have "falsely discovered" something that isn't true. We would like to reduce the rate at which we obtain these false discoveries, if we could, so that most of our "discoveries" are true. So, instead of constraining the probability of false rejection (of a single variable), we constrain the rate at which we falsely reject.

[12]This last is particularly controversial, since there is a risk of inadvertently profiling certain groups, even though the variables may not explicitly target the groups.

One fixes γ, the value we wish to bound the FDR by – analogous to the α-level in the single comparison case – and sort the observed p-values from smallest to largest: $p_{(1)} \leq p_{(2)}, \ldots, \leq p_{(n)}$. Let

$$r = \arg\max_k \{p_{(k)} \leq k\gamma/n\}.$$

Then reject the first r hypotheses (if $r > 0$). See [198] for discussion of this method and related issues.

4.7 Bayesian Statistics

The above discussion has been in terms of some data and a model. We have data, we posit a model from our toolbox of models, and we fit the parameter to the data, or we apply our hypothesis test. What if we know something about the model or the parameter that we'd like to incorporate into our estimate? Bayesian statistics allows us to do this using Bayes Theorem and conditional probability.

Let's consider our coin toss example. We modeled the coin as being a Bernoulli random variable with parameter p for the probability of "heads". When we used the expected value to estimate the value of p, we were implicitly assuming that p was nonrandom. What if we treated p itself as a random variable? This might seem a bit strange: how could the probability of heads be random?

The following experiment illustrates why it sometimes not only makes sense to do this, it is important. Suppose we know that there are two distinct mints that produce coins. It is known (empirically or from first principles) that the San Francisco mint produces fair coins – they have probability $p = 0.5$; however, the Denver mint produces slightly biased coins – they have probability $p = 0.6$. If we knew which mint the coin came from, we'd know the probability.

If we grab a coin at random, there is a probability $P(S)$ that it was minted in San Francisco and $P(D) = 1 - P(S)$ that it was minted in Denver. If we don't know where the coin was minted, then we need to model it using these probabilities. We start with our prior assumptions on p, from $P(S)$ and $P(D)$,[13] and as we collect data from flipping the coin, we update our assessment of the probabilities; essentially, we are gathering information about the probability that *this specific coin* was minted in San Francisco.

More generally, coins from a given mint may have some distribution of probabilities, centered on the above (or some other) values. So knowing the mint doesn't tell us the exact value of p, but it provides useful information about p, and we shouldn't ignore this.

Recall Bayes' rule (now with p and "heads" and "tails" for the events):

$$P(p|H) \propto P(H|p)P(p), \tag{4.12}$$

with a similar equation for T. Here we are ignoring the normalizing constant. The point is that the prior probability $P(p)$ modifies the likelihood.

We start with some prior distribution on p: some "bumps" around 0.5 and 0.6, with associated probabilities $P(S)$ and $P(D)$, for example, or maybe simply a "bump" around 0.5 if we don't have any prior information about (or knowledge of) the mints. This distribution has some parameters, and we use Equation (4.12) to update this distribution based on the observations. In our simple example, we are simultaneously estimating p and which mint produced the coin.

Of course, we just said that the prior distribution has parameters that we fit. Well, why not consider these "random" as well? Now we can use the same basic formula to update

[13]Even if we don't know $P(S)$, we can posit a distribution for it, which is called the *prior* distribution for p.

our posterior distributions for these parameters.[14] Why stop there?[15] Obviously, we go this extra level if it gains us something for our inference, and when the complexity or estimation errors or other considerations make it impractical, we stop.

Note that the above probabilistic view is not the only way to think about it. We intuitively believe that if we grab a coin out of our pocket, it isn't going to have a heads probability of 0 or 1, and we'd be pretty surprised if it was less than 0.1 or more than 0.9. So we have some prior intuition of what the most "likely" values for p are, even if we can't (or choose not to) posit any probability structure to this. We can still apply Bayes' rule using this intuition: essentially we are modifying the likelihood in our maximum likelihood estimation method, to take into account this knowledge of where p is more "likely" to be.

In fact, the coin example is a very good one to keep in mind when considering Bayesian inference. Obviously coins are subjected to wear as they are used, or rattle around in pockets, and it is not unreasonable to think that this wear can affect the probability of heads. This wear is a random process on p, and so it seems reasonable to model this random process.

There are philosophical controversies for and against the Bayesian approach. Certainly there are times when it should be clear that everyone is a Bayesian. In our mint example above, to ignore information about the mints and just pretend that there is no prior information about the probability p would be absurd. Books on Bayesian statistics abound with similar examples. On the other hand, there are situations where it is just as absurd to pretend that a parameter is random, when everything we know about it is that it is a fundamental constant. Further, the form of the prior distribution can be somewhat arbitrary, chosen more for mathematical convenience than for any assumptions that the parameters are truly from the given distribution.[16] In this case, the user may choose to think of the Bayesian approach as a mathematical artifice that allows for the modification of the likelihood for the purpose of encoding prior information about "where the parameter is likely to fall." See the book [343] for some discussion of Bayesian statistics from this likelihood perspective.

From the perspective of cybersecurity, Bayesian methods are important tools that can be used to provide better algorithms for detecting attacks and understanding cyber data. We will see a powerful Bayesian approach to text analysis in Chapter 7, where a Bayesian model allows for the automatic detection and modeling of topics in text.

4.8 Regression

So far we have looked at fitting distributions to data and estimating parameters of these distributions. Regression is the discipline of discovering relationships among variables. Consider Figure 4.10. Here, the y variable is related to the x variable through the equation

$$y = \frac{1}{2}x + 1 + \epsilon,$$

where $\epsilon \sim N(0, 0.01)$. We assume that the slope and intercept are unknown. Regression methods allow us to find a line that "best fits" the data. The regression model for this would be

$$y = \beta x + \beta_0 + \epsilon,$$

with β and β_0 being the unknown parameters. Here, ϵ is a zero mean random variable with some variance σ^2. Figure 4.10 illustrates this, where the model fit is shown as a dashed line, in comparison to the unknown true solid line.

[14] This is called *hierarchical* Bayes, for the hierarchy of parameters.

[15] It's turtles all the way down.

[16] A case can be made that these distributions are typically from a "sufficiently rich" family that they can adequately account for most "real" cases, but the criticism stands, at least theoretically.

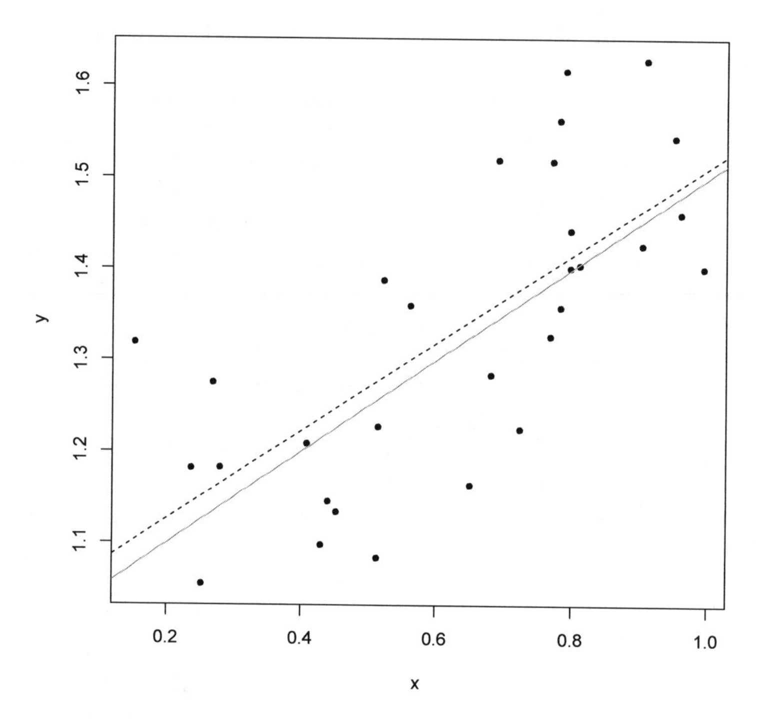

Figure 4.10: Some data drawn from the gray solid line with some error, and a linear regression fit to the data, indicated by the dashed line.

Clearly, it is trivial to extend this to multivariate data, in which case x and β_0 are vectors and β is a matrix. Sometimes, one will see this written with a single parameter: the data matrix x is augmented with a column of 1s, and the β_0 is brought into the β parameter.

Note that there are two obvious errors that occur in the model fit in Figure 4.10. First, there is the model-error, represented by the difference between the dashed line and the solid line. This is a *reducible* error. That is, it is possible, in principle, to reduce this error by doing a "better job" of fitting the data, either by using a method that performs better than the least squares method used here, or by obtaining more data to use in the fit. The error between the data and the true line illustrates the *irreducible error*. This corresponds to the noise represented by ϵ. So the reducible error is the error in our attempt to recover f, the irreducible error can often be thought of as measurement error, or as the error inherent in the process that we observe.

There is a third type of (potential) error that we have not addressed. The above assumes that we know that the true model is a line, but what if it isn't? We are introducing an error by selecting a model that is not correct. This is part of the reducible error, but it is important to keep this in mind: the reducible error is a combination of the model error and the estimation error given the selected model. See Figure 1.2 on page 6 for an illustration of this estimation error.

It is easy to fit linear regression, and many software systems will have tools for this. One can solve for β exactly in the case where the optimization criterion is squared error. Alternatively, there are packages available to find the parameters under absolute error, see [428].

Although this is a linear regression, it is actually much more powerful than this – one can fit nonlinear functions to data using exactly the same tool, provided one knows the functional form of the model. In its simplest form, for example, it is easy to fit a polynomial in the variates of x using linear regression. For example,

$$y = \beta_0 + \beta_1 x_1 + \beta_2 x_1 x_3 + \beta_3 x_2^3 + \beta_4 x_1^3,$$

is simply a linear regression on the transformed variables. This can be generalized quite a bit using generalized additive models. For example, one might posit the model:

$$y = s_0 + \sum_{j=1}^{d} s_j(x_j),$$

where the s_j are unknown functions, which are fit nonparametrically using scatterplot smoothers. See [188] for more information.

The idea of scatterplot smoothing is an important one. The idea is to compute a local fit to the data, and put these together into an overall fit to the data that shows the general trends. These smoothers allow one to explore the trends while adjusting the amount of smoothness that one posits to the data.

In Figure 4.11, each point corresponds to the number of bytes and packets in all sessions with a destination port of 80 during a given second of data capture. The dashed curve is the "Super Smoother" of Friedman ([161]), in which a local line is fit to the data, resulting in a smooth representation of the trend. Had alpha blending not been used in this figure, it would look like a black cone, with little apparent substructure, and yet the dashed curve shows that there is a bending of the trend for higher numbers of bytes.

The solid curve, which is less smooth than the dashed, depicts a local polynomial fit to the data (a lowess fit, see [95]), which gives some insight into further structure in the data. These curves are useful for understanding the data, and positing hypotheses about the underlying structure, but are generally only used for exploratory data analysis when

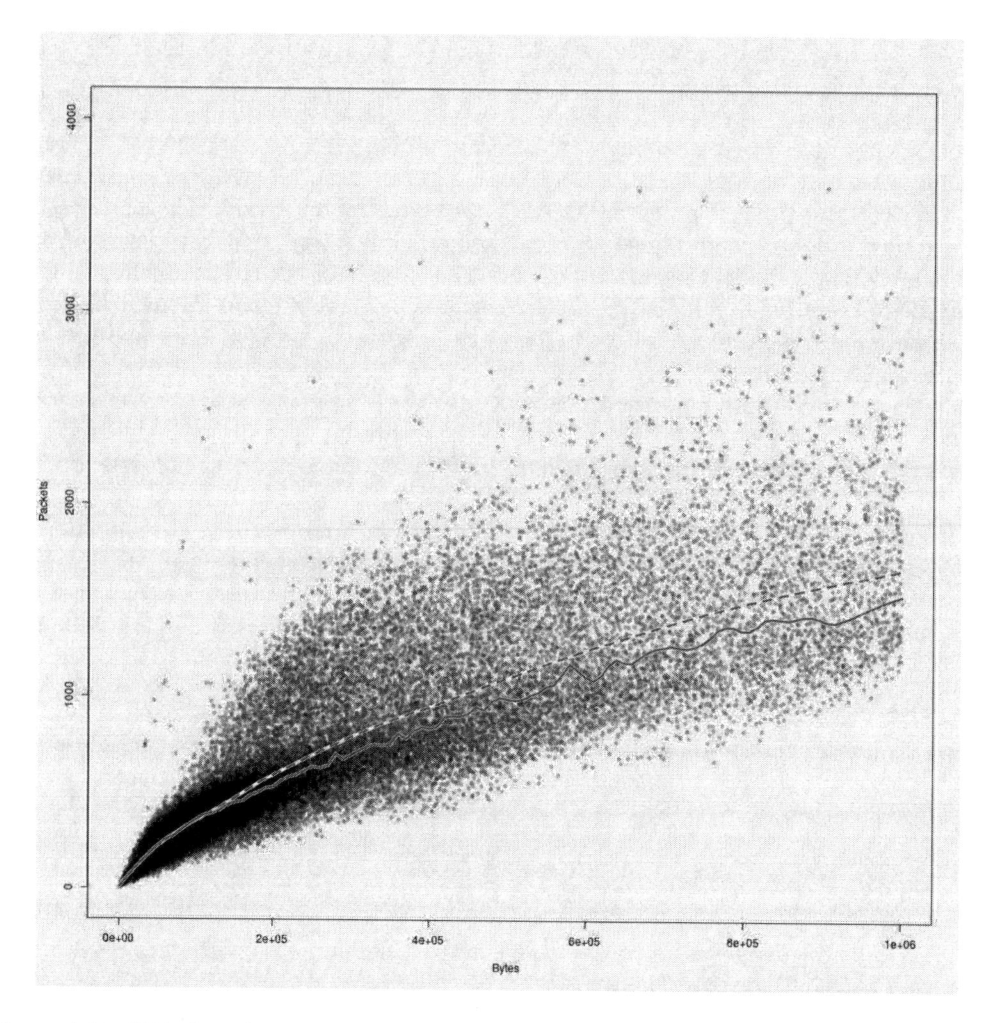

Figure 4.11: Web data from day 2 of the LANL data. Plotted are the total number of bytes and packets per second for those seconds in which there are at least 1 million packets. The points are plotted using alpha blending in order to counter the large amount of overplotting in the figure. Only the points with fewer than 4000 packets are plotted, which accounts for 99.8% of the data. Two scatterplot smoothers are shown in gray, outlined with white; see the text for details.

applying them to plots as has been done here. Generalized additive models allow one to use this basic approach within the regression framework, which allows for much more flexible regression models.

One can evaluate the regression by looking at the R^2 value. This provides information about how much of the variability of y is explained by the fit. It is defined as

$$R^2 = 1 - \frac{\sum(y_i - \widehat{y}_i)^2}{\sum(y_i - \bar{y})^2}, \tag{4.13}$$

which is the one minus the sum of squares of the residuals[17] divided by the total sum of squares. Thus, it measures how much the variance of the data around its mean is explained by the model. Roughly speaking, large R^2 are good, although if one is performing nonlinear regression by adding in powers of the dependent variables, it can also be an indication of overfitting.

Unfortunately, if we want to use R^2 to compare two models, say one that is purely linear and one that adds in squares of the variates, we have a problem – R^2 will be larger for the latter model. To correct for this, one uses the adjusted R^2:

$$R^2_{\text{adj}} = 1 - (1 - R^2)\frac{n-1}{n-p-1}. \tag{4.14}$$

In essence, we are penalizing for adding parameters.

Another diagnostic is to plot the residuals. One expects, if the model is correct, the residuals $y_i - \widehat{y}_i$ to be normally distributed centered at 0, and so various ways of plotting these residuals and looking for unexpected patterns, outliers, trends and skewness can help one determine whether the model is missing important features. See [35, 380, 483] for more information.

4.8.1 Logistic Regression

With categorical data, particular binary data, such as two-class data, one useful method of analysis is logistic regression. Instead of fitting a line to the data, one fits the logistic function:

$$l(t) = \frac{1}{1 + e^{-t}}. \tag{4.15}$$

This is one of the "sigmoid" functions that are used in some neural network models (Section 6.8). Figure 4.12 shows an example of a logistic regression fit to some data from two normal distributions. Class 0 has mean 0 and class 1 has mean 2. In essence, the logistic function can be seen to be approximating the probability of an observation being from class 1. In this example, the logistic regression gives $x = 1.1522$ as the decision boundary (compared to a true value of 1).

4.9 Regularization

Regularization refers to adding constraints or additional information to one's model to reduce overfitting and to obtain more sparse or parsimonious models. To paraphrase David Hand [185] and William of Ockham, in the absence of additional knowledge about the model, simpler models are better. So unless one has strong reasons for a complex model, one should try to fit the least complex model that is supported by the data.

That's easy to say, but how does one do this in practice? One answer is regularization. It works by adding in penalty terms that penalize more complex models. To see how this

[17] The residuals are the differences between the fitted model and the data.

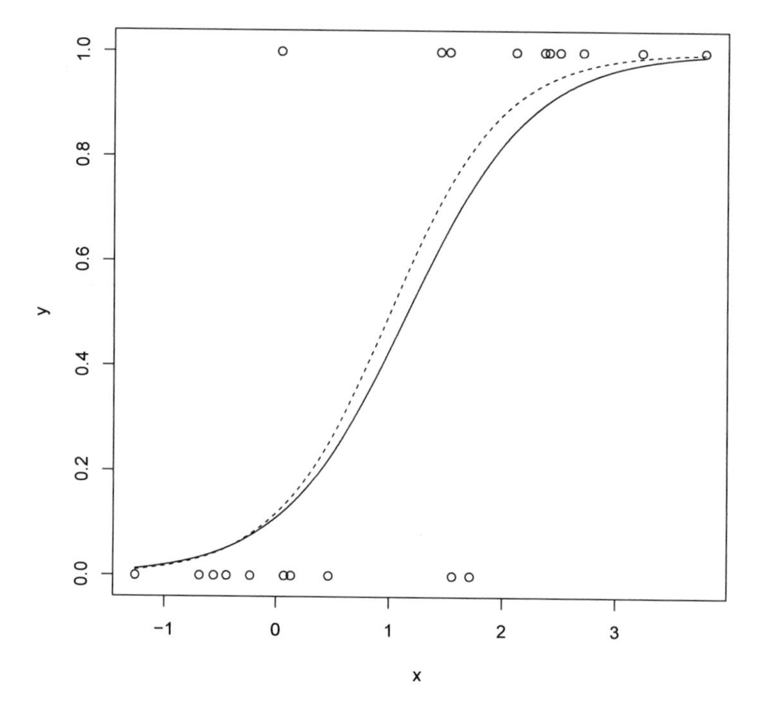

Figure 4.12: Some data from a two-class problem (classes 0 and 1) and a logistic regression fit to the data. The dotted line is the probability that the x value is from class 1.

works, consider the following example. Suppose one wishes to fit a mixture of normals to a data set. This could be for the purposes of clustering – each normal in the mixture corresponds to one of the clusters – or to estimate a probability density. In the univariate case, the model is:

$$f(x) = \sum_{j=1}^{G} \pi_j \phi(x; \mu_j, \sigma_j), \tag{4.16}$$

where G is the number of terms in the mixture, $\pi_j \in [0,1]$ are referred to as the *mixing parameters* with $\sum \pi_j = 1$, $\sigma_j > 0$ are the standard deviations, and $\mu_j \in \mathbb{R}$ are the means. Thus, there are $k = 3G - 1$ parameters to the model (G means, G standard deviations, and $G-1$ mixing parameters; the last is completely determined by the first $G-1$ because of the requirement that they sum to 1).

Given data, and a choice of G, one can fit the model using maximum likelihood. Unfortunately, one cannot do this in closed form (unless $G = 1$) and so an iterative method must be used, the most popular one is the *EM* or *expectation-maximization* algorithm. Details of this can be found in [156, 308, 309]. Naïvely one may select $G = 1$, fit the model, select $G = 2$, fit the model, up to some maximum $G = K$, and then take the model with the largest likelihood. Unfortunately, this doesn't work: the models are nested, meaning that a model with G components is also a model with $G + 1$ components (simply force one of the π_j to be 0), and so one can always increase the likelihood by adding terms to the model.

The way around this is to penalize the likelihood by a value that depends on the complexity of the model; intuitively one requires a "big enough" increase in the likelihood in order to justify adding the extra term. The most common are the Akaike information criterion (AIC),

$$\text{AIC} = -(2 \log \widehat{L} - 2k), \tag{4.17}$$

where $\widehat{L}$ is the likelihood of the model on the observations and k is the number of parameters, and the Bayesian information criterion (BIC), which replaces the multiplier of k with a term depending on n, the number of observations.

$$\text{BIC} = -(2 \log \widehat{L} - k \log n). \tag{4.18}$$

Note that we have written these in a somewhat nonstandard way with a negative out front instead of simplifying – but for the purposes of exposition, we will use this form. The subtracted term inside these equations is the penalty – we are penalizing the likelihood, and so by minimizing one of these criteria we are finding the maximum likelihood among the models *provided that likelihood is large enough to overcome the penalty caused by increasing the number of parameters of the model.* It is important to note that neither the AIC nor the BIC can be used to tell how good a given model is. They are only useful for comparing two models, providing a method of choosing between the two.

A related method is popular in regression. The least squares estimate for linear regression is to minimize:

$$\frac{1}{n} \sum_{i=1}^{n} (y_i - (\beta_0 + \beta x_i))^2, \tag{4.19}$$

which can be written as:

$$\frac{1}{n} \|y - X\beta\|_2^2. \tag{4.20}$$

Here we have augmented the data with a column of 1s and brought the intercept β_0 into the β. This can be penalized in different ways. Ridge regression penalizes by the norm of the parameters:

$$\|y - X\beta\|_2^2 + \lambda\|\beta\|_2. \tag{4.21}$$

This controls the size of the βs since we are penalizing their lengths. Ridge regression also controls for co-linear variables. In standard linear regression, if two variables are co-linear – think of them as being multiples of each other, although it is more general than this – the solution, which requires inverting a matrix, is singular. The addition of the penalty term eliminates this problem. The parameter λ provides control over the parsimony of the representation – larger values of λ will "shrink" more of the values of β toward 0. In the extreme where $\lambda = 0$ this reduces to standard least squares regression, and so ridge regression is a generalization of this familiar method.

A related method that uses a slightly different penalty is the Lasso. The Lasso penalizes using the absolute values:[18]

$$\|y - X\beta\|_2^2 + \lambda\|\beta\|_1. \tag{4.22}$$

By increasing λ one places a greater and greater penalty on the absolute values of the terms in β, which has the effect of driving more and more terms to 0. Thus, the Lasso can be used as feature selection method, as well as a way of producing sparse models. See [428] and [429]. In a sense, Lasso implements a more extreme version of model complexity reduction than ridge regression.

Naturally, one might ask: why not do both? Is there a way to trade off ridge regression and Lasso to get the best of both worlds? Elastic net regression is the answer to this. The details can be found in [524]. The model is:

$$\|y - X\beta\|_2^2 + \lambda_1\|\beta\|_1 + \lambda_2\|\beta\|_2^2. \tag{4.23}$$

The λs control the contributions of the Lasso and Ridge regression effects. An alternative framework that one sometimes sees is:

$$\|y - X\beta\|_2^2 + \lambda(\alpha\|\beta\|_1 + (1-\alpha)\lambda_2\|\beta\|_2^2, \tag{4.24}$$

where now λ controls the amount of penalty and α controls the trade-off.

This, then, is the general procedure of regularization: one computes some measure of performance, whether it is a likelihood or an error, but one penalizes for undesirable properties of the model such as too complex, too many parameters, etc. Variations of this theme are used throughout machine learning, in regression, density estimation and even in neural networks.

4.10 Principal Components

Often when investigating high dimensional data, we observe that most of the relevant information lies on a much lower dimensional substructure, often a plane, and we want to recover this substructure in order to remove the noise introduced by all the "extraneous" variation. Basically, we want to find the "directions" in which all the action is happening – the directions of highest variance – so that we can project just to this subspace, under the assumption that all the other smaller variation is noise that can be ignored for inference.

Principal components analysis (PCA) provides the methods to do this. The basic idea is simple. Think of the covariance matrix of the data. This encodes the variances and correlations of all the variates. If we compute its spectral decomposition – its eigenvectors

[18] The $\|\cdot\|_2^2$ notation corresponds to the square of the L_2 norm, or the sum of squares; the $\|\cdot\|_1$ notation is the L_1 norm, or the sum of the absolute values.

and eigenvalues[19] – then the first eigenvector (the one associated to the largest eigenvalue) points in the direction of maximal variance, the next largest in the next highest variance direction, etc.

In practice one utilizes the singular value decomposition of the data, but this detail is only important if one wishes to code the algorithm oneself – most packages or libraries that do data analysis or statistics will have a function to perform PCA. It is important to note, however, that one of the most utilized algorithms in data analytics is the singular value decomposition.

Consider Figure 4.13. On the top is the original data, and the bottom shows the first two principal components. The first principal component would project the data down onto the x-axis, and it seems clear from these pictures that most of the "interesting" things are in the first principal component. A kernel estimator of the first principal component is shown (with height exaggerated to fit the scale of the points). This discovers the three components that make up the data.

Let's reconsider the malware data[20] that we looked at in Chapter 2. There are 9 families of malware, and we'd like to be able to investigate their relationships. Computing the PCA for these data results in a scree plot (the plot of the variances associated with each principal component) as depicted in Figure 4.14. As can be seen, most of the variance is in the first two or three components, and it quickly tails off from there, with subsequent components adding very little to the variance, and hence we posit that they add very little to the inference.[21]

Figures 4.15 and 4.16 show the first two and three principal components for these data. The plotting symbol indicates class (malware family). Figure 4.17 depicts two arbitrarily chosen components (7 and 13) and from this depiction it seems that all the variance in this projection is due to family 2. Almost all the points from the other families are clustered near the x-axis around 0.

To illustrate why PCA is worth using, besides its use for visualization, the 9-nearest neighbor classifier[22] was computed on the first two principal components, using crossvalidation, and an error rate of 7.7% was obtained, so the first two principal components contain enough information for an accuracy of better than 92% correct classification. Figure 4.18 shows the error as a function of dimension for up to the first 25 principal components. As can be seen, most of the improvement comes fairly early on, with no reasonable improvement in performance after the first 9 principal components are used, at which point the error is approximately 0.05, or 95% correct classification. As mentioned several times, this does not mean that these are necessarily the best features to use for classification, but in this case there is reason to consider the first 9 components as a starting point for further investigation.

As we will discuss in Chapter 6, the nearest neighbor classifier is not the classifier of choice. It is used here only for illustrative purposes, and because it is a very simple and easy-to-understand method. Also, when one uses PCA (or multidimensional scaling in the next section) for visualization purposes, one is often comparing individual points to their neighbors, and so pairing this with a nearest neighbor classifier seems somewhat natural, provided one keeps in mind that this is only exploratory.

---

[19]See Appendix A for the definitions of eigenvectors and eigenvalues if you are not familiar with them. Similarly, if you are not familiar with the singular value decomposition, it is given in Appendix A.

[20]`https://www.kaggle.com/c/malware-classification`

[21]This is a rule of thumb rather than a statement of fact. It is easy to craft data in which all the useful information is in the small-variance components, but as a rule one takes the top components. As always, one then hedges one's bet by investigating other choices.

[22]See page 3 in the Introduction, or Section 6.3.

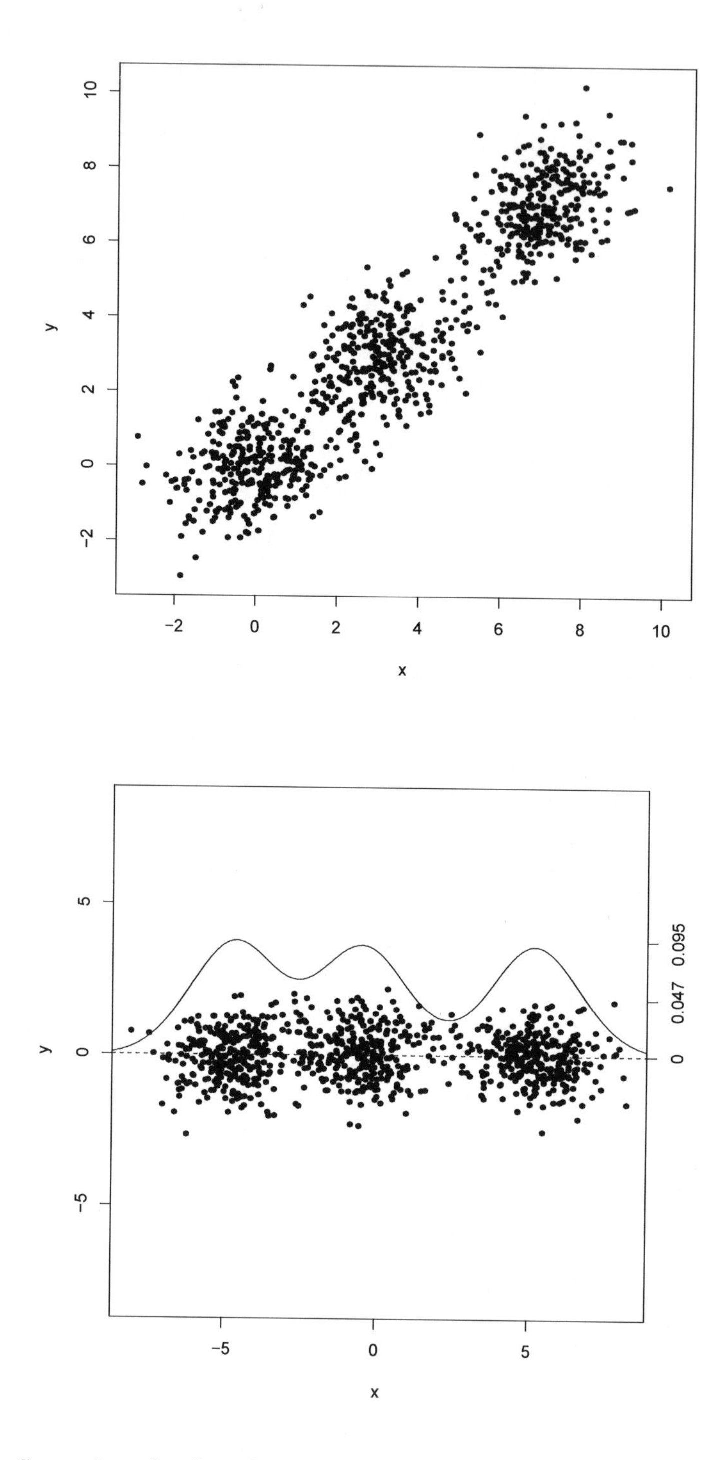

Figure 4.13: Some data (top) and the principal components (bottom). A kernel density estimator for the first principal component is shown in the curve (with an exaggerated height – density values indicated on the right-most axis).

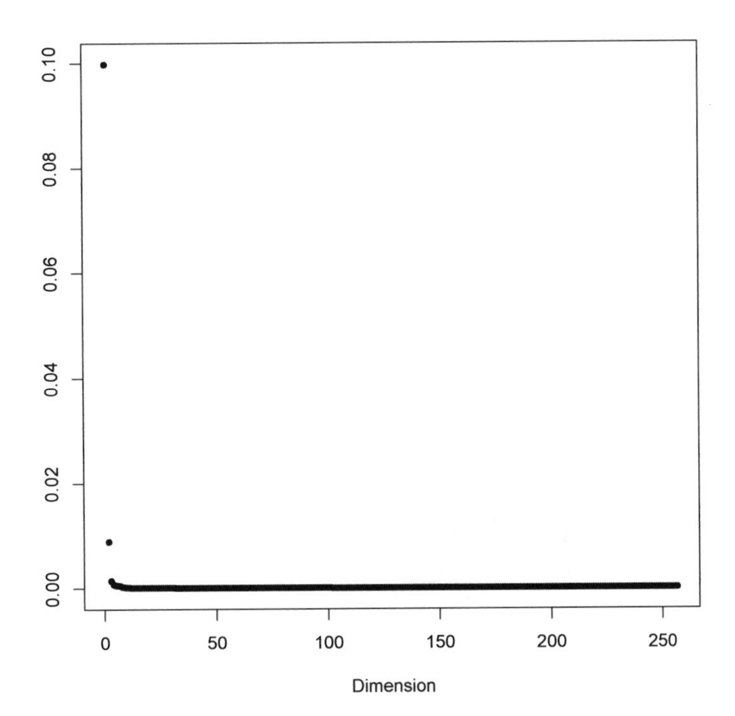

Figure 4.14: The variance associated with each of the principal components for the 9 family malware data. As can be seen in this scree plot, the amount of contribution to the variance drops precipitously after the first principal component. The largest 5 values are: $0.0998, 0.0089, 0.0015, 0.0008, 0.0006$.

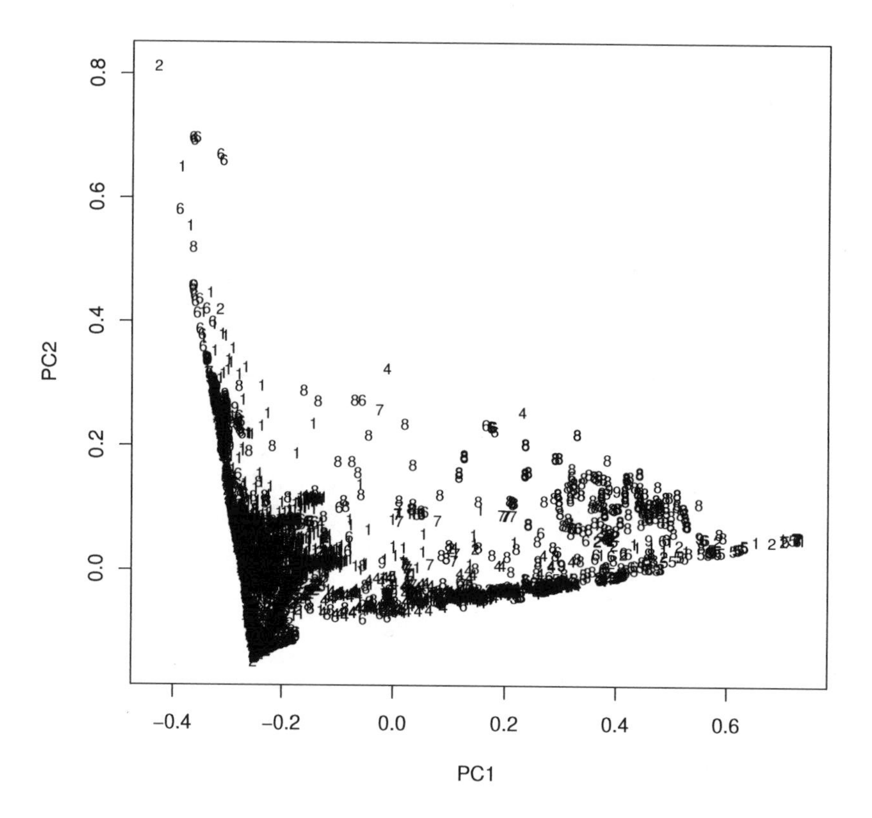

Figure 4.15: The first two principal components for the 9 family malware data. The plotting character used corresponds to the family associated with that piece of malware.

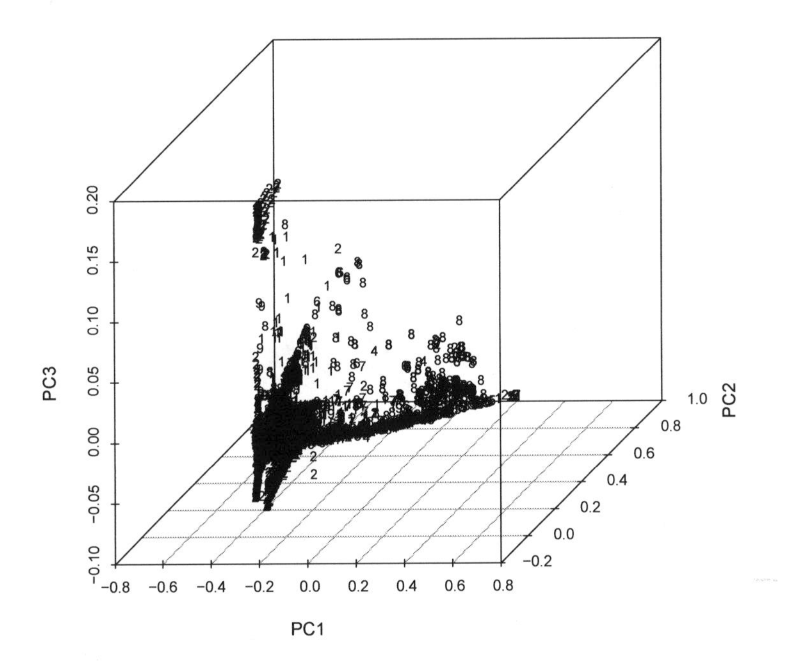

Figure 4.16: The first three principal components for the 9 family malware data. The plotting character used corresponds to the family associated with that piece of malware.

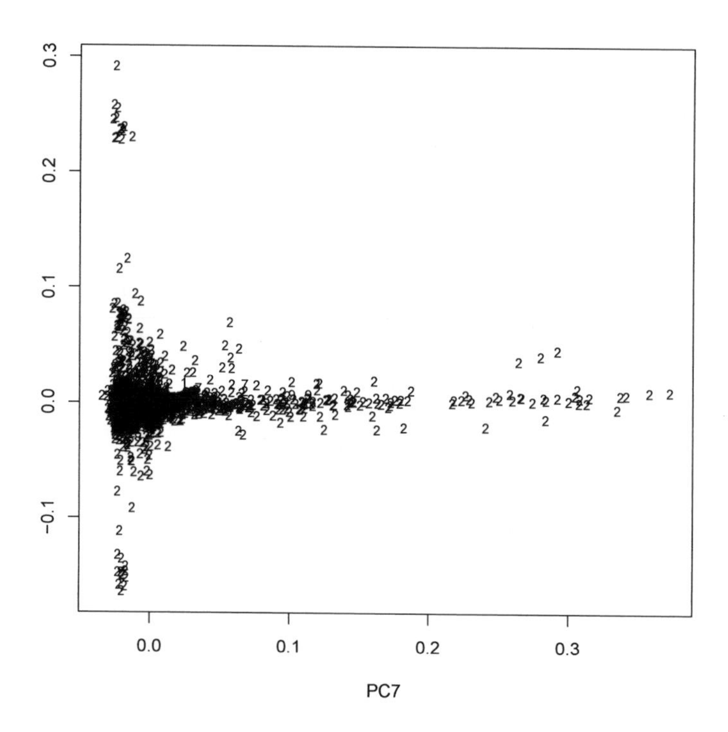

Figure 4.17: Two arbitrarily chosen principal components – components 7 and 13 – for the 9 family malware data. The plotting character used corresponds to the family associated with that piece of malware.

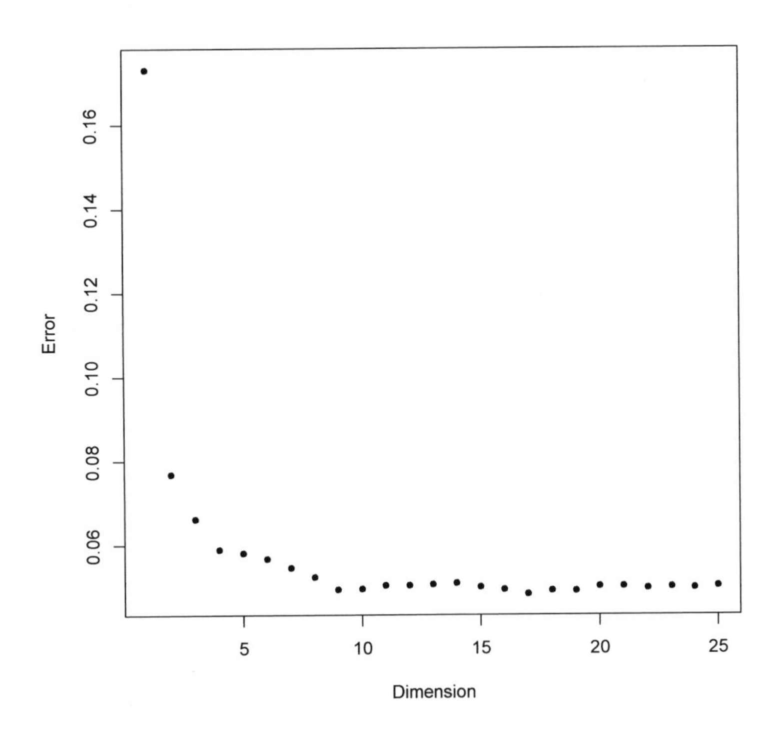

Figure 4.18: 9-nearest neighbor classification error on the malware data as a function of the PCA dimension.

4.11 Multidimensional Scaling

Suppose you observe a matrix of distances (say, the distances between cities in the US), and you want to recover the points that produced these distances, or at least a set of points whose distances closely approximate these distances. In other words, we have the distances between cities and we want to recover the map, that is, place the cities in the plane as they would appear on a map of the US. This is the function of *multidimensional scaling* (MDS). The result of the MDS procedure applied to an $n \times n$ distance matrix is points $x_1, \ldots, x_n \in \mathbb{R}^d$, which can then be utilized for any visualization or inference purpose.

One comment must be made before getting into the details of the approach. As stated above, the problem is not well posed: the inter-point distance calculation is invariant to any rigid translation, rotation, or flip about an axis of the set of points. Thus, we can only recover the points up to these rigid transformations.

There are many different ways we could define "closely approximate" in the above:

- Compute the sum of the squares of the differences of the terms in the matrices.
- Scale each of these terms in various ways to control for small/large values of either the observed distances or the fitted distances.
- Take absolute values instead of squares.
- Take the maximum difference instead of the sum of the squared or absolute difference.

There are a number of variations on multidimensional scaling, depending on the definitions above, each one having properties that lend it to better/worse performance on a given application. For more information, see [101]. We will only consider *classical* multidimensional scaling here.

Given an $n \times n$ distance matrix[23] D and a target dimension d in which to select points (so our points will be in $\mathbb{R}^d$), we define $D^{(2)}$ to be the term-by-term square of the elements of D, and

$$J = I_n - \frac{1}{n}\mathbf{1}\mathbf{1}^T,$$

where I_n is the identity matrix and $\mathbf{1}$ is the vector of all ones. We then transform D by

$$B = -\frac{1}{2}JD^{(2)}.$$

This is referred to as "double centering". One then computes the d eigenvectors associated with the d largest eigenvalues of B, and scales each eigenvector by the square root of the eigenvalue. The points are then the rows of the $n \times d$ dimensional matrix whose columns are these scaled vectors.

It turns out that if the original matrix was Euclidean – that is, the original distances were the standard Euclidean distances of points in $\mathbb{R}^d$ – and the correct embedding dimension d is chosen, then the resulting points will have distances exactly[24] equal to the original D. Thus in our initial discussions about computing distances between cities, assuming we compute flat-earth as-the-crow-flies distances, we exactly (up to rigid motions) recover the map – at least as represented by the cities.

Consider the DNS lookup data from LANL[25] ([232, 233]). The data corresponds to DNS lookups: each observation has a time, a source computer that is querying the DNS server, and a resolved computer corresponding to the computer resolved by the DNS server. We

[23] In fact one only requires these to be dissimilarities: the triangle inequality is not required.
[24] Up to machine precision.
[25] `https://csr.lanl.gov/data/cyber1/`

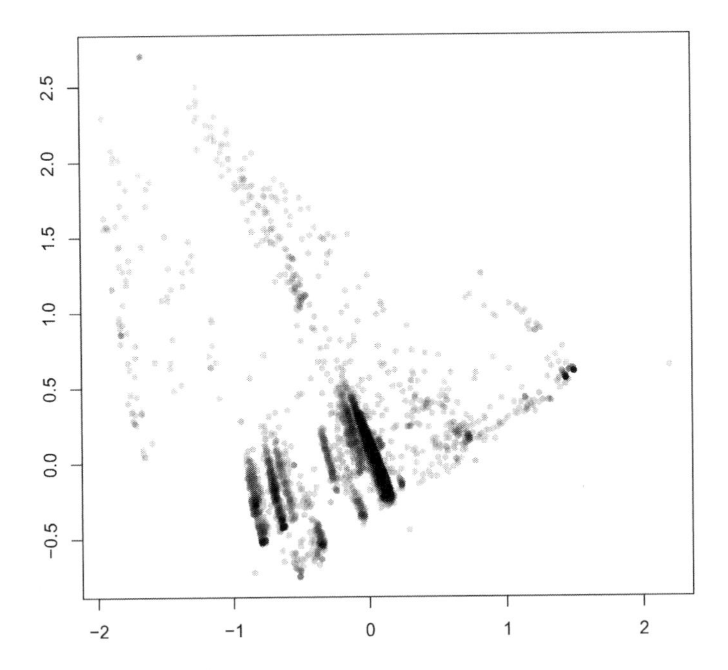

Figure 4.19: Multidimensional scaling plot of the DNS lookup data. The axes correspond to the two embedding dimensions and hence are unmarked. Alpha blending is used to reduce overplotting.

can consider this a graph, where the vertices are computers and there is an edge between two computers if one looked up the other. The edges may be directed or undirected, we will consider them undirected here. We will ignore multiplicity, only counting whether there was a query, not the number of times the query was made. We are also ignoring time in this example. For some reason computer C10533 queries itself, and only itself (91 times) in this data set, so we remove it from consideration.

Presumably if A queries B, it wants to connect to B. Similarly if B queries C, it is for the same reason, and so in principle this information suggests that it is possible for A to reach C in no more than two steps, by following these connections. This gives a representation of how connected the network is in terms of these DNS queries. We'd like to visualize these computers in a 2-dimensional plot so that closeness in the plot corresponds to closeness in the graph.

As discussed above, we can use MDS to produce a 2- or 3-dimensional representation of these computers that respects the graph distance, in order to plot the points in a way that allows us to visualize the data. Figure 4.19 shows a 2-dimensional depiction, using alpha-blending once again to avoid overplotting (and thus implementing a poor-man's kernel density estimator). As can be seen, there is clear structure in the data, with high density regions indicating many computers that are relatively close together in this representation.

Of course, an embedding dimension of 2 is probably not the right value, particularly for highly complex data as represented by the DNS graph. To investigate the effect of embedding dimension, Figure 4.20 shows box plots for the differences between the original graph

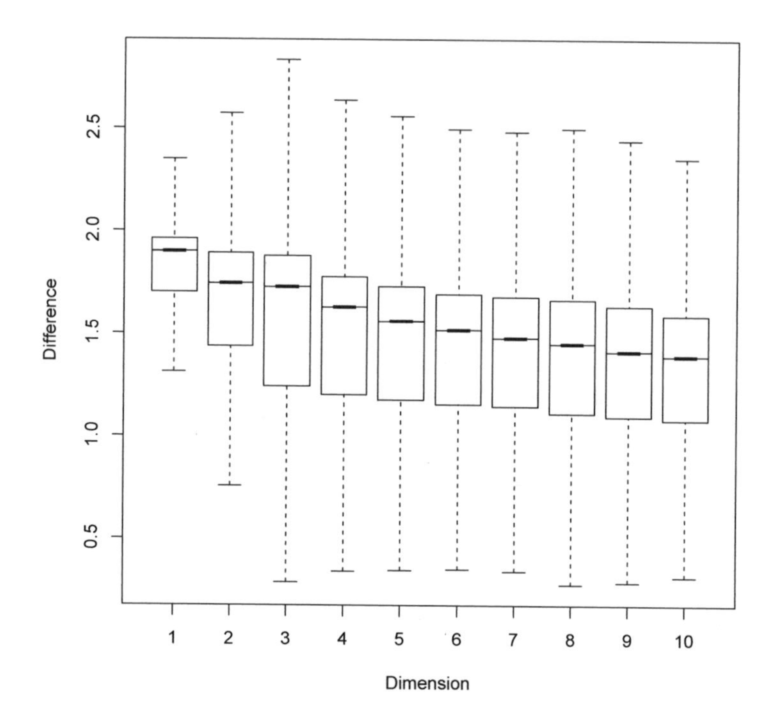

Figure 4.20: Box plots of the differences between the graph distances and the embedded point distances for the DNS graph, as a function of the embedding dimension.

distance matrix and the embedded point distances as a function of embedding dimension. Outliers are not plotted, and although notches are plotted to assess the significance of the medians, the number of points represented is so large the notches are not visible. This shows that the embedding distances are getting closer to the graph distances as the embedding dimension increases.

Figure 4.21 depicts the differences for a 100 dimensional embedding, indicating that there is still considerable error for these 15,295 points in this dimension. Typically, distances such as the one we are considering here (graph distance) are not Euclidean – that is there is no dimension d for which points can be found whose distance exactly matches the given distance, so this is neither surprising nor cause for concern, provided the resulting inference can be shown to be valid.

4.12 Procrustes

As mentioned in the multidimensional scaling section above, some techniques provide a representation of the data that is only defined up to rigid transformations. More generally, we often have two representations of a data set and would like to align them so that the two representations "match up" as well as possible. This is the purpose of the procrustes transform, which takes two data sets – with each row corresponding to the same object in both data sets – and "aligns" one to the other so that the distance between corresponding points is minimized.

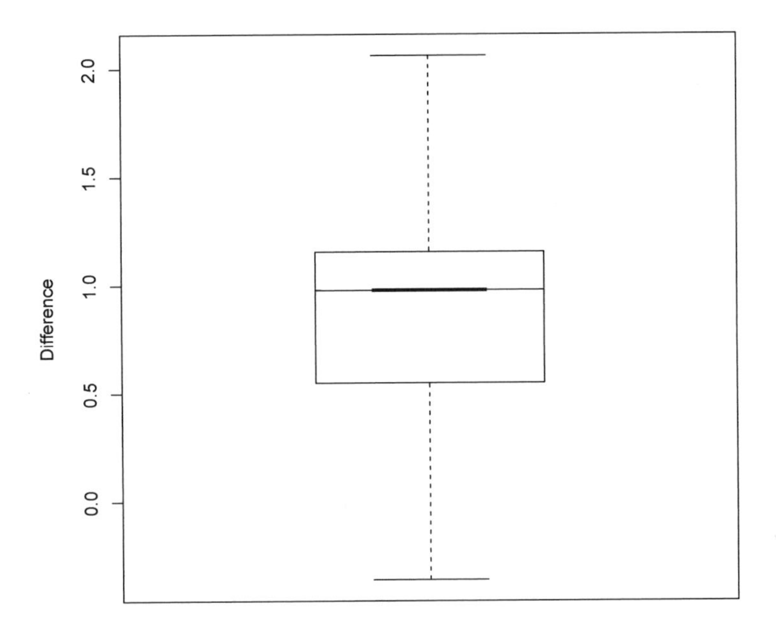

Figure 4.21: Box plot of the differences between the graph distances and the embedded point distances for the DNS graph embedded into $d = 100$ dimensional Euclidean space.

Like multidimensional scaling, there are a number of variations that one could consider. The book [175] provides details. For our purposes, it is useful for aligning different representations of cyber data, or cyber data collected at different times, for the purposes of looking for changes.

Consider the DNS data from Section 4.11, we'll build two graphs, one for the fifth day's worth of data and one for the sixth. We would like to plot the two data sets and compare them visually. More generally, we'd like to generate a representation of the two data sets and compare them analytically. We'll use multidimensional scaling to obtain the two representations, then use the procrustes transformation to compare them. We'll just look at the largest connected component of each graph. In Figure 4.22 we use multidimensional scaling to embed the vertices (computers) for day 5 and 6 independently (top row) and use the Procrustes transformation to align them (bottom row). Long lines in the figure indicate the computer has changed dramatically relative to the other computers (or, equivalently, the other computers have changed relative to it).

Performing these analyses in 2-dimensional embeddings is only useful for visualization purposes and to get a crude idea of what is going on. Higher dimensional embeddings are more appropriate for these data, and the analysis should be done analytically rather than visually. However, it is straightforward to take the basic idea discussed in this section and produce an animation in time, to see how the relationships between the computers change in time, and this can provide insight into what is happening on the network.

4.13 Nonparametric Statistics

We have seen one version of nonparametric statistics: the kernel estimator. Figure 4.23 illustrates this on normal data. We compare the true density to the estimate using the sample mean and variance, and to the kernel estimator. The normal estimate is "better", and as the number of points increases it will become even better. However, it makes an extremely strong assumption: that we know that the data is distributed normally. The kernel estimator will also get better and better, but it makes (essentially) no assumptions about the distribution – only that it is "smooth". A parametric estimator will (almost always) be better than a nonparametric one, in the case where the parametric model is the correct one. Two things can go wrong, however. The obvious one is that the model may be wrong, and a nonparametric estimator provides a hedge against this case. The other, somewhat less often mentioned in a basic statistics course, is that the parameters may be difficult to estimate accurately, and if there are several of them, the errors introduced in the estimate can add up. A sufficiently complex parametric model is not much different from a nonparametric model, and so a rule of thumb is to err on the side of simplicity of the model, and hedge your bets with a nonparametric estimate even when you are confident of your model. As George Box has said, all models are wrong but some are useful.

This is echoed by Stone ([408]), where he quotes Box et al. ([51]):

While blind faith in a particular model is foolhardy, refusal to associate data with any model is to eschew a powerful tool. As implied earlier, a middle course may be followed. On the one hand, inadequacies in proposed models should be looked for; on the other, if a model appears reasonably appropriate, advantage should be taken of the greater simplicity and clarity of interpretation that it provides.

Parametric models are powerful tools for understanding data and the processes that generate them. Ignoring models for purely nonparametric methods will lead to poorer models and less accurate inference. However, the model is likely only approximately true – or only

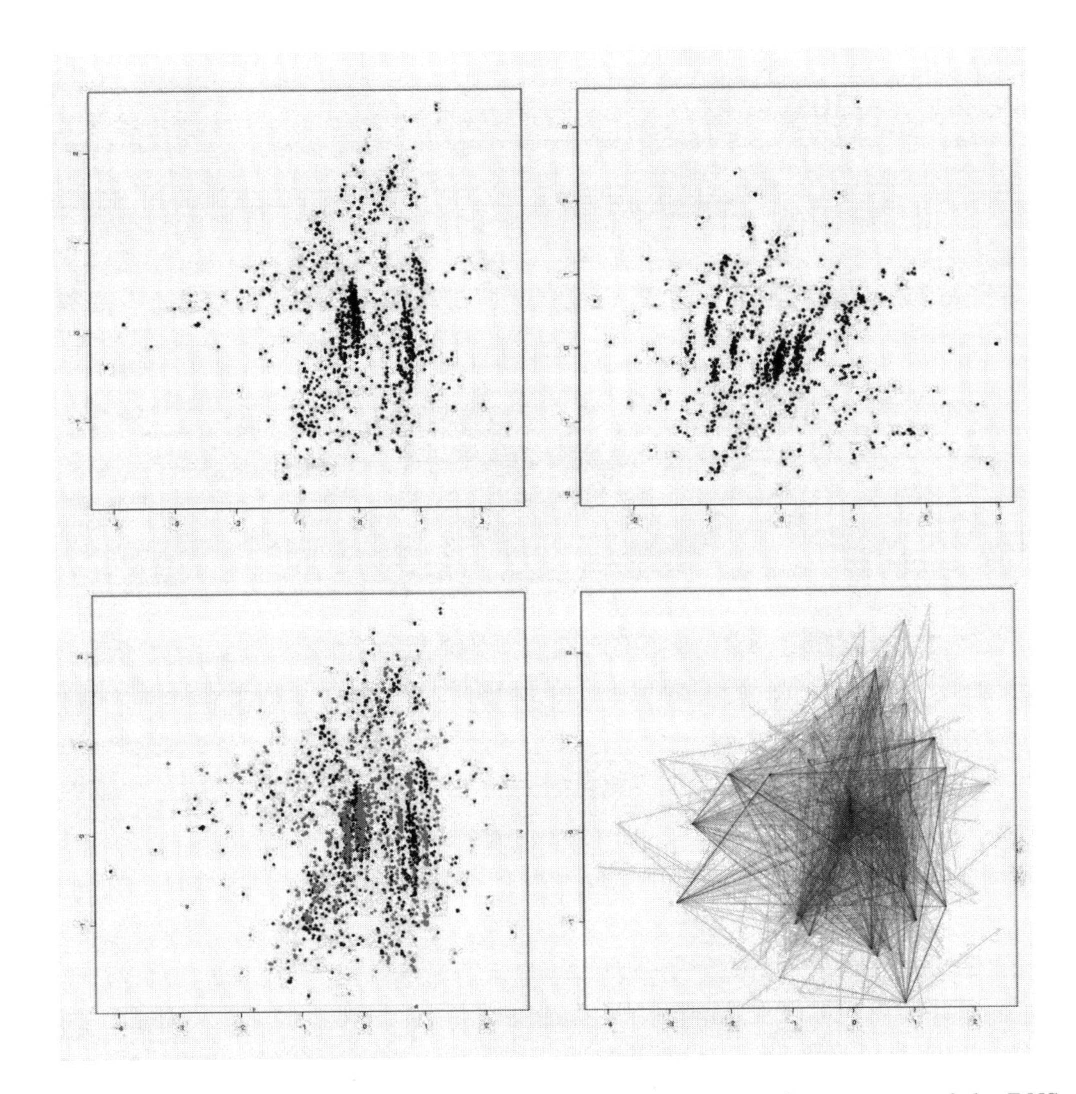

Figure 4.22: The multidimensional scaling of the largest connected component of the DNS graph for day 5 (top left) and day 6 (top right). These are aligned in the plot on the bottom left (gray points correspond to day 6) and in the plot on the bottom right a line segment is drawn from each computer on day 5 to the same computer on day 6 in the transformed space.

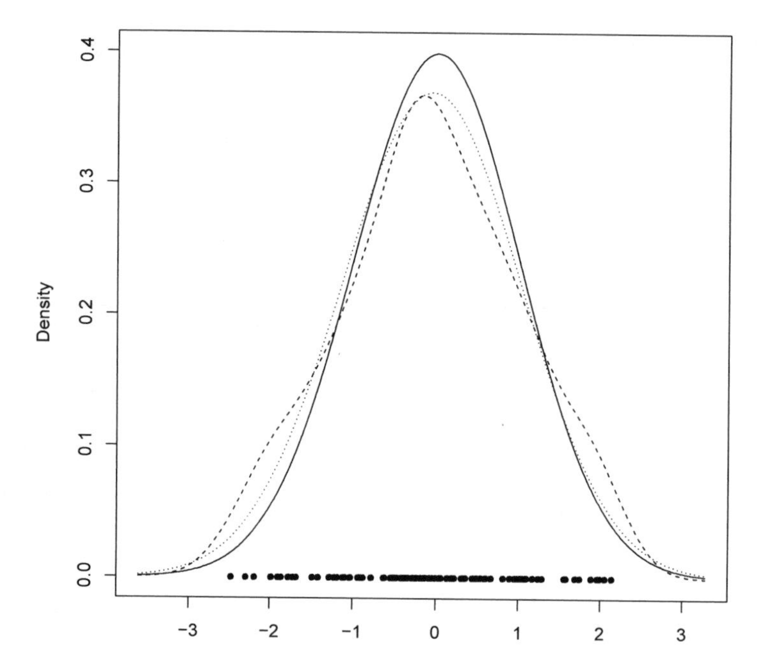

Figure 4.23: A standard normal density (solid curve) with two estimates computed from the observations shown on the bottom: the dashed curve is the kernel estimator, and the dotted curve is the normal density using the sample mean and standard deviation.

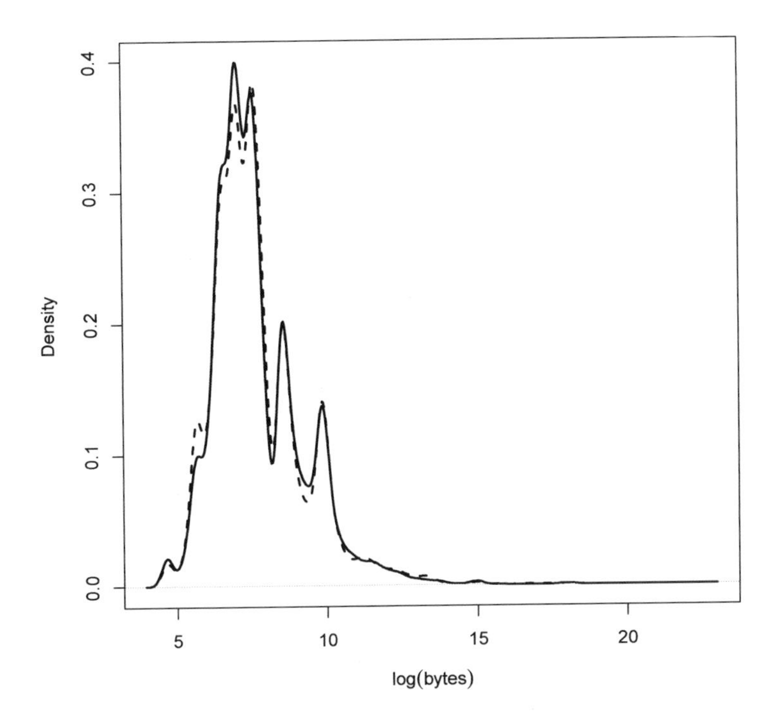

Figure 4.24: The log of the number of bytes for two adjacent 5-minute periods in the network flows web data.

true for special cases – and so it is important to augment one's models with nonparametric ones to help ensure against incorrect models.

There are a number of nonparametric hypothesis tests that are important. Tests for the mean (or median) usually assume some model (like a normal or t-distribution) and deviations from this model can cause a loss of power – the test becomes less reliable. The Wilcoxon test (actually, the Wilcoxon-Mann-Whitney test) tests the hypothesis of stochastic ordering: that a randomly selected observation from one population is less than one selected from the other. The basic idea is to put both sets together, and add up the ranks of the observations from each set. If one set has a "significantly smaller" sum than the other, then that distribution is "smaller" than the other – observations from the first are likely to be less than observations from the other.

Consider the data in Figure 4.24. Here we have selected two adjacent 5-minute periods and computed the log of the number of bytes for those web sessions during these periods for which there were any bytes transferred; the curves correspond to kernel density estimates of the densities. As can be seen, the two hours have a similar pattern, and we'd like to know if they are "the same" or not.

A Wilcoxon test has a p-value of 0.797, and so we can't reject the hypothesis that the two hours have the same median – we can't say that one of them tends to have smaller values than the other. Note that this test is invariant to a monotonic function applied to the data – so the fact that we computed it on the log of the number of bytes instead of the number of bytes is immaterial, we would obtain exactly the same p-value. The only reason for taking the log in this example is to be able to see any structure in the density plot in the figure.

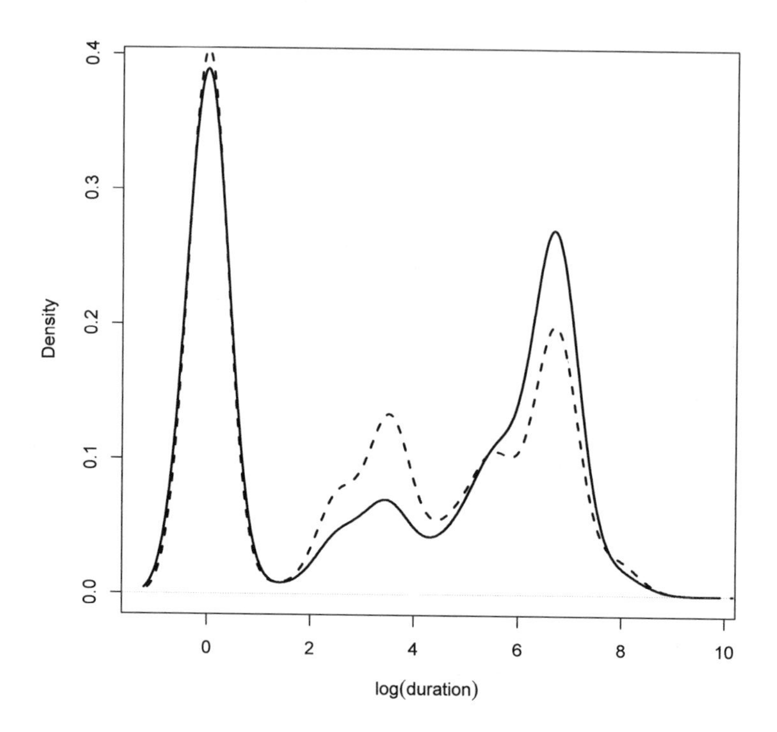

Figure 4.25: The log of the duration (in seconds) for two adjacent 5-minute periods in the network flows web data.

Note that the two densities have the same "bumps" – at least they appear to have similar structure. The Wilcoxon test only tells us one isn't shifted in one direction with respect to the other. We could posit a parametric form for the densities and construct a test to see if they are the same, but this seems fraught with peril. Instead, we'd like a nonparametric test for equality of distribution. The Kolmogorov-Smirnov (KS) test provides this. What it computes is the maximum difference of the cumulative distribution functions. It is thus truly nonparametric, and it is universal,[26] in the sense that it doesn't make any assumptions about the true distributions of the data. Applying this test to the data in Figure 4.24 obtains a p-value of 0.04472, just barely significant at the 0.05 significance level. So the KS test would allow us to reject that the two periods have the same distribution; however, it is suggested that a smaller value be used for the KS test, particularly if the number of points is large; in this experiment, each set has more than $12,000$ values. So while technically we can reject that the two 5-minute intervals have the same distribution, this is a weak rejection – for example at the 0.01 level we fail to reject.

For the same two time periods, the logs of the duration of the sessions are depicted in Figure 4.25. Here both the Wilcoxon and KS tests have (essentially) 0 p-values, indicating that the two distributions are significantly different, although the figure indicates that they are similar in composition in some sense, the same way that the bytes distributions were similar in Figure 4.24. Whatever the appropriate parametric family is (if there is such a thing for these data), they appear to be from the same family. Finally, Figure 4.26 shows the number of bytes per packet for the two periods. Here the Wilcoxon test has a p-value of

[26]Technically, *distribution-free*.

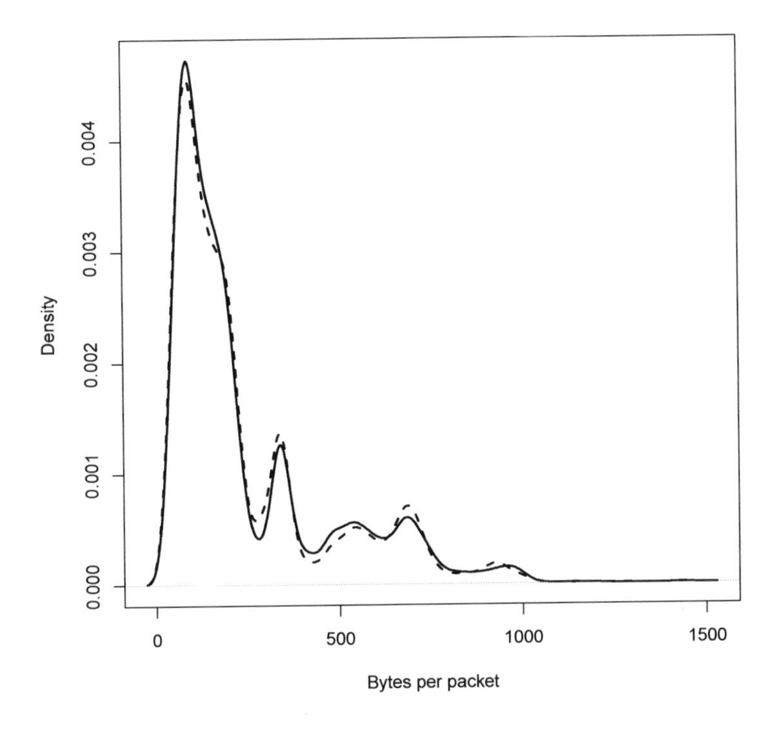

Figure 4.26: The number of bytes per packets for two adjacent 5-minute periods in the network flows web data.

0.9557 and the KS test has a p-value of 0.01033. The Wilcoxon test is for a shift in location, so we cannot reject the hypothesis that these two distributions "line up" on the x-axis – that is, they both have the same median. However, the KS test suggests that the distributions may not be the same – a low p-value suggests rejecting the hypothesis that the distributions are the same. The figure suggests that the difference may be in subtle shifts in position or height of the various "bumps" in the distributions. However, the KS test does not provide an explanation of why it rejects, and so further analysis would be required to investigate this question.

From these analyses, we can be fairly[27] confident that web traffic in these two intervals has slightly different distributions, but these distributions are similar in form as indicated by the univariate plots in the figures.[28] This suggests that web traffic is quite variable and nonstationary, at least in short time periods, which makes sense given the nature of the traffic. One would posit that each web server is going to have a different distribution; however, one might posit that similarly configured servers would have similar distributions, at least at a large scale. Further, one might investigate the "bumps" in these distributions – do they correspond to certain web pages, images versus html pages, on different servers?

Utilizing both parametric and nonparametric models is important, as is investigating different methods of each type. Throughout this book, we emphasize that "all models are

[27]We say "fairly" here because we failed to set our α level prior to running the tests. If we are using an α of 0.05, then the KS rejects, but it fails to reject at the 0.01 level.

[28]This analysis should be taken as tentative and suggestive, since we are only looking at univariate densities and are not investigating the multivariate densities. It is clear that the variables we are investigating are correlated, and we have not investigated this correlation structure.

wrong" while accepting that they can provide useful information even when they are wrong. In fact, nonparametric and parametric models can complement each other in interesting ways. See for example [355] in which a kernel estimator is paired with a Gaussian mixture model to help improve the model and the model fit. This approach is referred to as *alternation kernel and mixture* density estimation. Theoretical results for this estimator can be found in [214].

4.14 Time Series

For many computer security data sets, there is a temporal component that impacts the type of models that are appropriate. For example, consider the network flow data depicted in Figure 4.27. Here we have counted the number of TCP sessions initiated within each 5-minute interval over a day (day 3). There are two obvious patterns: there is a large scale cycle with a period of about 24 hours, and a smaller scale pattern – a dip, followed by a spike, every hour, roughly on the hour.

Figure 4.28 shows a zoom into the data, so that we can observe the pattern. The power spectral density is shown on the bottom of the plot, showing that there are three very strong frequency components to these data, at roughly $0.9, 1.75, 2.6$ hours.

There are a number of time series models and methods that might be considered for the analysis of cybersecurity data. One of the most flexible models is the ARIMA model. This is the autoregressive integrated moving average model that works as follows. The model requires three parameters, the number of autoregressive (AR) terms, the number

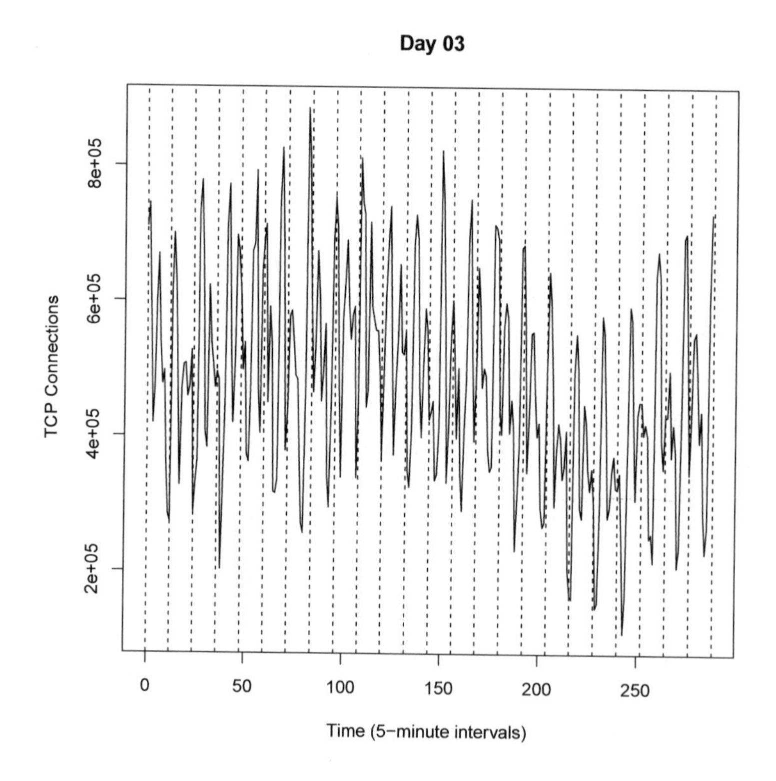

Figure 4.27: The number of TCP sessions, initiated within 5-minute intervals, for one day in the LANL flows data. The hours are indicated by vertical dotted lines.

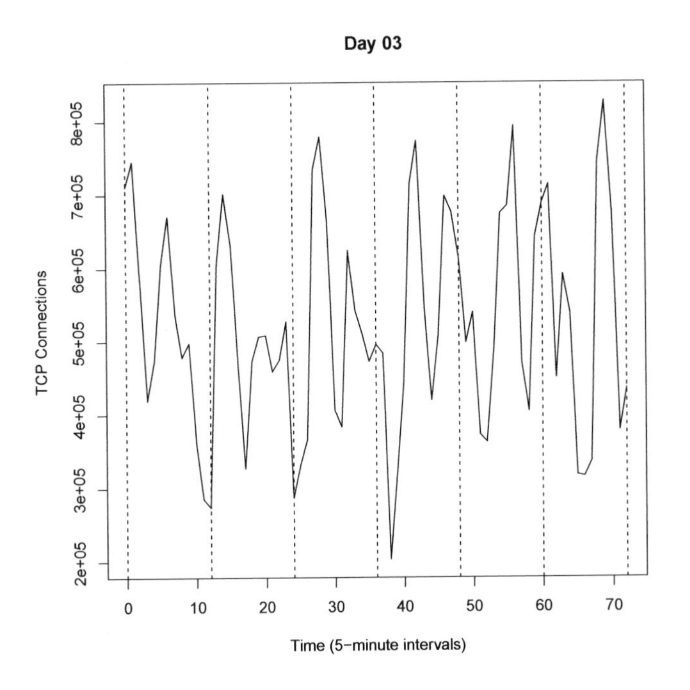

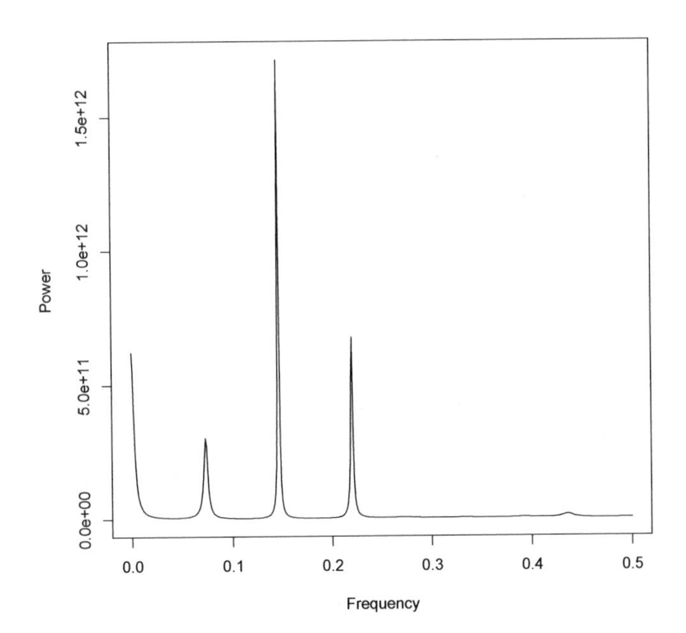

Figure 4.28: The number of sessions, as in Figure 4.27, zoomed into the first few hours of the day (top). A power spectrum is depicted in the bottom plot.

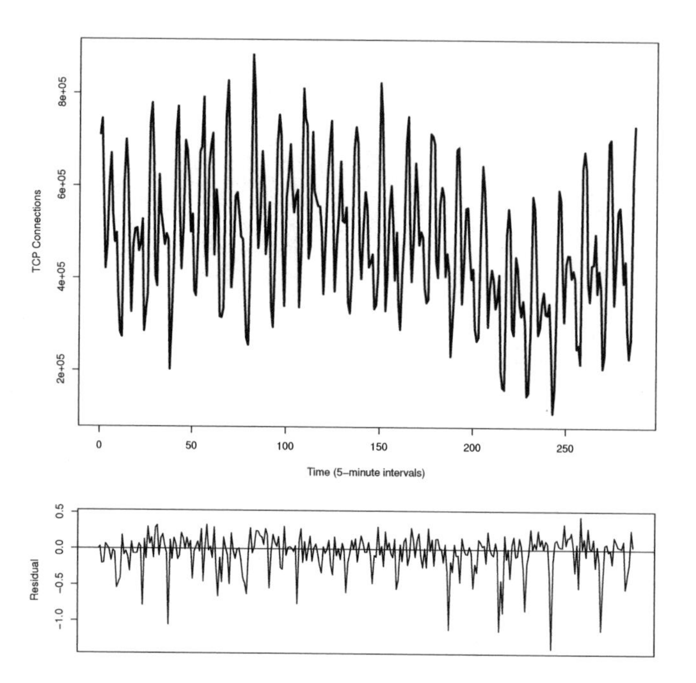

Figure 4.29: An ARIMA model for Day 3 of the LANL flows data: the number of TCP connections per 5-minute period is shown in the top plot. A (3,1,5)-ARIMA model was fitted to these data and the residual – the difference between the true value and the predicted value – is shown in the bottom plot. In this case, the residual is computed on the same data as was used to fit the model, and is normalized by the true value. In essence, this is a measure of the percentage of the error.

of integrated (differencing) times, and the number of moving average (MA) terms. The "integrated" part of the model corresponds to taking successive differences of the series, which removes trends. Once the differencing has occurred, the (p,0,q)-model[29] on this transformed series is:

$$x_t = \alpha_1 x_{t-1} + \ldots + \alpha_p x_{t-p} + \epsilon_t + \beta_1 \epsilon_{t-1} + \ldots + \epsilon_{t-q}, \tag{4.25}$$

where ϵ are error terms.

The terminology is suggestive, but may seem a bit confusing at first. A *moving average* is just what the phrase suggests: an average of things in a moving window of the time series. Thus we are averaging the last p observations in the time series. The *autoregressive* part comes from linear regression; if we think of linear regression as the line "in the middle of" (i.e., the average of) the observations, then the regression is "averaging out" the noise. So, the terms involving ϵs in Equation (4.25) are performing a "windowed" regression.

To see how one might use this, consider again the flows data discussed above. Figure 4.29 shows a model fit to the first day of the flows data from Figure 4.27. In this case, the model was chosen using an information criterion method using the `R` ([358]) package `forecast` ([207]).

Once a model is constructed, it can be applied to new data, and an error between the model prediction and the data is computed. This is standardized by dividing by the

[29]The middle term corresponds to the number of differences.

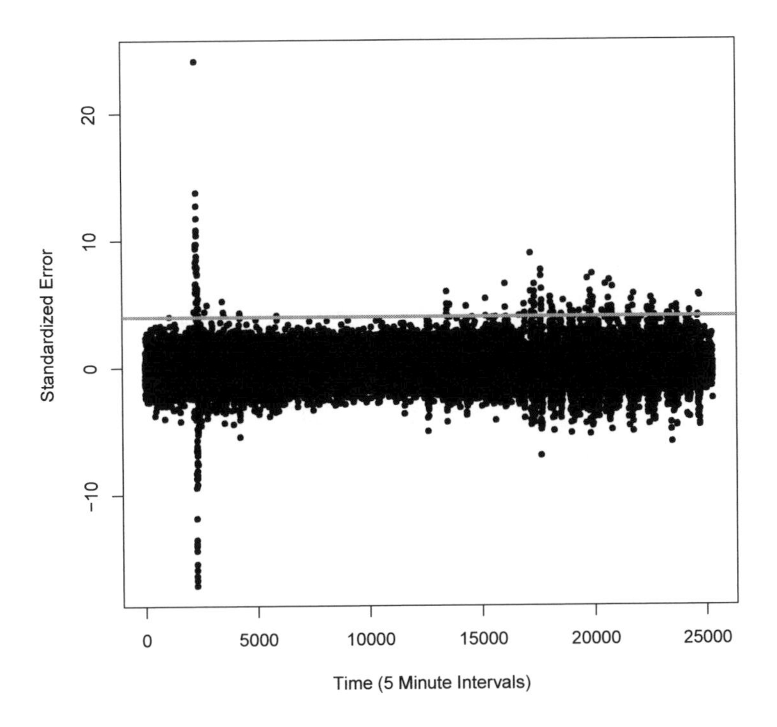

Figure 4.30: The standardized errors of the ARIMA model fit on day 3 for all 88 days. The gray horizontal line corresponds to 4-standard deviations.

standard deviation of the error for the model applied to the original data, so that we can set a threshold to detect deviations. Figure 4.30 shows this error. There are a number of deviations above 4 standard deviations, and these are candidates for unusually large numbers of connections on the network.

Note that there seems to be a change in the later days of these data, where there are many more deviations above the threshold. This may be because the model is no longer valid, and so one may wish to update the model periodically to account for overall increases in use or other "normal" changes in the network.

There are many other methods that a researcher should be familiar with. These include in particular hidden Markov models ([529] – also [132] is a nice short introduction) and state space models ([206, 207]). We discuss the hidden Markov model in the natural language processing chapter, Chapter 8.

This chapter has only discussed univariate time series. There are multivariate versions of these models, see for example [197] and [181] and other books. To see an example of why one might consider both TCP and UPD connection counts for the LANL flows data. Figure 4.31 illustrates that these are moderately correlated (the correlation coefficient is 0.46). In the top plot, because of the differences in the raw counts for these variables, we plot the standardized counts[30] for a small period on day 3. The bottom plot shows all the data, showing a strong correlation between the variables.

[30] These are z-scores: the mean has been subtracted and then the observations are divided by the standard deviation.

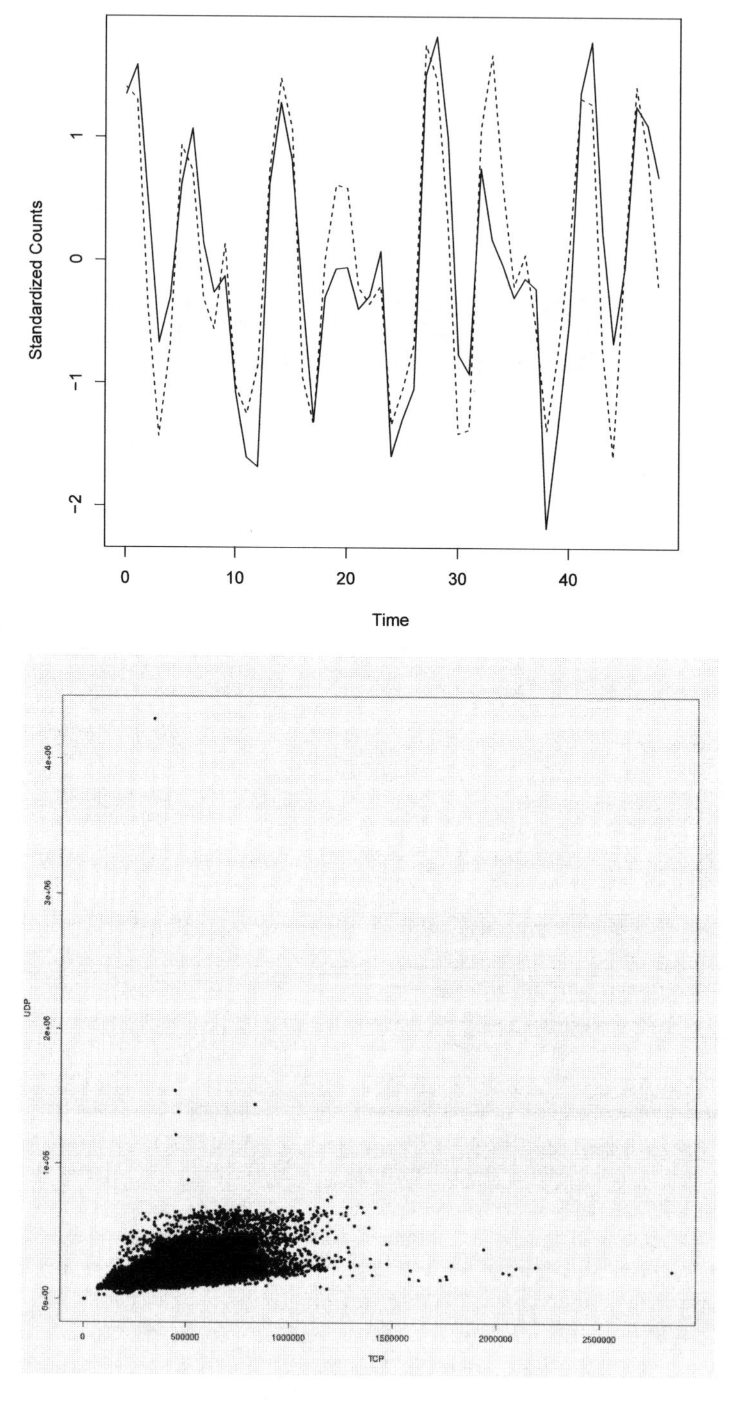

Figure 4.31: The number of TCP and UDP connections on the LANL network. The top plot shows standardized counts for the number of connections in 5-minute periods for TCP (solid curve) and UDP (dotted curve) for a short time on day 3. The bottom plot shows the raw values of TCP connections against UPD connections for the full data set.

This chapter (with the appendix) covers the basics of probability and statistics required for research in cybersecurity, and for the understanding of cybersecurity data and analysis methods. This is not by any means a complete or exhaustive coverage of the necessary methods and ideas; it is only meant as an introduction to provide some of the basic ideas necessary to understand some of the other algorithms throughout this book. A course in applied statistics is a good place to start to learn more about the methods discussed here, and there are a number of good books, for example [435], as well as the other books cited in the chapter, that cover most of the topics that have been discussed.

Chapter 5

Data Mining – Unsupervised Learning

Data mining is the science of extracting patterns from data. The goal is to extract interesting, nontrivial and useful patterns. However, such a task is very challenging since coming up with objective definitions for these terms itself is difficult, if not impossible. We give some examples to illustrate what we mean. For example, a *Times of India* news article in 2017 claimed that the change in the Greenland sea ice area affects the monsoon in India. We would consider this to be an example of a nontrivial and interesting pattern. Another example is that participants in a survey exhibited higher confidence in their answers when they were wrong, and a similar pattern was exhibited by participants who had lower conscientiousness [27]. On the other hand, as we have seen in Section 2.8, patterns are everywhere and we need tools for determining which patterns are important and which are spurious or not reliable.

Data mining draws upon techniques from statistics, artificial intelligence, databases and machine learning. Data mining techniques can be supervised or unsupervised. Supervised techniques use labeled training data to build a model for classification or prediction. The labels denote the output values or classes for the data instances. Unsupervised techniques can deal with data that is unlabeled.

In this chapter, we discuss mainly unsupervised data mining techniques, which include association rule mining, clustering and anomaly detection, with their security applications and adaptations. These techniques have been applied in a number of different security challenges, for example, credit card fraud detection, malware detection, intrusion detection, and stepping-stone detection. They are also important data mining techniques, which have found applications in many areas besides security.

However, we must take care to ensure that we adapt these techniques properly for security challenges, and not just routinely apply these techniques while losing sight of the unique needs of the security domain. Another issue is preserving privacy. The use of data mining can lead to unexpected violations of privacy, since it can find relationships between data items, which can defeat simple anonymization techniques.

The process of knowledge discovery from data involves a number of steps and there may be several iteration of these steps before satisfactory results are obtained. The list of steps includes: data collection, data preprocessing/wrangling and visualization, an optional data integration step in case data is obtained from multiple heterogeneous sources, feature extraction, the data mining algorithm(s), postprocessing and visualization, and presentation of the results. In cybersecurity, we may also need a feedback loop to improve the model over time. Concept drift or nonstationarity is a serious issue, especially in cybersecurity, mainly because of an active adversary. It also occurs naturally, due to changes in the environment – the computers on the network, new applications utilizing the resources, old applications falling out of use, changes in the way people use the resources, and so on.

5.1 Data Collection

Data may be collected from very different types of sources, e.g., sensors or databases, or it may be scraped from objects such as webpages, spreadsheets, documents, JSON files, CSV files, and XML files. Sometimes APIs may be available, and in other cases explicit and periodic scraping may have to be done. Explicit scraping is usually quite fragile. Hence, tools such as `import.io`[1] should be considered carefully before any custom code is written.

Three aspects of data collection need special attention for the data scientist. First, whether data is collected from a single source or multiple sources. Second, how much data should be collected. Third, how representative or diverse the data are. The first aspect determines the type and multitude of data collection and integration problems.[2] The second and third aspects are crucial to the validity of the patterns extracted from the data.

We will touch upon the first aspect below in the preprocessing section. For the second aspect, we should consider the number of independent factors. Some researchers recommend a data set size proportional to $\Pi_{i=1}^{|F|} N_i$, where N_i are the number of different possible values of the ith factor and $|F|$ is the number of different factors or features. We see that as the number of features increases, there is an exponential growth in the size of the data set required, since each feature will have at least two possible values.

As we have seen in Chapter 2, this is the curse of dimensionality, which has implications for both this issue of the number of observations necessary, and the structure of the high dimensional space in which the observations live. The third aspect is the most difficult to achieve, since there is no formal definition of a representative data set. A domain expert may be able to alleviate this problem to some extent. However, this problem is especially acute in security, since the attacker is constantly trying to devise new types of attacks.

Next, we consider the different types of data and associated operations, since a good understanding of these issues is essential to data preprocessing and analysis.

5.2 Types of Data and Operations

Data can be classified based on its many different dimensions, such as the source of data, the problem area, or whether it is public or private. For example, we may classify data as text, graph, spatial, record, time-series, etc. In another data classification scheme, we focus on the possible operations that can be done on the data. When we focus on the operations, then data can be broadly classified as categorical or numeric data.

Categorical data, sometimes also called nominal data,[3] is data that can be grouped into a number of groups or categories. We further distinguish between nominal categorical data and ordinal categorical data. Examples of nominal categorical data are gender, race, ethnicity, eye or hair color, etc. The only possible operators for nominal data are "equal to" and "not equal to." In ordinal data, there is an ordering on the values. For example, {small, medium, large} is ordinal categorical data. Other examples are student letter grades in courses and ranks in an organization hierarchy. Numerical data, also called quantitative data, is data that corresponds to measurements and is expressed in numbers. Numeric data can be

[1] https://www.import.io/

[2] Data integration is a vast problem in itself. Manual approaches do not scale. Machine learning solutions are being designed for this problem, but there is a long way to go, e.g., [409, 410].

[3] We will not use this term as synonymous with categorical data in this book.

discrete or continuous. Discrete refers to either finite numeric data or countably infinite set of numeric values. Continuous refers to data that is real valued (uncountable), but, of course, can only be represented with finite precision on a computer. In numerical data, we also distinguish between interval data and ratio-scale data. Interval data is measured on a scale of equal-sized units and the values are ordered. However, there may not be a true zero point. Examples of interval data include temperature measurements in Fahrenheit or Centigrade units, or calendar data. In numerical ratio-scale data, there is a true zero point and we can speak about values being an order of magnitude larger or smaller. Examples include temperatures in Kelvin, lengths, counts or monetary quantities.

5.2.1 Properties of Data Sets

There are four important properties of data sets, which require special care and consideration by the data miner. They are:

1. Dimensionality. This is the number of attributes or features of each instance or item in the data set.

2. Sparsity. Certain combinations of feature values may be missing from the data set, i.e., there may be no items or instances with those feature combinations. Sparsity is a measure of how often these combinations of features are missing and whether their absence is important.

3. Resolution. At what scale of precision are the values measured? Resolution is the difference between an overview of something from "30,000 feet above the ground," when one cannot make the finer details below, to the case when one is standing close to the same area.

4. Distribution. This refers to the similarity of the data items to each other or the diversity of the data set. When class labels are available, we can also check the distribution of the class labels.

5.3 Data Exploration and Preprocessing

5.3.1 Data Exploration

Data exploration is the process of getting to know the data and assessing its quality. The first step in exploration is to generate a data quality report [231]. It should include separate tables or plots for continuous and categorical features. The report should include the time period of data collection for each subset of the data set and the sources. Often, the classification process becomes easier if subsets of the data set are gathered at different times. For instance, in phishing email classification, if legitimate emails are gathered from recent years, say less than five years from now, and phishing emails are gathered from more than a decade ago, then a classifier can pick up on this artifact of data collection and get a very high accuracy.

For continuous features, we should include the minimum value, the maximum value, the 1st and 3rd quartile, the mean and median, and standard deviation. The number of distinct values, i.e., the cardinality, of each feature, and the percentage of data instances that are missing a value for each feature should also be reported.

These two items, cardinality and percentage of data instances that are missing a value for each feature, should also be reported for categorical features. In addition, we should report the mode, the second mode, and their raw as well as relative frequencies as a percentage of the total instances.

After we prepare a data quality report, we need to determine any issues such as: missing values, irregular cardinalities, and outliers. Once we determine the issues, we need to consider the source of the issue, i.e, whether the issue is due to invalid data or valid data. If we determine that the source of the problem is invalid data, then we need to regenerate the data and create a fresh data quality report. If the data is valid, then we must examine the issues more closely and take steps to deal with them.

1. Missing values. For missing values, we may disregard the feature, if more than half of the data instances are missing values for the feature, and we believe that the feature is not important nor significant. If that is not the case, then we may use imputation, which is the process of replacing missing feature values with plausible estimates for them. Generally, the mean or median is used for continuous features and the mode for categorical features. More advanced imputation techniques are possible and involve model building. A third way to handle missing values is to delete the data instance, which is generally not recommended, but occasionally unavoidable.

2. Irregular cardinality. This refers to a mismatch between our expectation for the cardinality of a feature and its actual cardinality. This may be due to errors in coding the feature values. For instance, if multiple people code the feature values for different subsets of the data, they may independently choose different encodings for the same value.

3. Outliers. Feature values may contain outliers due to noise, mistakes or valid reasons. For instance, typing mistakes could produce outliers. Sensors may pick up noise. One way to handle outliers in feature values is to use clamping techniques, unless outliers is what you are after. In this technique, all values above a certain threshold may be replaced by the value *upper*, and all values below a certain threshold may be replaced by the value *lower*.

The next step in getting to know the data is studying the distributions of the feature values for each feature, e.g., whether it is uniformly, normally, or exponentially distributed, is it skewed, or multimodal, etc. A final step is examining the correlations between pairs of features. For continuous features, scatter plot matrices can be used for this purpose initially, and if patterns are observed in the scatter plots, then calculations of various correlation coefficients, e.g., Pearson correlation coefficient or Spearman, can give more precise understanding of the degree of correlation. For categorical features, we may use a collection of bar plots. A simple bar plot is drawn that depicts the densities (*not frequencies*) of the different values of one feature, say f. Next we draw a bar plot of f, for each value v of the second feature, using only the data instances in which the second feature value is v. Now we look for any noticeable differences in these bar plots. If there are significant differences in these bar plots, then we may use rank-based correlation coefficients, such as Spearman, if ranking is possible for the pair of categorical features. This method is called *small multiples visualization*. This method is also appropriate for the case of a continuous feature and a categorical feature. We can plot the densities of the continuous feature for each value of the categorical feature.

Small Multiples Visualization Example. The following data set of eye color and height was obtained for nine astronauts. See Table 5.1. From the table, for the blue eye color, we have the following proportions for height: 2/3 Short, 1/3 Medium, and 0 Tall. Whereas for brown and black eye colors, we have 0 Short, 1/3 Medium and 2/3 Tall. A clear – but may be spurious – pattern emerges for this small data set of astronauts: the astronauts with darker eye color tend to be taller. □

Table 5.1: Eye Color and Height of Nine Astronauts

Name	Eye Color	Height
Bob	Blue	Short
Mike	Blue	Medium
Sam	Brown	Tall
Dan	Black	Medium
Blake	Blue	Short
Noah	Black	Tall
Pete	Brown	Medium
Stan	Black	Tall
Marc	Brown	Tall

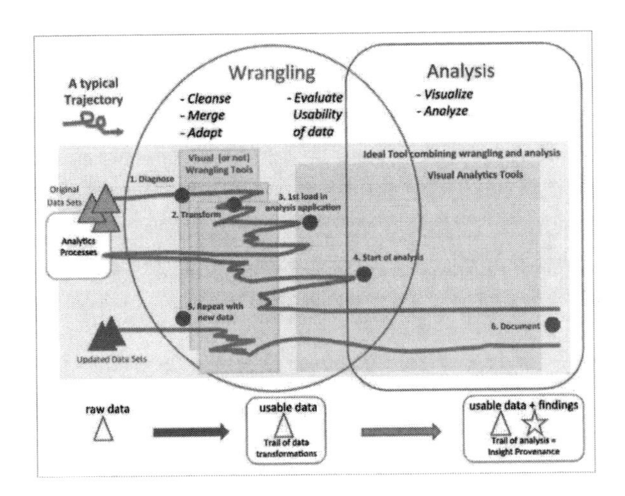

Figure 5.1: The process of wrangling and analysis.

5.3.2 Data Preprocessing/Wrangling

Data needs to be preprocessed before it can be analyzed. The goal of preprocessing is to put the data into a format that is appropriate for analysis. Preprocessing involves data cleaning, deciding what to do with any missing values or noise/corruption, identifying potential outliers, and data transformation/normalization. The process of data wrangling and analysis is shown in Figure 5.1, which is from `https://github.com/umddb/datascience-fall14/blob/master/lecture-notes/multimedia/wrangling.png`. For data cleaning and problems with data, we refer the reader to [361]. Two big challenges in the data preprocessing phase are: entity resolution (different "names" referring to the same object or entity) and conflict resolution (e.g., different values for the same attribute of a single object), especially when the data is obtained from multiple sources.

An optional step is data discretization. The final step before analysis is feature extraction. Some data mining algorithms such as decision trees cannot directly handle data that is continuous. In this case, we discretize the data into a finite number of ranges. For example, the temperature attribute of a place may be discretized into low, medium, and high by taking two values $T_1 < T_2$, low denotes all values less than or equal to T_1, medium all values in the interval $(T_1, T_2]$ and high is all values that exceed T_2.

Data normalization is a very important step in the preprocessing phase. Normalization ensures that an attribute does not dominate just because it is measured on a different scale. For example, a person's height is measured in feet (or meters) and the age is measured in years. A basic normalization step is to replace the value x by x_{norm}, where

$$x_{norm} = \frac{x - x_{min}}{x_{max} - x_{min}} \tag{5.1}$$

Attributes that obey the normal distribution can be normalized by scaling and centering as follows. Assume $x_1, x_2, \cdots, x_n$ are distributed normally with mean μ and standard deviation σ. Replace each x_i by

$$\frac{x_i - \mu}{\sigma} \tag{5.2}$$

Then, the new values have mean 0 and variance 1. Such a variable is called a standard normal variable.

Data preprocessing also includes encoding data values to make them suitable for analysis. For example, neural networks can take only data that is presented as numbers. Categorical data can be encoded by using what is called a one-hot encoding scheme. In this encoding scheme, we use as many new features as there are values in a category, with every new feature set to 0, except for one of the features, which is set to a 1. The feature that is set to 1 is different for different values. For example, the category $\{red, blue, green\}$ would be encoded as three new features, A, B and C, as shown in Table 5.2 below. Why do we do

Table 5.2: One-hot Encoding for Categorical Data

Value	**A**	**B**	**C**
Red	1	0	0
Blue	0	1	0
Green	0	0	1

this? Why not use the much simpler encoding scheme, Red - 1, Blue - 2, Green - 3? The reason is that the latter encoding scheme introduces new relationships between the features that did not exist before, e.g., it now imposes an ordering and also relationships based on magnitude. For example, we are saying that Blue is twice as large as Red and Green three times as large. The golden rule for a faithful encoding scheme is to preserve the *meaningful* relationships that exist in the data set, and to avoid creating new relationships that did not exist before.

An important issue about such encoding schemes is known as the *dummy variable trap.* We do not want to create a situation where independent variables become so tightly correlated that one can be predicted from the other. This would be the case, for example, if we used two variables to encode gender for a data set in which there are only two possibilities: male and female. If we introduce two variables: one to encode male and another to encode female, then either of them can be predicted from the other. The solution is to drop one of the categories and encode the rest. For example, we only encode female with 1 and then 0 for this category means male, or vice versa.

In cybersecurity, another step is also needed fairly often: data de-obfuscation and/or data de-decryption. To see this, recall the example of obfuscated command from Section 3.4.1.

5.4 Data Representation

How to represent the data plays a central role in the quality of the output from the analysis. A dramatic example of this is given in [210]. We have touched upon this issue in Chapter 2 under feature selection. We now discuss some of the key unsupervised data mining techniques and algorithms such as association rule mining, clustering and anomaly detection. Anomaly detection can be both unsupervised and supervised. In this chapter, we will cover mostly unsupervised anomaly detection.

5.5 Association Rule Mining

When you buy a book such as this one at the Amazon.com web site, as you check out, the web site gives recommendations, telling you that people who bought this book also bought books A, B and C. Did you ever wonder, how does the web site make these recommendations and how effective are these recommendations? The machinery behind these recommendations is the subject of our study in this section.

In association rule mining, deriving correlations or associations between data items is the goal. A word of caution is essential to keep in mind, however. Correlation is not causation. Just because two variables or features or attributes are correlated does not mean that one of them causes the other. Association rule mining is also sometimes called frequent itemset mining. Association rule mining has been applied widely to many different domains including text mining, market-basket analysis, disease data sets and biomedical data sets.

A popular algorithm for association rule mining is called Apriori algorithm. A few definitions are needed to make further progress. The input to the Apriori algorithm is what is called market-basket data. Basically, we may assume a database of transactions $DB = \{T_1, T_2, \cdots, T_m\}$, where each transaction, T_i is a subset of a universe $U = \{x_1, x_2, \cdots, x_n\}$ of items. An *itemset* is any nonempty subset of U. The *support* of an itemset I is the fraction of transactions that contain all the items in I, i.e., the number of transactions that are supersets of I.

$$support(I) = \frac{|\{T \in DB \mid I \subseteq T\}|}{m} \tag{5.3}$$

Given a minimum support threshold (called min-support), σ, we say that an itemset I is *frequent* provided its support is at least σ, $support(I) \geq \sigma$. We now give an example to illustrate these definitions and the Apriori algorithm.

In Table 5.3, consider the following transaction database of customers at a book-seller's web site. Assuming a support threshold of 0.5 or 50%, we see that there is a frequent

Table 5.3: Five Different Customers' Book Purchases

Trans. ID	**Items**
1	Goodnight Moon (GM), A Murder is Announced (AMiA), Security Analytics (SA)
2	AMiA, Right-ho Jeeves (RJ), Pigs Have Wings, Hard Times (HT), SA
3	AMiA, SA, Machine Learning with R, HT
4	AMiA, Nemesis, RJ
5	SA, GM, HT

itemset of two items {A Murder is Announced, Security Analytics}, whose support is 0.6, since its elements appear in three transactions each, and each of the two books in this set is also a frequent singleton itemset with the same support, 0.8. There is no frequent itemset containing more than two items.

Now we observe that if an itemset is not frequent, then none of its supersets can be frequent. This downward closure property of frequent items is called the Apriori principle.

Based on this principle, we can describe the Apriori algorithm for finding all the frequent itemsets as follows. The extend operation in Algorithm 1 does a union of two itemsets and checks the support threshold to ensure the resulting itemset meets the support threshold before adding it to the set of frequent itemsets, FI.

Algorithm 1: Apriori

Input: A Transaction Database and a support threshold σ
Output: Frequent Itemsets

1 i = 1;
2 extended := false;
3 1-FI = FI = All frequent itemsets of size 1;
4 **If** 1-FI is empty **then** return(FI);
5 **repeat**;
6 Extend every i-itemset in FI by 1-FI;
7 **If** any set was added to FI **then** extended := true **else** extended := false;
8 i := i + 1;
9 **until** extended == false;
10 return(FI);

Apriori Example. We run the Apriori Algorithm on our example transaction DB. Initially,

FI = 1-FI = {AMiA}, {SA}, {HT}

Next, we try to extend every 1-itemset in FI by 1-FI, getting three possible extensions given below:

{AMiA, SA}
{AMiA, HT}
{SA, HT}

Two of them meet the support threshold:

{AMiA, SA}
{SA, HT}.

In the next iteration, only one candidate is produced:

{AMiA, SA, HT}

This does not meet the support threshold and Apriori terminates. □

Besides generating the frequent itemsets, the Apriori association rule mining algorithm can also produce "actionable" rules or recommendations. For this, it needs a confidence threshold. The confidence of a rule $X \rightarrow Y$, where X and Y are both itemsets, and $X \cap Y = \emptyset$, is defined as its conditional probability, $P(Y|X)$ i.e.,

$$confidence(X \rightarrow Y) = \frac{support(X \cup Y)}{support(X)} \tag{5.4}$$

Typically, we insist upon rules for which $X \cup Y$ is a frequent itemset, which means both X and Y are also frequent itemsets by the Apriori principle mentioned above. So we can extend the Apriori algorithm to take pairs of frequent itemsets (X, Y) and check whether the confidence of either of the two rules that can be formed from them, viz. $X \rightarrow Y$ and

$Y \to X$, exceeds the confidence threshold or not. Alternatively, we can take a frequent itemset containing at least two items and decompose it systematically into a left-side and a right-side for a potential rule.

The *lift* of an association rule can be used to determine its importance. Lift was introduced in [225]. It is defined as:

$$lift(X \to Y) = \frac{confidence(X \to Y)}{expected\text{-}confidence(X \to Y)} \tag{5.5}$$

Expected confidence of a rule is defined as:

$$Expected\text{-}confidence(X \to Y) = support(X) \tag{5.6}$$

Combining the two equations, we get that

$$lift(X \to Y) = \frac{confidence(X \to Y)}{support(X)} \tag{5.7}$$

In the definition of lift, we assume that there is no statistical relation between the rule head X and the rule body Y. In other words, they are independent. Lift measures how much the rule deviates from this assumption of independence. The range of lift values is $[0, \infty]$.

1. A lift value close to 1 means that the independence assumption holds.
2. A lift value greater than 1 indicates that the rule body and rule head appear together more frequently than expected.
3. Conversely, a lift value smaller than 1 indicates that the rule body and rule head appear together less frequently than expected.

Consider the rule $AMiA \to SA$ where AMiA denotes A Murder is Announced and SA denotes Security Analytics. The support of AMiA is 0.8 and the confidence of the rule is $0.6/0.8 = 0.75$, so the lift of this rule is $0.75/0.8 = 0.9375$, which is close to 1. So we believe that the independence assumption holds.

We illustrate the association rule mining algorithm on a phishing URL data set,[4] in .arff format (a format for WEKA[5]), from the UCI Machine Learning repository [116]. The data set was contributed by [320]. The data set is relatively balanced containing 11055 instances, 4898 phishing and 6157 legitimate, each instance has 30 features and the class attribute: phishing/legitimate.

```
@attribute having_IP_Address  { -1,1 }
@attribute URL_Length   { 1,0,-1 }
@attribute Shortining_Service { 1,-1 }
@attribute having_At_Symbol   { 1,-1 }
@attribute double_slash_redirecting { -1,1 }
@attribute Prefix_Suffix  { -1,1 }
@attribute having_Sub_Domain  { -1,0,1 }
@attribute SSLfinal_State  { -1,1,0 }
@attribute Domain_registeration_length { -1,1 }
@attribute Favicon { 1,-1 }
@attribute port { 1,-1 }
```

[4] https://archive.ics.uci.edu/ml/datasets/phishing+websites

[5] We used the stable version 3.8 in our experiments.

```
@attribute HTTPS_token { -1,1 }
@attribute Request_URL  { 1,-1 }
@attribute URL_of_Anchor { -1,0,1 }
@attribute Links_in_tags { 1,-1,0 }
@attribute SFH  { -1,1,0 }
@attribute Submitting_to_email { -1,1 }
@attribute Abnormal_URL { -1,1 }
@attribute Redirect  { 0,1 }
@attribute on_mouseover  { 1,-1 }
@attribute RightClick  { 1,-1 }
@attribute popUpWidnow  { 1,-1 }
@attribute Iframe { 1,-1 }
@attribute age_of_domain  { -1,1 }
@attribute DNSRecord   { -1,1 }
@attribute web_traffic  { -1,0,1 }
@attribute Page_Rank { -1,1 }
@attribute Google_Index { 1,-1 }
@attribute Links_pointing_to_page { 1,0,-1 }
@attribute Statistical_report { -1,1 }
@attribute Result  { -1,1 }
```

Here are the first few data instances:

```
-1,1,1,1,-1,-1,-1,-1,-1,1,1,-1,1,-1,1,-1,-1,-1,0,1,1,1,1,-1,-1,-1,-1,1,1,-1,-1
1,1,1,1,1,-1,0,1,-1,1,1,-1,1,0,-1,-1,1,1,0,1,1,1,1,-1,-1,0,-1,1,1,1,-1
1,0,1,1,1,-1,-1,-1,-1,1,1,-1,1,0,-1,-1,-1,-1,0,1,1,1,1,1,-1,1,-1,1,0,-1,-1
```

Apriori ignores the class attribute, which is called Result above. The min-support threshold used is 0.85 and the min-confidence threshold is 0.9. WEKA reports the 5 best rules as follows:

```
1. port=1 Iframe=1 9429 ==> RightClick=1 9423
<conf:(1)> lift:(1.04) lev:(0.04) [399] conv:(58)
2. Iframe=1 10043 ==> RightClick=1 10035
<conf:(1)> lift:(1.04) lev:(0.04) [424] conv:(48.05)
3. on_mouseover=1 Iframe=1 9529 ==> RightClick=1 9521
<conf:(1)> lift:(1.04) lev:(0.04) [402] conv:(45.59)
4. port=1 9553 ==> RightClick=1 9512
<conf:(1)> lift:(1.04) lev:(0.03) [370] conv:(9.79)
5. on_mouseover=1 9740 ==> RightClick=1 9665
<conf:(0.99)> lift:(1.04) lev:(0.03) [344] conv:(5.52)
```

WEKA reports confidence and lift, as well as a few other measures: leverage and conviction. The conviction, conv, of a rule $X \to Y$ is defined as:

$$conv(X \to Y) = \frac{1 - support(Y)}{1 - confidence(X \to Y)} \tag{5.8}$$

The conviction is not so easy to interpret, but the leverage of a rule $X \to Y$, which is defined below, is much easier. It measures how much the probability of X and Y together exceeds the probability we would obtain, if they were independent.

$$lev(X \to Y) = support(X \cup Y) - support(X)support(Y) \tag{5.9}$$

WEKA also has an option to run classification and association rule mining, which uses the algorithm proposed in [278]. We show the best rules obtained by this algorithm below.

```
1. SSLfinal_State=1 web_traffic=1 4183 ==> Result=1 3968    conf:(0.95)
2. SSLfinal_State=1 Domain_registeration_length=-1 Google_Index=1 4244
==> Result=1 3895    conf:(0.92)
3. SSLfinal_State=1 Request_URL=1 4328 ==> Result=1 3958    conf:(0.91)
4. SSLfinal_State=1 Domain_registeration_length=-1 4757 ==> Result=1 4314
conf:(0.91)
5. SSLfinal_State=1 Domain_registeration_length=-1 RightClick=1 4547
==> Result=1 4107    conf:(0.9)
```

In addition to lift, information-theoretic measure [42], statistical hypothesis tests [478], and other measures have been introduced to control the rules generated by Apriori algorithm. To reduce the need for manual analysis, rules are summarized [294], generalized [36], clustered [14], or evaluated as a compression of the original database [30]. These methods investigate relationships among the rules to generate concise rule sets.

Practical considerations: On a reasonable sized transaction data set, setting a low value for the support threshold produces a large set of association rules and setting too high a value would produce few, if any, association rules. So how should we set the support threshold in practice? We can experiment with different values of the support threshold and then decide the best value to use based on an examination of the frequent itemsets or rules generated. A domain expert would need to be closely involved in the process to determine the quality of the rules.

We can offer a rule of thumb for the support threshold. To start with, we can set the minimum support threshold to

$$\frac{\sum_{i=1}^{m} |T_i|}{m * n} \tag{5.10}$$

Observe that the numerator denotes the total size of all the transactions, and the denominator is the number of transaction times the number of unique items in the transactions. The ratio $\sum_i |T_i|/n$ denotes the expected number of repetitions of each item, if all items are equally likely. Since the support of an n-itemset, where $n > 1$, cannot exceed the support of its least frequent itemset, it seems reasonable to start with the expected support of singleton itemsets. We have converted this to a fraction between 0 and 1, a support threshold.

Even if we set the support threshold to a reasonable value, we may still get a lot of rules. This happens frequently for medical data sets in which the number of transactions (the patients usually) may be small, usually a few hundreds to a few thousands, but the number of their attributes (observations, tests, etc.) may be many. An example of such a data set is the Heartfelt data set from [80]. The Heartfelt study collected data on adolescent health. The target population for the study was African, European, and Hispanic American adolescents, between 11 and 16 years of age, "residing in a large metropolitan city in southeast Texas with an ethnically diverse population. There were 383 adolescents recruited, and the collected data included totally 105 attributes and 16,912 records. The attributes included age, gender, ethnic/racial group, physical maturity, resting blood pressure and heart rate, ambulatory blood pressure, heart rate and moods reported at 30-minute intervals, body mass index, fat free mass, and psychological characteristics such as anger and hostility" [80].[6]

For such data sets, a different approach to control the number of rules generated by the Apriori algorithm is proposed in [80]. They use a semantic network based on the UMLS (Unified Medical Language System) to specify which rules are interesting, nontrivial and useful. Their method reduced the original input of 1.2 million rules to about 26,000 rules, which is a reduction by a factor over 46. Comparison with the research published by the

[6]Amazon customer transactions on the other hand probably number in the few millions daily, with the average customer purchasing less than 10 products - our estimate.

researchers who participated in the Heartfelt study showed that 8 out of 10 associations found by the researchers were captured in the final set of rules.

Efficiency considerations: To improve the efficiency of the Apriori algorithm, two important concepts are closed itemsets and maximal frequent itemsets. Given a transaction database, an itemset I is closed provided no superset of I has the same support as I. One could look for closed itemsets and then check if any of them are frequent. Such a procedure can be more efficient than the basic Apriori algorithm given above. Furthermore, I is a maximal frequent itemset, if I is frequent and none of its immediate supersets are frequent. In our example database, we can see that the last 2-itemset is both closed and maximal.

A number of other useful alterations can be made to the pseudo-code above, e.g., adding efficient data structures, for increasing the efficiency of the algorithm, but these are beyond the scope of this book and we will refer the reader to any good book on Data Mining, e.g., [182].

5.5.1 Variations on the Apriori Algorithm

A serious limitation of association rule mining is the constraint of binary data. A transaction either contains an item (1) or does not contain an item (0). Another issue is that some attributes may be categorical, interval, or continuous. We have already mentioned the issue of efficiency above. A final issue is the large number of rules obtained in typical applications. Researchers have addressed many of these issues.

One approach for categorical variables is the one-hot encoding discussed above. Intervals can be turned into Boolean attributes using membership in a interval. For continuous attributes, we can discretize first and then encode each interval with a Boolean attribute. However, how to discretize is not so clear. Small intervals can cause problems with the min support threshold and large intervals cause problems with the min confidence threshold.

Efficiency has been tackled with hashing and other data structures, e.g., the FP tree [183]. Another idea is to partition the database so that each partition fits into main memory. First, candidate frequent itemsets are computed for each partition, "locally" frequent itemsets, and then these are checked to find the globally frequent itemsets.

For the large rule set problem, researchers have introduced various notions of "interestingness" for rules. Some researchers have also proposed directly finding discriminative patterns [85] rather than finding frequent itemsets and then filtering them using notions of interestingness.

Related to association rule mining is correlated pattern mining. It discovers the correlation between items and reduces the number of rules generated. However, it still generates a lot of rules and so researchers have proposed closed correlated pattern mining [239].

An incremental association rule mining algorithm has also been proposed for the case where new itemsets are added to the data set.

We next consider clustering, which is a very important data mining technique.

5.6 Clustering

Clustering is a popular, exploratory, and unsupervised technique whose goal is to group similar objects together into a cluster and keep dissimilar objects apart in different clusters. Some motivating applications of clustering in cybersecurity include:

- Detecting traffic similarities: some examples where this could help are: botnet detection, alpha flows, traffic characterization, etc.
- Clustering of commands into malicious or benign, malware types, attack types.

- Clustering of logs to monitor employee behavior over time. This could help to determine an anomalous remote login, i.e., intrusion detection.

However, clustering is easier said than done. First of all, a big challenge is that there is no good definition of a cluster. Second, as we have observed before in Chapter 2, even randomly generated data can be clustered by a clustered algorithm. Most clustering algorithms do not check whether the data has any clustering tendency or not. They assume it does and act according to this assumption. In this section, we review six important clustering algorithms. Some of them, e.g., K-means and hierarchical clustering are already used widely, while some of the others, we believe, will gain in popularity in the future.

In the subsequent discussion, it will be convenient to imagine a concrete spatial data set of points in two-dimensional space. Later, we shall see how to adapt the clustering algorithms for nonspatial data sets such as malware and text data.

Broadly speaking, there are two kinds of clustering algorithms: partitional and hierarchical. We start with partitional algorithms first.

5.6.1 Partitional Clustering

A very popular partitional clustering algorithm is K-means. K-means assumes that the number of clusters, K, is given as an input to the algorithm. It is a very simple iterative algorithm with two main steps: assigning points to cluster centers, or centroids, and recomputing the centroids after the assignment. In the following algorithm, a clustering denotes an assignment, or label, for each item indicating its membership in a cluster. K-means requires a distance function D, which is a metric, along with the number of clusters. We refer the reader to [423] for the definition of a metric and for a list of distance functions. How is the centroid of a cluster calculated? Assuming, a coordinate system for the items, we just take the mean of each coordinate separately for all the items in a cluster.

Algorithm 2: K-means

Input: A data set equipped with a distance function D, which is a metric, and a whole number $K > 1$

Output: A clustering with K clusters

1 Initialize: Randomly choose K items as the cluster centers;
2 **while** not done ;
3 Clustering: Assign each data item i to closest cluster center c, according to distance $D(i, c)$;
4 Update cluster centers: For each cluster center, c := centroid of all items in its cluster ;
5 **If** assignment of any item to cluster changed **then** done := false **else** done := true;
6 **end-while** ;
7 return(Clustering);

What is the goal of the K-means algorithm? The algorithm tries to minimize the residual sum of squares (RSS), i.e., the sum of the squared distances from the points to their respective centroids. However, the algorithm does not guarantee a global minima at termination since the optimization function is not convex.

The algorithm runs in $O(NKtd)$ time, where t is the number of iterations, d is the time for each distance calculation which varies with the dimensionality of the data items, and N is the number of items in the data set. However, one problem is that there is no guarantee that t will be small. In [19], researchers have shown that, in the worst-case, t can be a super-polynomial function of N. Hence, in practice, an upper bound is placed on the number of iterations. K-means can be considered as a special case of the expectation-maximization algorithm. The K-means algorithm can be proved to terminate with a locally optimal solution under certain conditions.

How to choose the initial centroids? It has been shown that there are better methods than random choice, which has a high chance of assigning many centroids to the same cluster. In particular, if the algorithm selects a point with probability proportional to the square of its distance to the nearest preceding centroid, then it can be shown that, in expectation, the clustering is suboptimal by at most a log(K) factor [20].

5.6.2 Choosing K

Note that increasing K will reduce the sum of the squared distances. To see this, let each point be in its own cluster. In this case, the centroid of the cluster will be the point itself and the objective function will achieve its minimum value, 0. But, this obviously defeats the purpose of the algorithm, which is to group the points. Therefore, we can vary the number of clusters and plot the objective function of K-means against the number of clusters, finding the point of diminishing returns and taking that as the number of clusters. This is called the "knee" or "elbow" method.

Another possibility is an empirical rule of thumb, which states that we should choose $\mathrm{K} = \sqrt{\mathrm{N}/2}$.

A third method is called the m-fold crossvalidation method. In this method, we divide the data set into m parts. We build a clustering on $m - 1$ parts and test the quality of the clustering on the mth part.

A number of internal cluster validation indices are compared in the paper [178]. Internal cluster validation indices try to estimate the number of clusters without using any external information. Not mentioned in [178] is the G-means method of Hamerly and Elkan [180], which uses a statistical test for deciding whether to split a cluster into two clusters. The idea is that one center should be used for one Gaussian cluster. The researchers claim that their method is even better than the X-means method of [344]. The latter uses regularization for learning K, basically a penalty for using a large K. They score each model using the Bayesian Information Criterion (BIC): $BIC(C|X) = Log\text{-}likelihood(X|C) - 0.5\mathrm{K}(d+1)log\ N$, where C is the model, X is the data set, $\mathrm{K}(d + 1)$ is the number of parameters in model C with dimensionality d and K centers, and N is the number of data instances as above. $Log\text{-}likelihood(X|C)$ can be calculated as $nln(RSS/n)$, where RSS is the residual sum of squares. Of course, many other criteria are possible instead of BIC.

5.6.3 Variations on K-means Algorithm

Two important variations of the K-means algorithm are bisecting K-means and the K-medoids algorithm. In bisecting K-means, all the data items are in a single cluster initially. The algorithm is iterative. In each iteration, it picks a cluster, finds two subclusters using the basic K-means algorithm and bisects it into clusters.

In some situations, bisecting K-means performs better than the K-means algorithm. It consists of the following steps: (1) pick a cluster, (2) find two-subclusters using the basic K-Means algorithm, (bisecting step), (3) repeat step 2, the bisecting step, for ITER times and take the split that produces the clustering with the highest overall similarity, (4) repeat steps 1, 2, 3 until the desired number of clusters is reached.

In the K-medoids algorithm, we use the median of a cluster instead of the centroid to represent the cluster center. The rest of the algorithm is unchanged. In case of categorical data, we may use the modes instead of the means.

A third important variation is constrained K-means [466]. Sometimes, background knowledge is available from the application domain, which can be used to generate constraints on the clustering. Given a set of must-link (two data instances must be in the same cluster) and cannot-link (two data instances must not be in the same cluster) constraints, researchers presented a constrained K-means clustering algorithm and studied its performance. The idea is to assign a point to the closest cluster that does *not* violate the constraints. We refer the reader to the original paper for more details.

We ran K-means algorithm with two clusters on phishing URL data set mentioned above in Section 5.5 with Euclidean distance. The within cluster sum of squared errors is 83290 after three iterations and random initialization. With K-means++ initialization the within cluster sum of squared errors is 83821 after four iterations, which is slightly worse than random. WEKA has two other ways of initializing cluster centers, but only two distance functions for K-means, Euclidean and Manhattan are implemented. Since 11,055 instances with 30 attributes each are hard to visualize, we give a toy example consisting of long jumps in feet of ten athletes $\{21, 22, 24, 22, 21, 21, 22, 22, 24, 24\}$. If we group them all together in one cluster, the within cluster sum of squares is:

$$\sum_{i=1}^{10} (x_i - \overline{x})^2 = (21 - 22.3)^2 * 3 + (22 - 22.3)^2 * 4 + (22.3 - 24)^2 * 3 = 14.1 \qquad (5.11)$$

But if we choose the three cluster centroids as 21, 22 and 24, then the within cluster sum of squares is 0. We run SimpleKMeans from WEKA with random initialization and K=3, and sure enough it comes back with the right centroids in two iterations. The WEKA file for this example is really simple and given below.

```
@relation LongJumps
@attribute abcissa numeric
@data
21
22
24
22
21
21
22
22
24
24
```

5.6.4 Hierarchical Clustering

Hierarchical clustering algorithms are also iterative, but more expensive computationally than partitional algorithms. There are two versions: *agglomerative*, in which in each iteration one combines two clusters, and *divisive*, in which in each iteration one splits a cluster. We describe the agglomerative method.

A basic step in these algorithms is merging or combining two clusters with the highest similarity (closest distance) into one cluster. Ties may be broken arbitrarily. Initially, each data item is in a singleton cluster by itself. In each iteration, the algorithm chooses two clusters to be merged based on a selection criterion. The cycle then repeats, until only one cluster is left. The sequence of merges gives rise to what is called a cluster dendrogram. By cutting the dendrogram at any desired level, we can get the desired number of clusters. Because the algorithms must calculate the similarity (or distance) of each pair of clusters in every iteration, each iteration costs $O(N^2)$ time and there are exactly $N-1$ iterations until we reach a single cluster, giving an overall $O(N^3)$ time complexity for hierarchical clustering algorithms. This means that when N is large, these algorithms take an enormous amount of time. Choosing to stop after $k > 1$ clusters have been found, rather than going all the way to 1 cluster, doesn't significantly reduce the run time unless k is very large.

A number of different selection criteria are available.

1. Single linkage. In single link, the distance between two clusters c and c' is the minimum distance between any element of c with any element of c'.

2. Complete linkage. In complete link, we take the maximum distance instead of minimum.

3. Average linkage. In this method, we compute the average of all the pairwise distances between an element of c and an element of c'.

4. Centroid method. Here we take the distance between the centroid of cluster c and the centroid of cluster c' as the distance between c and c'. A variant of this is to use the Mahalanobis distance, which utilizes the covariance in the distance calculation.

5. Medoid method. Instead of the two centroids, in this case we take the distance between the medoids of the two clusters as the distance between them.

There are several other criteria that are used. See [128, 213, 230] and similar references.

Complete linkage tends to find "round" clusters – "clumps" – (as do most of the other approaches), while single linkage tends to find chains – "snakes". To see this, consider Figures 5.2 and 5.3. Here we have generated data from two clusters, in one case (the left column) two normal densities, separated on the x-axis, and in the other two concentric circles. Hierarchical cluster trees were constructed on the data sets using the two linkage methods, and these were cut at $k = 2$ to produce two clusters. Figure 5.2 shows the data, with plotting symbol indicating the clusters identified by the cut in the tree. Figure 5.3 shows the cluster trees with the symbols at the leaves indicating the true clustering.

As indicated in the figures, single linkage clustering tends to perform poorly with "clumps" – particularly if there are outliers; it tends to put outliers in their own cluster. Complete linkage operates similarly to K-means, and will tend to work well with well-separated groups that are round or elliptic in shape. Single linkage does very well with groups consisting of disjoint lines or curves, provided the within-curve distances are smaller than the between-curve distances.

An example of a (complete linkage) hierarchical clustering of a subset of the KDD-Cup data is given in Figure 5.4. The data correspond to a random sample of the Satan, Neptune and nmap attacks; however, the clustering algorithm does not use the class labels. For each case, 200 samples were selected, for a total of 600 observations.

Satan and nmap are scanning attacks (probes) looking to determine the hosts and applications on the network and potential vulnerabilities. Neptune is a denial-of-service attack; it is a SYN-flood, where a large number of connection requests are sent to a server with spoofed source IPs. This results in half-open connections being maintained, since the spoofed IPs don't respond to the SYN/ACK from the server, filling the connection table of the server and denying access to legitimate requesters.

As seen from the cluster tree, there are a number of clusterings at various levels; one would select the level to cut the tree and this results in a set of clusters. A heat map for these data is shown in Figure 5.5. In this, the data matrix is shown as an image, with gray scale indicating the value of the variate in that cell of the matrix. The columns correspond to the variables and the rows correspond to the observations. Hierarchical cluster trees are

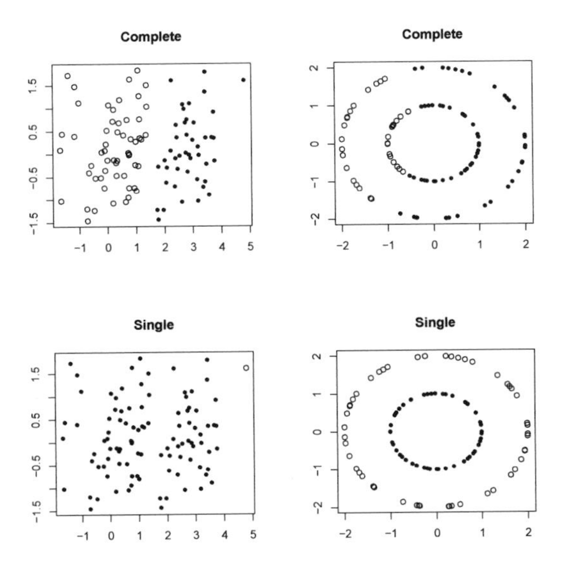

Figure 5.2: A comparison of hierarchical (agglomerative) clustering on two different data sets using complete linkage and single linkage methods. The symbols in the plots correspond to the clusters identified by the algorithm.

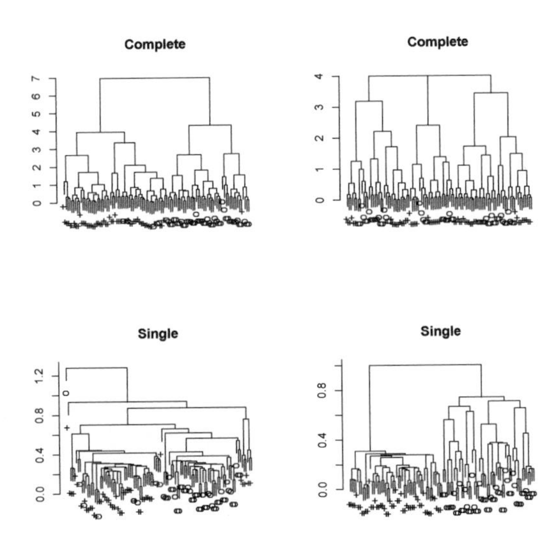

Figure 5.3: A comparison of hierarchical (agglomerative) clustering on two different data sets using complete linkage and single linkage methods. The symbols at the leaves of the cluster tree indicate the "true" grouping of the data.

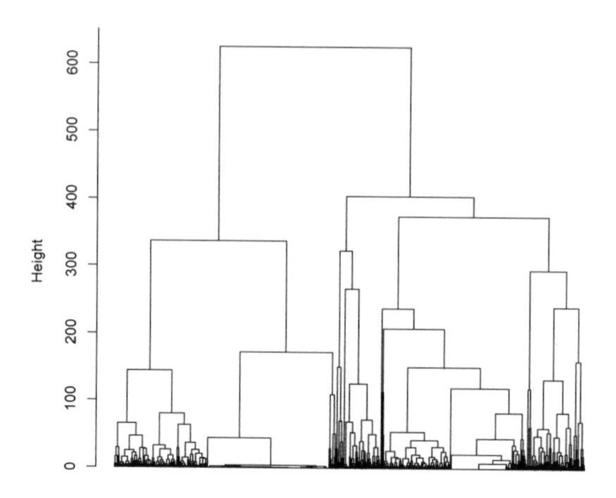

Figure 5.4: A complete linkage clustering of the KDD-Cup data corresponding to the Neptune, Satan and nmap attacks.

Table 5.4: The Number of Observations from Each of the Three Attack Classes in the Hierarchical Clustering Depicted in Figure 5.4. The Tree was Cut to Produce 7 Clusters

cluster	neptune	nmap	satan
1	89	72	24
2	111	0	9
3	0	46	1
4	0	66	2
5	0	10	0
6	0	6	0
7	0	0	164

shown as computed on the rows and the columns. The row tree corresponds to the tree shown in Figure 5.4.

The hierarchical clustering can be analyzed by considering how the three classes are distributed throughout a given choice of cut. For an example, Table 5.4 shows a cut at $k = 7$, producing 7 clusters, and the number of observations of each attack class within each cluster are shown in the table. As can be seen, the first cluster is a bit of a mix, but cluster 2 is nearly purely Neptune, clusters 3–6 are pure, or nearly pure, nmap, and cluster 7 is pure Satan. Thus, the clusters are clearly separating certain aspects of the different attacks, even though the class labels were unknown to the algorithm. This indicates that the clusters do have some intrinsic meaning related to the problem of distinguishing the attacks, and gives an indication that the problem of distinguishing the attacks from each other, using these data, may be a relatively easy one.

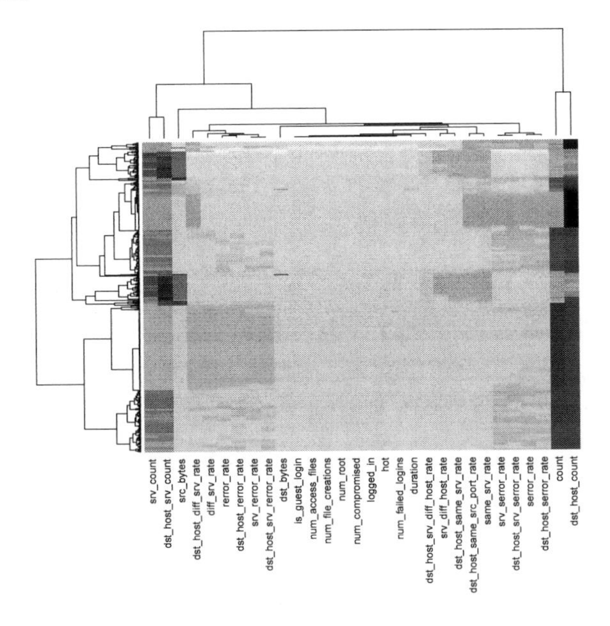

Figure 5.5: A heat map of the KDD-Cup data corresponding to the Neptune, Satan and nmap attacks.

5.6.5 Other Clustering Algorithms

Many other clustering algorithms have been proposed in the literature. We summarize some of the most effective ones.

DBSCAN and LDBSCAN

The DBSCAN paper [142, 377] has two key ideas: a density-based clustering model and the clustering algorithm.

The density-based clustering model. Given a radius ϵ, a distance measure, and a threshold on the number of neighbors, *minPts*, a core point is defined as a point with more than minPts neighbors within the radius ϵ. The idea of DBSCAN is to find areas of minimum density that are separated by areas of lower density. All the neighbors of a core point, within the radius, are considered to be part of the same cluster and called *direct density reachable*. If such a neighbor is also a core point, then its neighborhood is included as well and called density reachable. Noncore points are called *border points*, and all points within the same cluster are called density connected. Points that are not density reachable from any core point are called *noise* and not included in any cluster. Note that a border point can be a border point for more than one cluster, but it is included in only one cluster. This choice of cluster is made arbitrarily.

We now present an abstract DBSCAN algorithm whose actual implementation requires the use of efficient data structures for handling the operations. The complexity of a straightforward implementation of this algorithm would be $O(n^2 f(m))$, where n is the number of points and $f(m)$ is the cost of computing the distance function for a pair of m-dimensional points. DBSCAN inspired considerable research, e.g., HDBSCAN, LDBSCAN, etc. We mention one extension of DBSCAN called LDBSCAN.

DBSCAN is based on a "global view" of density, which may not be suitable for some data sets or applications. LDBSCAN tries to correct this with a local density based algorithm [127].

Algorithm 3: Abstract DBSCAN Algorithm

Input: Data set, distance function, Minpts threshold, ϵ
Output: Clustering
1 For each point, determine whether it is a core point;
2 Merge neighboring core points into clusters;
3 **For** each noncore point nc **do**;
4 **If** $NC = \{c \mid$ c is core neighbor of $nc\} \neq \emptyset$ **then** ;
5 Assign nc to an arbitrary $c \in NC$;
6 **else** mark nc as noise ;

Model Based Clustering

Model based clustering is similar to K-means, except in three aspects: it utilizes both the means and the covariance (shape) of the clusters, and provides methods for determining (or specifying) the properties of the shape; it provides a natural way to assign a probability vector to an observation, rather than a hard cluster threshold; and, through an information theory criterion, it provides an estimate of the number of clusters (see Section 4.9 for some detail on this).

The basic tool of model based clustering is the mixture model. For example, if one models the data as a mixture of normal densities, the model is:

$$f(x) = \sum_{j=1}^{k} \pi_j \phi(x; \mu_j, \Sigma_j). \tag{5.12}$$

Here ϕ denotes the (multivariate) normal, π_j corresponds to the mixing proportion – the proportion of the overall density accounted for by the jth component – μ_j corresponds to the mean, and Σ is the covariance matrix. By requiring a certain structure to the covariance matrix, such as diagonal matrices – the axes of variation are parallel to the coordinate axes – or requiring all covariances to be equal, etc., one can specify a model structure to the clusters.

Once the model is fit, a new observation x then has probability:

$$\frac{\pi_j \phi(x; \mu_j, \Sigma_j)}{f(x)} \tag{5.13}$$

for cluster j. One can threshold on these probabilities to produce a hard clustering, as in K-means, or one can use the probabilities as a "soft" threshold. Further details can be found in [156, 308, 309].

Fitting the model is via the EM (Expectation Maximization) algorithm. This is a general algorithm that is appropriate for any mixture of densities (and for many other models as well). See [309, 310].

We ran EM clustering in WEKA on our long jumps data set (Section 5.6.1) with the crossvalidation method of choosing cluster numbers and it returns with two clusters in two iterations. Cluster 0 with three instances and Cluster 1 contains 7 instances. The Log likelihood is -0.97698. The cluster information is shown in Figure 5.6.

Unfortunately, WEKA did not finish in more than one hour with EM clustering and crossvalidation ($-X10$ option) method on the phishing URL data set. We terminated the run.

Heuristics for faster convergence of the EM algorithm are explored in [304]. They combine Markov Chain Monte Carlo (MCMC) simulation technique to escape local minima.

Algorithm 4: Normal Mixture EM Algorithm

Input: Data set, initial parameters for the model, maxIters, ϵ

Output: Parameters of the model

1. For each point, determine the probability for each component, using Equation (5.13). This is the E step;
2. Using these probabilities for all points, compute the new parameters by computing the appropriate weighted sums; for example, the proportions are the summed probabilities divided by n; the means are the weighted averages; the covariances are the weighted version of the sample covariances. This is the M step;
3. Stop when the parameters change by less than ϵ or the maximum iterations have been used.

```
              Cluster
Attribute           0       1
                (0.3)   (0.7)
=============================
abcissa
  mean        23.9677 21.5684
  std. dev.    0.2525  0.4953
```

Figure 5.6: EM clustering with crossvalidation to select number of clusters on the long jump data set.

MCMC techniques can slow down the algorithm, so for faster convergence, they use several ideas: sufficient statistics, simplified model parameter priors, fixing covariance matrices and iterative sampling from small blocks of the data set. Promising results are shown in preliminary experiments.

Affinity Propagation

Our next clustering algorithm, which was proposed in [160], is called "affinity propagation." The main problem with K-means type algorithms, which choose a set of random initial points as cluster centers, is that this initial choice can have a big effect on the final clustering returned. In affinity propagation, on the other hand, the data points exchange messages until a good set of cluster representatives and corresponding clusters emerge. The number of clusters is not pre-specified as in K-means.

Suppose that 1 through n are the data points and s is a similarity function, such that $s(i, j) > s(i, k)$ iff i is more similar to j than to k. The diagonal of s (i.e., $s(i, i)$) represents "preferences," which means how likely a point is chosen to be a representative. When it is set to the same value for all points, it controls how many clusters the algorithm produces. A value close to the minimum possible similarity produces fewer clusters, while a value close to or larger than the maximum possible similarity, produces many clusters. It may be initialized to the median similarity of all pairs of points.

The algorithm updates two matrices in alternating message passing steps: the responsibility matrix R and the availability matrix A. Responsibility $r(i, k)$ measures the accumulated evidence for k to represent i taking into account other potential representatives, and is sent from i to k. Availability $a(i, k)$ measures the accumulated evidence for how appropriate it would be for i to choose k as its representative taking into account the support from other points for k, and is sent from k to i.

1. Responsibilities are updated using the rule: $r(i,k) \leftarrow s(i,k) - max_{k' \neq k}\{a(i,k') + s(i,k')\}$.

2. Availabilities are updated using the rule ($i \neq k$):

$$a(i,k) \leftarrow min\{0, r(k,k) + \sum_{i' \notin \{i,k\}} max\{0, r(i',k)\}\}.$$

Self-availability is updated differently: $a(k,k) \leftarrow \sum_{i' \neq k} max\{0, r(i',k)\}$.

These update rules require only "local" computations. Iterations are performed until either the cluster boundaries remain unchanged over a number of iterations, or some predetermined number of iterations has occurred. The representatives are extracted from the final matrices as those whose 'responsibility + availability' for themselves is positive, i.e., $r(i,i)+a(i,i)) > 0$. Numerical oscillations are possible in these updates, so a damping factor $0 < \lambda < 1$ is used to prevent them. The new update value is λ times the old value + (1 - λ) times the computed update value.

Spectral clustering

The idea of spectral clustering is to partition a graph. Assume that we have a set of n data items $1, 2, \ldots n$ or "points," and we have a pairwise similarity function s_{ij} (or distance function d_{ij}). Then we can turn this into an undirected graph with the points as vertices in several different ways.

1. The ϵ neighborhood graph, an unweighted graph in which we connect all points whose pairwise distance is at most ϵ.

2. k nearest neighbor graph. Here we have two choices: (i) we connect vertices i and j provided either i is among the k-nearest neighbors of j or vice versa. This is called the k nearest neighbor graph. (ii) we connect vertices i and j provided both i is among the k-nearest neighbors of j and vice versa. This is called the mutual k nearest neighbor graph. In either approach, the weight of the edge $\{i,j\}$ is $w_{ij} = s_{ij}$.

3. The fully connected graph, a weighted graph in which we connect every pair of points i, j, again $w_{ij} = s_{ij}$.

Now, assume that we have a undirected, weighted graph G on n vertices $1, 2, \ldots n$ with nonnegative weights $W = (w_{ij})_{n \times n}$. Define the degree of a vertex i as $d_i = \sum_{j=1}^{n} w_{ij}$. Construct a degree matrix D as a diagonal matrix with $D_{ii} = d_i$. Now construct the unnormalized Laplacian of a graph G as $D - W$. The spectral clustering algorithm works as follows. We refer the reader to [463], for properties of the Laplacian matrix, two normalized

Algorithm 5: Unnormalized Spectral Clustering

Input: Laplacian matrix L and desired number of clusters K
Output: K clusters
1 Compute the first K eigenvectors $e_1, \ldots, e_K$ of L;
2 Let E be an $n \times K$ matrix containing e_i as columns;
3 Clustering $\{C_1, \ldots C_K\}$ = K-means(E, K) /* each row of E is a point */;
4 Output clusters $B_1, \ldots B_K$, with $B_i = \{j \mid y_j \in C_i\}$;

spectral clustering algorithms, and an exploration of why spectral clustering works better than other clustering methods. See also [288]. In [63], a generalized version of the spectral clustering algorithm using the p-Laplacian of a graph is given. The p-Laplacian is a nonlinear generalization of the Laplacian. Authors show that the clustering obtained is often better than the one obtained from the Laplacian. We should also note that the results of the clustering depend significantly on the structure of the graph, see [292].

Hybrid Clustering Algorithms

There are also quite a few proposals that combine clustering techniques. An interesting proposal, called MOSAIC, that combines K-means with hierarchical agglomerative clustering is in [91]. It first runs a representative-based clustering algorithm such as K-means and then builds a certain kind of proximity graph, called Gabriel graph, on the cluster representatives. The final step is an iteration of merging clusters using a fitness function and updating the graph. The researchers provide some experimental evidence that this method combines the strong points of the two basic clustering algorithms that are being combined. Authors claim that Gabriel graphs are more efficiently constructed using approximation algorithms as opposed to the popular Delaunay graphs.

5.6.6 Measuring the Clustering Quality

There are two different approaches to measuring the quality of a clustering: extrinsic versus intrinsic. In the extrinsic approach there is external information available, perhaps in the form of a labeling of the data items into classes. In this case, we can compute the purity of a cluster c by taking the percentage of items in c that belong to the most frequent class of c, which is considered as the label of the cluster. We can then calculate the purity of the clustering as the average purity over all the clusters. Besides purity, we can use maximum matching, F-measure, conditional entropy, normalized mutual information, and variation of information. The latter three measures are referred to as entropy-based measures. There are also correlation-based measures: discretized Huber statistic [204] and its normalized version.

A number of internal cluster validation indices are compared in the paper [178]. Internal cluster validation indices try to estimate the number of clusters without using any external information.

Another possibility is to consult domain experts, when they are available. It is generally a good idea to do this whenever possible; however, note that experts don't always agree, and the existence of clusters that are not predicted by the experts may be indications of either a novel discovery, or an error in the clustering solution.

In the intrinsic approach, we use an optimization function that is calculated from the given data set without using any external information. The sum of square errors is one such criterion. A second possibility is to evaluate how compact is each cluster and how well-separated are the clusters. Many different indices have been developed to determine the quality of a clustering. See [250, 305] for definitions of several, and illustrations of their use.

Another general method for determining the quality of a clustering method for a given data set is to repeat the clustering several times and determine how consistent it is. There are several ways to do this. One can split the data into two sets, cluster each, then see how well each clustering "predicts" the other data. Alternatively, one can perform crossvalidation (bootstrap resampling) and see how often withheld points are clustered together or together with other points. There are several variations on these themes. See [321, 430] for examples and discussion.

5.6.7 Clustering Miscellany: Clusterability, Robustness, Incremental, ...

So far, we have not discussed an important issue, i.e., whether the data has any clusters to begin with. We refer the reader to two important works for this issue [4, 90]. For characterization and detection of noise in clustering, see [109]. For a unified view of robust clustering methods, we point to [110]. For incremental, graph-based clustering of dynamic data streams, we refer the reader to [286]. A density-based clustering over an evolving data stream with noise is given in [67]. A good survey of clustering algorithms can be found in [502]. For a taxonomy of clustering methods, see [212]. An intriguing graph-based algorithm for clustering is the normalized cut algorithm of [387].

5.7 Manifold Discovery

As mentioned in Chapter 2 high dimensional data has complex, counter-intuitive properties. A folk-theorem[7] is that "real" data "lie on" a low dimensional substructure in the high dimensional feature space that we observe. The purpose of *manifold discovery* is to find this low dimensional structure – to reduce the dimensionality of the data, removing noisy and uninformative dimensions so that the lower dimensional structure can be used for inference.

In fact, as Martin Wainwright tells us ([467]), finding a low dimensional structure on which our data lie is our only hope! In high dimensions, all of the methods designed to work in an asymptotic, "high n, low d" world fail, and they fail catastrophically. Only by reducing the dimension – through feature selection, sparsity, manifold discovery or model constraints – can we hope to obtain good and reliable inference. The paper [124] provides a nice discussion of some of the issues related to the curse of dimensionality and the ways to counter, and even embrace, the strange behavior of high dimensional data.

Feature selection methods as discussed in Chapter 2 reduce the dimensionality, but they do so by removing all the information in the features, retaining only the "most important" features. Manifold discovery seeks to find an embedding into a lower dimensional space that (potentially) uses all the features, but reduces the noise and irrelevance in these features by finding the "true" structure that holds the most inferentially relevant portion of the data.

One method (cf. Section 4.11) is called *multidimensional scaling*. It takes an inter-point distance (or dissimilarity) matrix and produces a set of points in $\mathbb{R}^d$ whose inter-point distance is (approximately) the same as the original. This produces an embedding into a lower dimensional space, and the "manifold" is then the support of the data in this lower dimensional space.

A related method is the *isomap* algorithm ([426]. Here one computes a graph on the data using proximity to define the edges. That is, the vertices of the graph corresponds to the points, and there is an edge between two vertices if their distance is less than ϵ, where $\epsilon > 0$ is a parameter of the model. Alternatively, one may connect two vertices if one of them is one of the k-nearest neighbors of the other, where again k is a parameter of the data. Other distance-based graphs could be utilized in this step, see [300] for some examples. See [21, 38] for methods to handle large data sets where computing all inter-point distances is impractical. Once the graph has been obtained, the shortest path distances are computed in the graph, and it is these distances which are used with multidimensional scaling to embed the data. In mathematical terms, the graph distances are approximating the intrinsic metric of the "manifold".[8]

Most other approaches are similar: one constructs a graph on the data, then using some properties of the graph (distances, the adjacency matrix, etc.) one constructs an embedding of the data. For examples, see [33, 34, 270]. A quite different method appears in the neural network literature. This method, called *autoencoding*, is discussed in Section 6.8.

Manifold learning has been used for analyzing and visualizing network flows data ([342]), detecting bots in gaming ([79]), and correlating attacks to reduce false alarms ([339]). Anomaly detection using manifold learning approaches is discussed in [139].

[7] A pretty much universally accepted truism, used more for guidance in modeling than as a provable fact.
[8] Quotes signify the lack of assumptions about the mathematical structure of the support of the data.

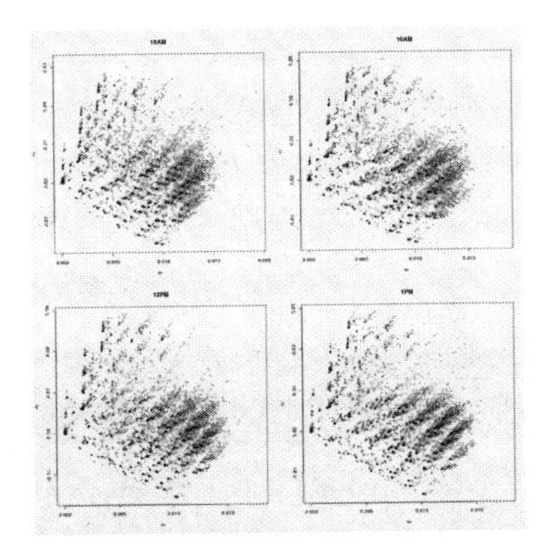

Figure 5.7: Four hourly graphs, embedded into $\mathbb{R}^2$.

5.7.1 Spectral Embedding

Spectral embedding corresponds to the first part of the spectral clustering algorithm discussed above. One constructs (or is given) a graph (directed or undirected, as appropriate) and utilizes either eigenvectors or singular vectors of the adjacency matrix of the graph (in the case of *adjacency spectral embedding*) or the Laplacian – or some scaled or otherwise modified version – to embed the vertices of the graph in $\mathbb{R}^d$ for some value of d.

Given a graph – such as the graph of connections between computers on a network within a given hour – the largest connected component of the graph is extracted, and the graph is embedded using adjacency spectral embedding ([287, 288, 418]).

1. Extract the adjacency matrix, the binary matrix with a 1 in position ij if there is an edge from vertex i to vertex j.

2. Compute the singular value decomposition of this matrix, retaining the top 2 singular vectors. Thus, $A = UDV^T$, with D being the diagonal matrix of (nonnegative) singular values.

3. Use the vectors in U to embed the vertices. In the case of an undirected graph, one can use the eigenvectors rather than the singular vectors.

The resulting embedding (into $\mathbb{R}^2$) is depicted in Figure 5.7. This data comes from Los Alamos and is described in [441]. The hourly graphs for 4 contiguous hours on day 3 were constructed and embedded using the above procedure. In practice, one would select the embedding dimension either through analysis of the singular values, as is done in principal components, or through some analysis utilizing the intended inference task. However, for the purposes of visualization and illustration, we have embedded into $\mathbb{R}^2$.

The most interesting aspect of the plots in Figure 5.7 is the obvious similarity of the patterns in the four plots. Clearly there is structure in these graphs that is similar across time, and we can use this structure to analyze the graphs or look for anomalies. The obvious clusters in the figure would be investigated to determine whether they represent specific types of behavior. Also, one could construct measurements from these embeddings (such as the sum or other statistic of the distances between the points from one hour to the next, after a Procrustes transform has been performed) and tracked in time, to look for times that have a very different pattern than their predecessor.

Quite a bit is known about spectral embedding of certain models of random graphs. A popular method for finding communities in graphs is called spectral clustering. One embeds the vertices using spectral embedding and then performs clustering, usually a k-means clustering, although model-based clustering can sometimes be superior. See [288].

5.8 Anomaly Detection

Anomaly detection is a well-studied challenging topic, which is very useful for security challenges, since it is the most capable of detecting so-called "zero-day attacks." In anomaly detection, we build a model or tests for normal behaviors and anything that is not classified as normal is called an anomaly or an outlier. The main difficulty lies in defining normal behavior. In the security field, this problem is even more difficult, since attackers are constantly trying to make anomalous behavior look normal.

A common assumption underlying anomaly detection methods is that the number of outliers is significantly smaller than the number of normal instances. There are quite a few surveys on anomaly detection, e.g., see the latest one by [72] and references therein. There is also a book on Outlier Analysis [5].[9] Because of these excellent sources, after discussing popular anomaly detection approaches, we shall briefly review some exciting recent work.

In [72], the researchers define point anomalies, contextual or conditional anomalies and collective anomalies. Point anomalies are the simplest type of anomalies, where a single data instance can be considered an anomaly. For example, a single credit card transaction can be fraudulent. Contextual and collective anomalies are more complex. In contextual anomalies, the context determines whether something is an anomaly or not. The attributes are divided into two categories: contextual attributes and behavioral attributes. In collective anomalies, a group of related data instances is anomalous.

We shall focus on the following approaches to anomaly detection:

1. Statistical methods
2. Distance-based approaches
3. Nearest neighbor based approaches
4. Density-based approaches
5. Clustering-based approaches
6. One class learning

5.8.1 Statistical Methods

A simple test for outliers is statistical and was proposed by the famous statistician Tukey. The first quartile Q_1 is the value $\geq$ one-fourth of the data, the second quartile Q_2, also called the median, is the value $\geq$ half of the data and the third quartile Q_3 is the value $\geq$ three-fourths of the data. The interquartile range (IQR) is calculated as $Q_3 - Q_1$. Tukey proposed that the values that are above the lower quartile and the upper quartile by 1.5 times IQR can be considered as outliers, i.e., the values that exceed $Q_3 + 1.5IQR$ or the values below $Q_1 - 1.5IQR$.

Assuming that data is distributed normally, one can use the deviation from the mean to determine outliers.

More sophisticated anomaly detection algorithms are based on distance, density or clustering. The distance and density methods are commonly referred to as proximity-based approaches. In proximity-based approaches, a common assumption is that outliers are significantly dissimilar than the rest of the data set.

[9]In our discussion, the terms outlier and anomaly are used synonymously.

5.8.2 Distance-based Outlier Detection

In one distance-based approach, we need two parameters as inputs: a radius ϵ and a percentage threshold π. A point is an outlier provided at most π percent of all other points are closer than ϵ. Mathematically,

$$Outlierset(\epsilon, \pi) = \{p \mid \frac{|\{q \in DB \mid dist(q,p) < \epsilon\}|}{|DB|} \leq \pi\} \tag{5.14}$$

Here $|DB|$ denotes the size of the entire data set or database. The calculations of outliers can be done more efficiently using an index or a grid-based method.

Another distance-based approach uses the Mahalanobis-distance, MD. If Σ is the covariance matrix of the data set, then

$$MD(d, \mu) = \sqrt{(d - \mu)^T.\Sigma^{-1}.(d - \mu)},$$

where μ is the mean of the data and d is any data instance. We then set a threshold on the set of Mahalanobis distances and declare all points as outliers, whose MD value exceeds the threshold.

5.8.3 kNN Based Approach

This method requires a user specified parameter k. Once k is given, we take as outliers those objects with the largest kNN distance. Here *kNN distance* is defined as the distance to the kth nearest neighbor. This is a simple method that requires $O(k)$ additional computation per point, once the kNN sets for all the objects have been identified.

5.8.4 Density-based Outlier Detection

In density based schemes, the key idea is that the density around an outlier is significantly different than the density around its neighbors. One density-based scheme, which we call local outlier factor approach, assigns to each point its degree of being an outlier. Like the kNN based approach, this algorithm requires an input parameter k as well. We calculate the local outlier factor (LOF) of each data instance d as follows:

1. Find kNN distance of d and its kNN distance neighborhood, which is the set of all points within d's kNN distance. Note that this can be more than k because of ties.

2. Find reachability distance of d with respect to another data instance d' as maximum of kNN distance of d' and distance of d to d'.

3. Find local reachability density (LRD) of d as inverse of the average reachability distance based on the k nearest neighbors of d.

4. Calculate the LOF of d as follows:

$$LOF(d) = \frac{\sum_{d' \in kNN(d)} \frac{LRD(d)}{LRD(d')}}{k} \tag{5.15}$$

The advantage of this approach is that it detects certain data instances as outliers that other methods can't. However, the disadvantage is that it is not clear what value of LOF should be used to determine outliers. Hence, we need a second parameter, a threshold $t > 1$, to declare outliers. There are many variants of and attempts to improve the LOF approach, which led to a large body of research.

5.8.5 Clustering-based Outlier Detection

Many clustering algorithms have a natural definition of "noise" or outliers, so they can be also used for outlier detection. For example, DBSCAN, includes the concept of noise points, which can be regarded as outliers. There are at least three different kinds of clustering-based anomaly detection methods:

1. Techniques assuming that outliers do not belong in any cluster.
2. Techniques based on the notion that the normal data instances are closer to their cluster centroids than the outliers, which are further away.
3. Techniques considering that normal data instances are in large dense clusters and outliers are in small or sparse clusters.

5.8.6 One-class Learning Based Outliers

This is an example of a classification based technique. The assumption is that we can learn a classifier to distinguish the normal and anomalous data instances. One-class learning methods can be used for outlier detection. We will discuss more about this in the next chapter, when we cover one-class learning. Of course, multiclass learning methods can also be used for anomaly detection. The problem in security is getting labeled data for both normal and anomalous data instances. Hence, we emphasize one-class anomaly detection.

We have briefly discussed the evaluation of anomaly detection schemes. The best evaluation method is to have ground truth available, but this may be difficult. In such a situation, a combination of simulations and/or manual analysis of the outliers may have to be used.

In addition to the above anomaly detection techniques, we mention two recent works on unsupervised anomaly detection, which we believe are quite promising. In [37], researchers have proposed an embedding technique that solves the problems of sequential clustering and detection of anomalous temporal data. Comparisons with four different methods, three of them based on dynamic time warping, show promising results. In [89], researchers design an unsupervised anomaly detection scheme for the online activity of users. They introduce a dispersion score for users, which is based on events, each event is a pair consisting of the volume and inter-arrival time. They model the normal behavior of an online user using a mixture of two log-logistic distributions. The parameters are estimated using EM. On a ground truth data set, they report a recall of 0.7 and a precision of at least 0.9. A recent application of manifold learning techniques to anomaly detection in image data is given in [335].

Much of security data consists of commands and file names, i.e., categorical data. Similarity measures for categorical data are evaluated in [48]. How to do anomaly detection on categorical data is explained in [9, 81, 105]. Finally, methods for anomaly detection in high-dimensional spaces can be based on subspace embedding techniques, such as manifold learning, or subspace sampling techniques, such as [223].

5.9 Security Applications and Adaptations

The techniques in this chapter have been employed for a number of different security challenges. For example, association rule mining (Section 5.5) has been used for intrusion detection, malware detection and stepping-stone detection; clustering (Section 5.6) for malware, and anomaly detection (Section 5.8) for intrusion detection and detecting spear phishing. We cover some of this work below, selecting work that *either extends some of the above techniques, or introduces new techniques for some classical data mining tasks.* We begin with the classic paper, [266].

5.9.1 Data Mining for Intrusion Detection

Host-based Intrusion Detection

In this paper, the researchers analyzed a sendmail data set that was used in previous experiments [267] and provided by [154]. This data set consists of a set of traces of the *sendmail* program. Normal traces include "a trace of the *sendmail* daemon and a concatenation of several invocations of the *sendmail* program." The normal traces were generated by 112 artificially constructed messages, with an attempt to include as many normal variations as possible. The messages produced a combined trace length of over 1.5 million system calls.

Abnormal traces include "three traces of sunsendmailcp attacks, two traces of the syslog-local attacks, two traces of the decode attacks, one trace of the sm5x attack and one trace of the sm565a attack." Each file of the trace data has the process ids and encodings of the system calls made by the program.

The first step was feature selection. A sliding window was used to scan the normal traces and a list of unique sequences of system calls was created. This list is called the normal list. Then they scanned the abnormal traces, and labeled every sequence of system calls (with the same sliding window) as normal if it appeared in the normal list and abnormal otherwise. The reason behind this procedure is the "exhaustive" nature of short sequences of system calls collected by [267]. Thus, each data item has n positional attributes, one for each system call, in a sequence of length n. Table 5.5 shows labeled sequences with a sliding window of length 7.

Table 5.5: Labeled System Call Sequences of Length 7

System Call Sequences (length 7)	Class Label
4 2 66 66 4 138 66	normal
...	...
5 5 5 4 59 105 104	attack
...	...

Next, the data set was split into training (80% of normal traces and abnormal sequences from 2 traces of the sunsendmailcp attacks, 1 trace of the syslog-local attack and 1 trace of the syslog-remote attack) and testing (remaining 20% of normal traces and all the abnormal traces not used in the training portion). Then researchers used the RIPPER algorithm [97], which is an inductive rule learning program, on the training data set. RIPPER outputs if-then rules for the minority class, and the "true" default rule for the remaining items. Here is an example of the kind of rule generated by RIPPER from [267]:

```
normal :- $p_2$ = 104, $p_7$ = 112.
```

This rules states that the sequence is normal if p_2 is 104 (the system command *vtimes* and p_7 is 112 (the system command *vtrace*). The RIPPER rules can be used to label test sequences as normal or attack. They still needed a strategy for labeling the traces as normal or intrusion. For this purpose, sliding windows of odd length $2m + 1$ starting from 7 were used with a sliding step of m, and predictions for each sequence are obtained from RIPPER. For each region of length $2m + 1$, if more than m predictions were attacks, then the region is marked as attack. If the percentage of attack regions is above a threshold determined empirically, then the trace is labeled as an intrusion.

Four experiments were conducted to learn the threshold. There were two parameters for the experiments: the normal:abnormal ratio and sequence lengths. Since RIPPER learns rules for the minority class, in two of the experiments, the authors made the normal class as minority class. The idea was to show that normal class patterns could be used to detect unseen attack traces (i.e., attacks not in the training data). In a second task, the system call prediction task, only the normal traces were used for training. Here again, four prediction experiments were conducted. We refer to [266] for the remaining details.

Network-based Intrusion Detection

In another experiment, the goal is to construct an anomaly detector using just the normal data. In some of the host-based intrusion detection work above, both intrusion and normal traces were used. For network-based intrusion detection, another *tcpdump* data set[10] was used to predict the destination of connection requests from the other connection features.

The data set had three runs of *tcpdump* on generated network intrusions and one run on normal network traffic. Now *tcpdump* is not designed specifically for security tasks, so the researchers went through multiple iterations of data preprocessing and feature extraction. Briefly, the authors studied TCP/IP and its security-related problems to understand the protocols and figure out the relevant features. After the first experiments on *tcpdump* data with RIPPER in which varying window sizes and five-fold classification were used, the authors added temporal and statistical features to the mix. Table 5.6 shows the data fields.

Table 5.6: Description of Data Fields

time	Converted to floating pt seconds ... hr*3600+min*60+secs
addr and port	The first two fields of the src and dest address make up the fake address, so the converted address was made as: x + y*256
flag	Added a 'U' for udp data (only has ulen) X - means packet was a DNS name server request or response. The ID and rest of data is in the 'op' field. XPE - means there were no ports... from "fragmented packets"
seq1	Packet's data sequence number
seq2	Data sequence number of data expected in return
buf	Number of bytes of receive buffer space available
ack	Sequence number of the next data expected from the other direction on this connection
win	Number of bytes of receive buffer space available from the other direction on this connection
ulen	Length, if a udp packet
op	Optional info such as (df) ... do not fragment

```
time,src_addr,src_port,dest_addr,dest_port,flag,seq1,seq2,ack,win,buf,ulen,op
38141.504694,1,7000,2,7001,U,,,,,,148,""
38141.510076,3,20,2,3421,.,1811081902,1811082414,366784001,9216,512,,""
38141.515159,3,20,2,3421,.,512,1024,1,9216,512,,""
38141.516172,4,80,2,2609,.,,,438528422,9112,,," (DF)"
38141.516647,5,25,2,1362,F,266688477,266688477,580609140,4096,0,,""
```

The data is stored in comma-separated-values format. The first few lines of one of the data sets is shown above. It shows that there are missing values for some fields in the records.

[10] http://ivpr.cs.uml.edu/shootout/netdoc2.htm - Accessed 30 July 2018. The paper's link is obsolete.

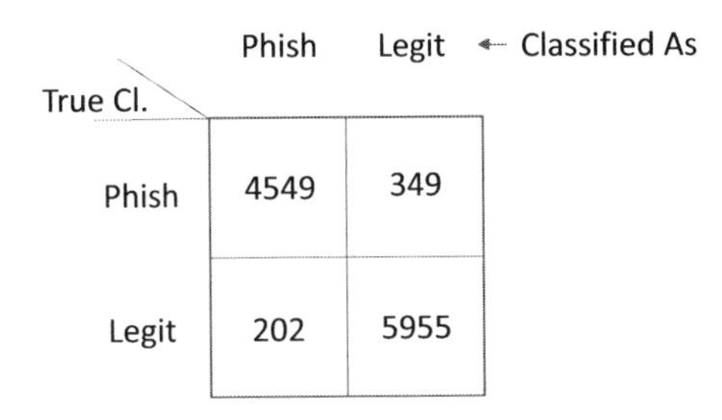

Figure 5.8: The confusion matrix for JRiP on the Phishing URL data set.

The final set of experiments use association-rule mining for intra-record patterns in the audit records and the frequent episodes algorithm [295] for inter-record patterns. These two are then applied again to the *tcpdump* data set. A frequent episode is defined as a set of events that occur frequently within a certain window of time. A certain minimum frequency, *min_fr* of simultaneous occurrence in sliding time window is specified. Events in a *serial* episode must follow a partial order in time, parallel episodes have no such constraints. The definitions of support and confidence are now based on frequencies of simultaneous occurrence. Notice the parallels with association rule mining. These two algorithms are then run on audit data under different settings to construct a rule knowledge base. This knowledge base can then be used for anomaly detection by designing scoring functions for deviations from normal profiles.

Note that attacks that have just a few traces will not show up in the frequent episodes algorithm. For these kinds of attacks, we refer the reader to [399]. Also, from the presentation of the researchers in [266], it is hard to determine whether the data set was balanced or unbalanced. No security analysis of the methods is attempted either, which means it is hard to determine their effectiveness against an active attacker.

We run JRip (WEKA's version of RIPPER) on the same phishing URL data set in Section 5.5. There are 29 rules, we show the top 5. The time to build the model was 2 seconds. These rules should be compared with the ones given above for the classification version of Apriori in WEKA.

```
(URL_of_Anchor = -1) and (SSLfinal_State = -1) => Result=-1 (2049.0/0.0)
(URL_of_Anchor = -1) and (SSLfinal_State = 0) => Result=-1 (969.0/0.0)
(SSLfinal_State = -1) and (Links_in_tags = -1) and
(Links_pointing_to_page = 0) and (Request_URL = -1) => Result=-1 (248.0/8.0)
(SSLfinal_State = -1) and (Links_in_tags = -1) and (having_IP_Address = -1)
=> Result=-1 (203.0/21.0)
(SSLfinal_State = -1) and (having_Sub_Domain = -1) and (DNSRecord = -1)
=> Result=-1 (171.0/16.0)
```

WEKA also gives the confusion matrix, shown in Figure 5.8, using JRip as a classifier. The accuracy is pretty good, 95.02%. We can compute a lot of metrics from the confusion matrix. If we regard phishing as the positive class, then there were 202 false positives, i.e., legitimate URLs classified as phishing, and 349 false negatives, phishing URLs classified as legitimate. The recall is 92.9%, the precision is 95.7%, and the F-score is 94.3%.

The work of [266] inspired scores of researchers who have applied data mining techniques to problems such as intrusion detection, malware detection, etc. For example, association rule mining has been used in stepping-stone detection, in profiling [17], and malware detection. Hierarchical clustering algorithms have been used for malware clustering. However, much of this work has ignored the unique needs of the security domain, and thus much remains to be done.

5.9.2 Malware Detection

In [500], researchers have presented a method for malware detection based on association rule mining. The hypothesis of their technique, as well as a lot of work on malware detection, is that malware exhibits its behavior through system calls. The unfortunate fact is that we can reduce the halting problem for Turing machines to the problem of determining a program's behavior, so the latter problem is also undecidable. However, we can still estimate program behavior, which is the basis for their method and other malware detection techniques.

Inspired by the work of [378], Xiao et al. [500] extract system calls from the headers of Windows Portable Execution (PE) format files. In [378], libBFD, a library in the GNU Bin-Utils toolset, was used to extract information from the object format. This information is included: the list of DLLs used by the executable, the list of DLL function calls made by the executable, and the number of different functions calls within each DLL.

After extracting system calls, a goal directed version of association rule mining called objective-oriented association rule mining algorithm is used. In this version of association rule mining, the attributes or features are split into objective attributes and nonobjective attributes.

The class labels of the software are added as new attributes. The malware class attribute is the only objective attribute and the rest are nonobjective attributes. The *utility* of a rule of the form

$$I_1, I_2, \ldots I_M \Rightarrow Obj$$

is defined as the degree to which the itemset $\{I_1, I_2, \ldots I_m\}$ supports the objective Obj. It can be calculated by dividing the difference of the positive utility and the negative utility of all the records that contain the itemset by the frequency of the itemset. This still leaves open the definition of positive utility and negative utility of a record. These are usually dependent on the domain and require domain expertise. In the malware detection application, the positive utility is defined as 1 for all records that have the class malware and negative utility as 0.

The researchers collected 9,216 programs, which included 1,694 hamwares and 7,522 malwares, which is the opposite of a realistic scenario. The hamware was "randomly selected from the system32 directory of a newly installed Windows XP system," and the malware was collected from two web sites: "vx.netlux.org" and "offensive.computing.net." The malware included mainly viruses, trojans, and worms. They unpacked and disassembled all of the malwares; 1,662 malwares were packed, and 21 malwares could not be unpacked. They extracted a total of 6,181 distinct window APIs from the samples (excluding the 21 packed malware instances). They calculated the program distribution for each API and sorted them in descending order. The program distribution of an API A is the ratio of the number of executables that definitely call A to the number of total executables. Only 9% of the API's have a ratio exceeding 35% and 538 had a ratio exceeding 2.5%. Information gain was applied to the APIs for selection. After that, the objective association mining algorithm was applied and rules were selected for classification. We refer the reader to [500] for the results. The authors report the confusion matrices as well as detection rate and accuracy. However, the times taken for association rule mining or the overall classification process are not reported.

5.9.3 Stepping-stone Detection

When an attacker is targeting a victim's computer, the attacker typically uses a long chain of intermediate hosts to hide behind. These intermediate hosts used by the attacker to create a long attack chain are called stepping stones. Association rule mining has been used for stepping stone detection by quite a few researchers including, e.g., [201].

The data set consisted of simulations on the university network. A transaction represents a fixed length time slot and consists of a number of connection patterns. Each connection pattern is an unordered pair of the form $\{IP, IP'\}$, where *IP* and *IP*' are IP addresses. They simulated three kinds of servers (SSH server, proxy server and SOCKS server) to simulate attacks. Using *tcpdump*, they collected a total of 4,029 traffic records with 1,667 stepping-stone transactions and the remaining were nonstepping stone. For transactions, they experimented with different length intervals ranging from 1s to about 25s. The association rule mining algorithm was run with different values for support and confidence. The association rules were then analyzed for potential stepping stones.

5.9.4 Malware Clustering

The techniques in this chapter have also been used for clustering malware [31], where the hierarchical clustering algorithm is used with a twist for efficiency.

To extract features, software execution is monitored and a behavioral profile is created by abstracting system calls, their dependencies, and the network activities to a generalized representation consisting of OS objects and OS operations. The authors extend taint tracking [379], which tracks the flow of information through a program, to better capture malware behaviors. System calls are the taint sources, i.e., the researchers taint the returned values and out-arguments of system calls. Also, the in-arguments of system calls are checked for taint information. The idea is to track the usage of information obtained from interactions with the OS. An example from [31] is tracking the output of the *GetDate* command in a subsequent *CreateFile* command. This leads to the determination that filenames created by the software change on every execution, a behavior more indicative of malware. The entire code of the program is also tainted to determine whether it reads its own code. Control flow decisions based on tainted data are also recorded. This helps them in finding out similarities between programs that perform the same kind of "date-checks" or try to disable "anti-virus" software. Network activity of a program cannot be tracked through system call traces, since there is a single API function called regardless of the activity. Hence, the researchers use *Bro*, a network analysis tool,[11] to recognize application-level protocols, e.g., SMTP, IRC, etc.

The second step is to build the behavioral profile. It consists of a set of operating system objects, their types, their dependencies, and the operations that were performed on them. The idea is to overcome the limitations of too rigid an attention to syntax, e.g., similar functions but different names, and also the different ways in which the operations may be structured by malware writers to defeat signature-based schemes.

The third step is clustering. To reduce time and improve scalability, since malware is exploding, authors use LSH, *Locality Sensitive Hashing*, to avoid calculation of $\Theta(N^2)$ pairwise distances for N items/clusters. LSH is an approximate, probabilistic approach, which solves the approximate nearest neighbor search problem in sublinear time. It can also be used to compute an approximate clustering.

LSH maps a set of points into a table so that close/similar points have a higher collision probability than distant/dissimilar points. The family H of hash functions is constructed to satisfy the property: $Pr[h(a) = h(b)] = similarity(a, b)$, for a hash function h chosen uniformly at random from H and points a, b. The researchers define $lsh(a) = h_1(a), \ldots, h_k(a)$ using k hash functions selected independently and uniformly at random from H so that $Pr[lsh(a) = lsh(b)] = similarity(a, b)^k$. They use the Jaccard similarity function,[12] since [61] constructs a family of hash functions with the needed properties for this function. For a given similarity threshold t, locality sensitive hashing computes an approximation of the

[11] `https://www.bro.org/`

[12] For two sets a and b, Jaccard(a, b) $= \frac{|a \cap b|}{|a \cup b|}$

true set of all close pairs $\{(a,b) \mid Jaccard(a,b) > t\}$. A final check is made to remove pairs whose Jaccard similarity is below the threshold. The remaining pairs are sorted by similarity and the single-link clustering algorithm (Section 5.6) is run. This gives part of the clustering, since it deals with only the pairs whose similarity is above the threshold. For the remaining details, we refer to [31].

The data set consisted of 14,212 malware samples from ANUBIS over a four-month period. These samples were given to six different anti-virus programs. Only if a majority of those six reported it as malware, the sample was retained. This resulted in 2,658 samples.[13] The quality of the clustering was measured using average recall and precision. The method achieved 87 clusters while the gold truth had 84. The precision was 0.984 and the recall 0.930 for $t = 0.7$.

We refer the reader to [145, 146] for a more comprehensive study of malware clustering in which traditional methods, such as K-means and hierarchical clustering, as well as the more recently clustering techniques, such as affinity propagation and spectral clustering, were used.

5.9.5 Directed Anomaly Scoring for Spear Phishing Detection

In [192], researchers present a new method for anomaly detection, called directed anomaly scoring (DAS). They applied this method to detect 19 spear phishing emails in a large data set of around 370 million emails, i.e., a highly unbalanced data set.

A taxonomy of spear phishing attacks or impersonation models is given [192]: address spoofer, name spoofer, previously unseen attacker, and lateral attacker.

An address spoofer uses the email address of a trusted sender in the From field of the header. The attacker could also spoof the name in the From field as well and try to match the header exactly as the header of an email from the trusted sender modulo the dates and times. Such an attacker could be caught by email authentication methods such as DKIM (Domain Keys Identified Mail protocol) and DMARC (Domain-based Message Authentication, Reporting & Conformance). However, not many organizations are currently using DKIM/DMARC. The name spoofer spoofs the name in the From field, but not necessarily the email address. Since the email address is not spoofed, DKIM/DMARC cannot catch this kind of attack.

A previously unseen attacker spoofs neither the name nor the email address of a trusted sender. However, such an attacker uses the name of a seemingly trusted authority that the victim would be familiar with, e.g., University of Houston IT Team with an email address: uh-help-desk@uh-it.edu. A lateral attacker sends an attack email from a compromised account, usually in the victim's organization.

The researchers consider only the last three impersonation models. Next, they list three common problems with typical anomaly detection approaches: parametric nature, direction agnostic, and signal combination.

1. Many approaches assume an underlying probability distribution, whose parameters must be determined, or they require proper setting of some input hyperparameters. For example, in distance-based approaches, we need two hyperparameters: the radius ϵ and the percentage threshold π.

2. For example, in the standard deviation based approach, a standard deviation of 3 would be considered just as anomalous as a standard deviation of -3 from the mean (assume a standard normal variable). However, in security settings, fewer visits to a domain would render the domain as suspicious, but a large number of visits to a domain may be harmless.

[13] This choice was criticized by [271].

3. They treat an event as anomalous even if only a few features are anomalous. Researchers feel that for spear phishing attacks all features would be anomalous.

Hence, they propose a new anomaly detection algorithm. In this algorithm, a relation "more suspicious than" is defined on pairs of events. An event is identified with its d-dimensional real-valued feature vector. Event E is more suspicious than event E' provided $E_i \leq E'_i$ for all $1 \leq i \leq d$. The $\leq$ is for convenience. For features where low values are more suspicious, they use $\leq$, and for features where high values are more suspicious, they use $\geq$. The score of an event E is the number of events E' such that E is more suspicious than E'. Once the score of all the events has been calculated, they just return the top N most suspicious events, where N needs to be chosen correctly.

They use information from three kinds of logs (SMTP logs, network intrusion detection system logs and LDAP logs) from the enterprise, the network traffic and an alert budget β for each sub-detector, of which there are three, one corresponding to each impersonation model introduced by the researchers. The researchers have outlined how to construct a real-time alert service based on these inputs.

The results are impressive, with some caveats. First, it is difficult to be certain that there are no false negatives, i.e., some attacks may have been missed by the team that identified attacks from the data set. Second, even one attack missed may be one too many, if it leads to significant damage. The DAS algorithm was able to detect 17 out of 19 spear phishing emails. However, as the authors observed, it is challenging to construct ground truth since there are 370 million emails. So they used a combination of successful attacks that were detected and the manual analysis of 15,521 alerts. These alerts were generated from a combination of running (1) an older version of their detector that used manually chosen thresholds instead of the DAS algorithm; and (2) a batched version of their anomaly scoring detector, which ran the full DAS scoring procedure over the click-in-email events in the evaluation window and selected the highest scoring alerts within the cumulative budget for the chosen timeframe.

We conclude this section with some observations: no method is foolproof. Hence, defense-in-depth and user education/training are very important in practice. Also, to our knowledge, some methods, such as spectral clustering and affinity propagation, have not been used in security challenges.

5.10 Concluding Remarks and Further Reading

We saw some interesting unsupervised techniques in this chapter, e.g., association rule mining, clustering, and anomaly detection, and their applications/adaptations to security challenges. We end with a word of caution on all data-mining and statistical analyses: the Bonferroni principle.

This principle states that the probability of finding a pattern increases with the number of comparisons that are performed on a data set (the multiple comparison problem), or with an increase in data, even if the data is random. We explained the remedy in the Statistics Chapter 4, viz., the Bonferroni correction.

Chapter 6

Machine Learning – Supervised Learning

The main topics in machine learning that are relevant to cybersecurity are classification and clustering. Classification is the assignment of a class label, such as spam email versus benign or "ham" email, distinguishing viruses from legitimate code, determining that a user is masquerading as another user, etc. Clustering is the automatic grouping of data. There are other applications of machine learning that do not fall into these categories. For example, the detection of unusual events – characterizing normal behavior and flagging abnormal behavior – is also important for cybersecurity. Applications to control and automation, such as self-driving cars and flying and coordinating drones are also areas of application for machine learning. Clustering was covered in Chapter 5 so we will focus on classification (supervised learning) in this chapter.

The classification task is as follows. We observe pairs $\{(x_1, y_1), \ldots, (x_n, y_n)\}$, where the x_i are features – measurements of some observable object we wish to classify, such as email or user activity – and the y_i are class labels, usually denoted by integers or strings such as "spam" vs. "ham"[1] or "virus" vs. "benign". If we denote by X the space of features, and Y the set of class labels, we seek to find a function, called the *predictor*:

$$g : X \to Y, \tag{6.1}$$

so that given an observation $x \in X$ with true (but unknown) class $y \in Y$, $g(x) = y$. Of course, since we don't know y, we can only hope that our classifier g is correct "most of the time", and we seek methods to construct classifiers from training data $\{(x_1, y_1), \ldots, (x_n, y_n)\}$ so that the resulting classifier will have a low error on new data. In other words, we want to find a function g that minimizes

$$L(x, y) = \sum_{i=1}^{n} l(g(x_i), y_i)$$

where l is any loss function. If we average this function over the n data instances, we get the empirical risk.

$$R_{emp}(g) = L(x, y)/n \tag{6.2}$$

The empirical risk minimization principle states that we should choose our function g, also sometimes called the hypothesis, to minimize this risk.

[1] We use the term *ham* to refer to emails that are not spam.

The distinction between supervised and unsupervised learning is in the variables y in the above formulation. These correspond to labels – we are providing *supervision* to the algorithm. In clustering or unsupervised learning, we provide no class labels and rely on the algorithm to "find" the groups. This distinction is somewhat arbitrary, since even in classification we provide some supervision in the form of clustering criteria or model selection criteria. In supervised learning, we give explicit information as to which group the observation belongs, and it is this relationship that we wish the algorithm to learn.

Machine learning algorithms can be classified along at least four dimensions. The first dimension is inductive versus transductive learning. In *induction* we derive general rules, or equivalently try to learn a function, from the observed training data, which will be applied to the test data by evaluation. In *transduction*, we learn models from training data, e.g., a separating or decision boundary, and apply them to test data. The second dimension is generative versus discriminative learning. *Generative* models try to learn the joint distribution $P(x, y)$, whereas discriminative models try to learn the conditional distribution $P(y|x)$ directly. In general, discriminative models perform better than generative ones. The third dimension is active versus passive learning. In *passive* learning, which can be supervised, semi-supervised or unsupervised, the learning algorithm is given some combination of labeled and unlabeled data (one of these could be empty) and the algorithm goes ahead and does its job of learning a good predictor. In *active* learning, the algorithm may not have any labeled data at all to start with, but it can issue queries strategically and sequentially to learn the labels of some data instances. The idea is that there is a large amount of unlabeled data available and it is costly to get annotated data. The whole trick is for the algorithm to make as few queries as possible and learn a good predictor. The fourth dimension is online, or streaming, versus offline, or batch learning. In *online* learning, a learner must make predictions "in real-time," and cannot wait to see more data. In *batch* learning, a learner gets to see the whole training data set to build the model and then makes predictions on the test data.

In this chapter we discuss several classification methods. An important question for the practitioner is – how to choose the correct classifier? The examples we give will suggest that random forest is the best classifier to use, since in our examples they are generally the best of the ones tested. This hypothesis was tested convincingly in [150], in which many classifiers were tested on many standard machine learning data sets, and random forests and support vector machines (SVMs) were consistently at the top in performance. The data used in these tests wasn't restricted to cybersecurity applications, but the overall performance of these two classifiers was impressive, and the breadth of applications suggests that they are each "best in breed."

However, saying that on average a given method outperforms other methods doesn't mean it is the best for every problem. As a rule, random forests (and SVMs) are good, reliable algorithms, and should be considered for any classification problem. They are not necessarily optimal for any given problem. Knowing other methods and understanding their strengths and weaknesses is important for designing solutions to given problems.

David Hand ([185]) argues that more time should be spent on addressing the real-world issues associated with the problem, and less time on coming up with new classifiers. The extra apparent performance that one can obtain by carefully crafting a classifier to the particular data can be illusory, given the nonstationary nature of the world and the fact that the training data is finite and may not be representative. Simpler models are often better, because they are less susceptible to over-fitting, and are easily understood and analyzed. A procedure that starts with simple models and only extends to more complex models when the data provides strong justification, particularly with new data, is a good one to follow.

With that said, there has been a lot of excitement about "deep learning" and "neural networks". See, for example, the success of AlphaGo.[2] Many articles have been written about the great strides that deep neural networks have taken to solve extremely hard problems. This comes with a cost, as has also been reported.[3] As with all new technologies, *caveat emptor*; i.e., make sure you know what you are doing. It is a good idea to try several methods, collect new data to test them, look for ways to break them – a sort of "penetration testing" for machine learning – and investigate ways to visualize them and the data to get a better idea of the scenarios.

6.1 Fundamentals of Supervised Learning

From the above preliminary remarks, we see that any empirical risk minimization machine learning algorithm has two basic ingredients: a loss function (sometimes also called a cost function or objective function), and an algorithm that searches for an optimum of the loss function using the training data set (optimization procedure). An optional, but fairly common ingredient, is the addition of a regularization component to the loss function, which penalizes for model complexity. Recall the principle that simpler models are to be preferred.

Popular loss functions include: 0-1, cross-entropy, hinge, logistic, least absolute deviation (L1), and least square errors (L2).

1. The 0-1 loss function is $I(g(x_i) = y_i)$, where I is the indicator function and g is as above.

2. The hinge loss function is $\max(0, 1 - g(x_i)y_i)$.

3. The L1 loss function is $|g(x_i) - y_i|$.

4. The L2 loss function is $(g(x_i) - y_i)^2$, which is more susceptible to outliers.

5. The logistic loss function is $\log(1 + \exp(g(x_i)y_i))$.

Cross-entropy is given by the following formula in which $\hat{y}$ is our predicted distribution versus the actual distribution y, i.e., in this case, we must associate probabilities with each prediction. As usual, we sum the cross-entropy function over all the samples to get the total cross-entropy loss.

$$H_{\text{entropy}} = -\sum_i y_i \; log \; \hat{y}_i \tag{6.3}$$

The KL-divergence (Kullback-Leibler divergence, also known as relative-entropy) of $\hat{y}$ to y, written $KL(\hat{y}||y)$ is the difference of the cross-entropy and the entropy of y, where the entropy of y is just $-\sum_i y_i \; log \; y_i$.

Popular optimization procedures include: gradient descent (descent, since we are minimizing a loss function) and linear/quadratic programming. In practice, gradient descent is too slow for large data sets, and stochastic gradient descent (SGD) is used instead. In SGD, a single data instance is used to estimate the gradient. A variation on SGD is SGD with a minibatch, where a subset of the data instances is selected.

---

[2] https://www.wired.com/2016/03/two-moves-alphago-lee-sedol-redefined-future/

[3] https://www.wired.com/story/greedy-brittle-opaque-and-shallow-the-downsides-to-deep-learning/

6.2 The Bayes Classifier

It is a fact of life that for many problems there is no perfect classifier. It is impossible to build a rule that will always give the correct class for new data. There are generally two reasons for this. First, we don't always have a way of determining the class perfectly. If we don't know for sure whether an activity is malicious or benign, we can't know whether a given classifier is correct when applied to the activity. For most real-world applications we can't develop a training set that is both complete in its coverage and correct in its class assignment. The second reason is that distributions often overlap. If you want to determine if a person is a good candidate for a loan or is likely to default, there will likely be individuals with the same scores on the features you measure who have different values for class labels – one will default and the other won't. Any classifier will get one of them wrong.

It is possible, however, to define the optimal classifier, called the Bayes Optimal Classifier, usually just referred to as the Bayes Classifier. Using the notation from Equation (6.1), this is:

$$g = \arg\max_{g} P(Y = g(x)|X = x). \tag{6.4}$$

That is, the classifier that gives the highest probability of being correct. In terms of probability densities, the idea is to use the true probability densities for the different classes, and report the class whose likelihood is highest for the observation x.

This is great in principle, but not so helpful in practice, since we don't know the true densities. Although we have many ways of estimating them, these methods either assume we know the form (family) of the density, or they have difficulty with high dimensional or complex distributions. Further, no matter how much data one has for training – to build the classifier – it is finite and we cannot correctly determine the true densities.

6.2.1 Naïve Bayes

As we have seen in Chapter 2, high dimensional data is complicated and confusing. Estimation in high dimensions is problematic and difficult. The Naïve Bayes classifier is a method to get around these problems by considering the case where the features are conditionally independent given the class. Consider a d-dimensional problem, with each observation consisting of d features $(x_1, \ldots, x_d)$. Writing the Bayes Classifier as the likelihood, the Naïve Bayes assumes the likelihood reduces to the product of the marginal densities:

$$g = \arg\max_{j} f^j((x_1, \ldots, x_d)) = f_1^j(x_1) \cdot f_2^j(x_2) \cdots f_i^j(x_i) \cdots f_d^j(x_d). \tag{6.5}$$

Here the superscript denotes the class label, so f^j is the density for class j.

The Naïve Bayes classifier trades bias for variance. We can be sure that the independence assumption is wrong, and therefore the estimate of the overall density is going to be biased. However, trying to estimate all of the relationships amongst all the variables is going to naturally increase the variance of the estimator, and so by making the simplifying assumption of independence the Naïve Bayes estimator radically reduces the variance. This trade-off – the bias-variance trade-off – is fundamental and occurs throughout statistics and machine learning. See Section 4.3.1 for more details on the bias-variance trade-off.

6.3 Nearest Neighbors Classifiers

The nearest neighbor classifier is one of the simplest nonparametric classifiers, both to understand and to implement. The basic idea is simple: take your new observation and find the training observation closest to it. Associate that observation's class to the new observation.

It is instructive to use the nearest neighbor classifier to consider the issue of testing the classifier. In our framework discussed above, we have training observations $(x_1, y_1), \ldots,$ (x_n, y_n). If we use these observations to build our nearest neighbor classifier, how are we going to test it? The obvious answer to this question in general is to use the training data: pass the training data through the model and see how well the model performs. This is called *resubstitution*, and it can be a reasonable method for obtaining an estimate of the classifier performance for many models;[4] however, it is never used in practice due to its inherent bias. To see this bias in an extreme case, note that for the nearest neighbor classifier it is essentially useless. To see this, assume that there are no duplicate observations, $x_i \neq x_j$ if $i \neq j$. Then the resubstitution error is 0! After all, the nearest observation to x_i is x_i and so we obtain the correct class for each of our training observations!

There are two ways we could solve this problem. First, we could obtain new observations to use for our test – or equivalently hold out a proportion of the observations to use to test the algorithm's performance. This is always preferable, if it is possible. One way to implement this is to split the training set into two sets: a subset for training and a subset for testing. Then the observations we use for testing are distinct from those of training, and we obtain an unbiased estimate of the errors – assuming the split is done randomly and the training and testing sets are from the same distribution and are also from the same distribution as future observations, see [185]. The second is called crossvalidation. The idea is to withhold one observation (or a small number of observations), train on the remaining, and test on the withheld. Repeat this until all of the training observations have had a chance to be withheld. See Section 2.7 for further discussion on crossvalidation. As discussed there, we use the term "crossvalidation" for the leave-one-out version.

One issue with crossvalidation is the question of which algorithm is being tested. In practice, one would perform crossvalidation to estimate the performance of one's algorithm, but before deploying the classifier one would retrain on all the training data. In essence, the crossvalidation technique provides an estimate of the performance of the classifier *methodology*, rather than the specific classifier that is deployed. We are trading off a tiny bias caused by the fact that we are fitting many models to the data (one for each of the withholdings) for the reduced variance from using almost all the data in the model fit.

Note that this is not unique to crossvalidation; even if one splits the data into training and testing sets, ultimately one will train on all the data before deploying the classifier. Thus, in the absence of sufficient data to support a split into training and testing data, one should always utilize crossvalidation to estimate performance.

The obvious extension to the nearest neighbor classifier is the k-nearest neighbor classifier. In this classifier one fixes k and for each new observation x determines the k training observations nearest x. The set of classes for these training observations is then used to assign a class label to x by a simple vote: whichever class appears most in the set.

In the two-class case, it is always advisable that k be chosen to be odd. This is to avoid ties amongst the votes. For problems with three or more classes, it is not possible to guarantee there will be no ties. In these cases, one must decide what the classifier should return. The obvious choices are:

- One of the tied classes chosen at random.
- Order the class labels in some a priori manner and select the label with lowest order from the ties.
- Report the tie and the set of tied classes.

[4] It is actually not too bad an estimate for linear classifiers, which we discuss below, and it can be used to provide a rough bound on the performance of algorithms that require an excessive amount of time to train, but there are almost always better ways to assess performance than using resubstitution. In fact, it is probably true that no journal will accept an article that uses resubstitution to evaluate the algorithms.

Table 6.1: The Features Used in the KDD-Cup Example

duration	src_bytes
dst_bytes	hot
num_failed_logins	logged_in
num_compromised	num_root
num_file_creations	num_access_files
is_guest_login	count
srv_count	serror_rate
srv_serror_rate	rerror_rate
srv_rerror_rate	same_srv_rate
diff_srv_rate	srv_diff_host_rate
dst_host_count	dst_host_srv_count
dst_host_same_srv_rate	dst_host_diff_srv_rate
dst_host_same_src_port_rate	dst_host_srv_diff_host_rate
dst_host_serror_rate	dst_host_srv_serror_rate
dst_host_rerror_rate	dst_host_srv_rerror_rate

Alternatively, one could return a vector of length equal to the number of classes with the counts of the number of neighbors from each class, and let the users determine what inference they will make from this information.

At this point, the question arises as to how to choose k. What is often done is that one tries a range of values of k and selects the one with the best performance. Care should be taken in this process, however. Suppose one has split the data into two sets, a training set R and a testing set E. Consider the following procedure:

1. Select a set K of values for k.
2. For each k in K, use the training set R to define a k-nearest neighbor classifier g_k.
3. Evaluate g_k on E.
4. Choose the value of k with the best performance.

The problem with the above strategy is that it chooses the best value of k *for the test data* E. Thus, it introduces a slight bias into the estimator. A better solution is to either use crossvalidation on R to evaluate g_k, or to split R into two sets: a training set T and a *validation set* V, and use V instead of E in the above procedure. This latter process is often used for classifiers that require extensive training time, such as neural networks (Section 6.8), and are thus not well suited for crossvalidation.

To illustrate the nearest neighbor classifier, consider the KDD-Cup 1999 data.[5] This is a fairly complicated data set, designed to be a benchmark for exploring machine learning and pattern recognition algorithms. For the following experiment, the data was reduced by selecting a subset of the features, listed in Table 6.1. Only those observations from classes "normal", "portsweep" and "ipsweep" were retained.

The data was then reduced by removing duplicates, and randomly split into a training set of 1,000 observations from each class, leaving 816,601 for the test class. The nearest neighbor classifier ($k = 1$) on these data has the confusion matrix:

[5] Available at `http://kdd.ics.uci.edu/databases/kddcup99/kddcup99.html`

	ipsweep	normal	portsweep	error
ipsweep	2700	15	2	0.63%
normal	8289	793721	9518	2.19%
portsweep	10	10	2336	0.85%

In particular, note that the two attacks are each misclassified as "normal" around 0.5% of the time. The 3-nearest neighbor classifier has comparable results:

	ipsweep	normal	portsweep	error
ipsweep	2711	6	0	0.22%
normal	12011	785912	13605	3.16%
portsweep	19	14	2323	1.40%

Whether the above performance is satisfactory depends on the intended use. If one needs to investigate every potential attack, a false alarm rate of 2–3% is unacceptable. On the other hand, if this is simply a filter to select observations for further, more intensive (but automated) processing, then it may not be so bad.

It can be shown that the 1-nearest neighbor classifier has an error that is asymptotically – as the number of training observations goes to infinity – twice the Bayes error rate. The k-nearest neighbor classifier is asymptotically optimal – the error goes to the Bayes error as the number of training observations goes to infinity – provided k also increases at a slow rate. These statements can be made precise, but for practical purposes they are not terribly useful. Since one only has a finite data set with which to train a classifier, asymptotic results such as these cannot be used to gauge the expected performance of the algorithm. Still, it is good to know that the algorithm is, in some sense, "moving in the right direction" as the number of training observations increases.

6.4 Linear Classifiers

The linear classifier is the geometrically simplest of all classifiers and the easiest to draw (at least in one or two dimensions). The classifier simply defines a line (a plane in three dimensions and a hyperplane in higher dimensions) that splits the data so that one side of the line corresponds to one class and the other corresponds to the other class.

Figure 6.1 depicts data from two classes, and a linear classifier defined by the diagonal line. The points above the line are classified as "triangles", and those below are classified as "circles". The line is referred to as the *decision boundary*, and defines the equal-probability region – points on the line are of undetermined class (that is, the classes are tied).

Several things are clear from the plot in Figure 6.1. There are clearly several possible (in fact, infinitely many) linear classifiers one could define for these data that are equivalent (in terms of performance on the training data). Also, the proposed classifier in the figure has errors for both classes, and in fact no linear classifier can be defined that would have zero error on these training observations.

The data were drawn from two multivariate normal distributions, and so the Bayes error is not zero, but in this case the Bayes classifier is a linear classifier. In fact, it is the bisector of the line segment between $(0, 0)$ and $(2, 2)$. This suggests one method for defining the linear classifier, as this bisector of the segment joining the two means of the data from the two classes. This works, provided the covariances are equal and a multiple of the identity matrix, and one knows (or estimates) the means. More generally, there are several algorithms for fitting a linear classifier. It is instructive to write the formulas for two normal distributions with equal covariances, equate them, and solve for the corresponding hyperplane. This reduces to a quadratic equation, and is relatively straightforward to solve. In fact, in the case of equal covariances, the quadratic terms cancel, and the resulting equation is linear.

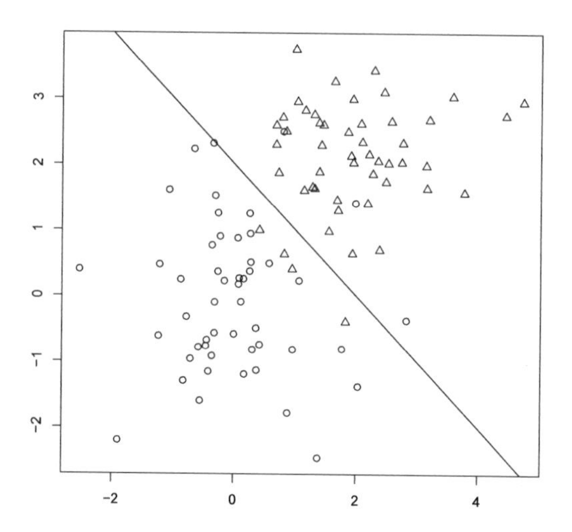

Figure 6.1: A two-class problem in two dimensions with a linear classifier indicated as a line. The two classes are depicted using different symbols, and the line is the decision boundary for the linear classifier.

Note that if we don't assume equal covariances, we still obtain a quadratic equation, and the resulting classifier is called a quadratic classifier.

Revisiting the KDD-Cup data from the previous section, the confusion matrix for a linear classifier is:

	ipsweep	normal	portsweep	error
ipsweep	2610	106	1	3.94%
normal	13903	770678	26947	5.03%
portsweep	4	36	2316	1.70%

This is considerably worse than the results for the nearest neighbor classifiers. This is not surprising. While it is always a good idea to investigate simple models like the linear classifier, for all the reasons mentioned in [185] and elsewhere in this book, it is often the case that nonparametric methods like the nearest neighbor classifier will be superior.

6.5 Decision Trees and Random Forests

In the 1970s and 1980s one of the most popular methods of artificial intelligence was expert systems. The idea was to query experts to determine how they performed a particular task, eliciting rules which could then be programmed into a computer, resulting in, it was hoped, a computer program that could solve the task as well as the human experts. These programs often consisted of a set of if-then statements, which encoded the expert's knowledge.

A field that showed some early successes was medicine, where expert systems were developed to diagnose illness. Figure 6.2 depicts such a decision tree. This is clearly not a serious attempt at diagnostics, merely an illustration of the ideas. At each node that is not a leaf, one or more variables are tested and a decision is made – in this case a simple binary decision – for further tests. At the leaves of the tree, the classification (or medical decision) is given.

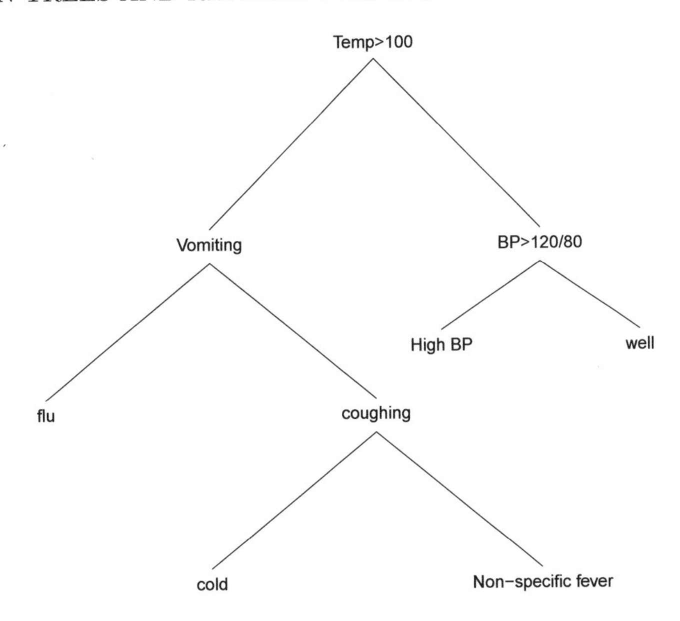

Figure 6.2: A decision tree for diagnosing illness. The left branches correspond to a true value for the node test, right branches correspond to a false node test.

In general, expert systems can be far more complicated than this simple example, but it illustrates a method that we will discuss in more detail below. In particular, expert systems are not restricted to trees of this type, and often involve domain specific features and algorithms in making the decisions. We will not discuss expert systems further, but the interested reader is encouraged to investigate references such as [130, 238, 275, 327].

Several problems with this approach soon surfaced. It is not always a simple process to get an expert to say how they solve a problem in ways that are translatable into computer code. While something like baking a cake seems like a simple enough process – one simply follows the recipe – imagine eliciting from a pastry chef how they come up with a new recipe.

Even something as simple as following a recipe has potential pitfalls; the phrase "add the eggs" requires the (usually unspoken) assumption that one cracks open the eggs and only adds the interior of the egg, not the shell.[6] This kind of "obvious" knowledge is not obvious to the computer, and must be extracted from the expert and encoded into the algorithm. In most real-world applications, there are many such unspoken and "obvious" facts that the experts don't necessarily consciously consider, of which the algorithm designer may be ignorant; thus the knowledge extraction can be extremely time consuming and tedious.

Another problem is that experts don't agree on how they do things. If one wishes to rely on a single expert this may not be a problem, although sometimes an expert may give different answers to the same question at different times. If one is interviewing several experts, a method of aggregating their answers needs to be developed. For a review of expert systems, see [275].

One possible solution to this problem is to design an algorithm that "learns" the expert system from data, thus eliminating the need for an expert. One such method was the classification and regression trees (CART) algorithm, described in [59]. The algorithm takes data such as the measurements a doctor would perform paired with a diagnosis and builds a tree to perform the classification. An example tree might look something like that in Figure 6.2.

[6]There is a visual joke in which a novice cook is told to "separate the eggs" and moves the uncracked eggs apart on the counter.

The CART algorithm is a two-stage algorithm. It recursively builds a large tree, in which the leaf nodes are pure – contain a single class. At each node, the data at that node is used to select a variable and threshold as in Figure 6.2. The variable is selected so that some measure of "goodness" of the split is satisfied – for example, splitting off a large bulk of one class from the others. The algorithm is then recursively applied to each set of data, those that satisfy the criterion and those that don't. It stops when the node is pure – contains only one class. Other stop-criteria might be to stop when the depth reaches a set maximum, or the node is "nearly pure" by some given definition. Variations allow for a validation set to determine whether there is an advantage to the split in classification performance, and to stop if there is none.

Once the tree is built, it is generally pruned back, to improve on generalization performance and reduce over-fitting. The details of the algorithm are in the book [59], and in the many papers that have been generated since.

As noted, the CART tree has a very important property that makes it quite useful – it is (relatively) easy to understand and to explain to a practitioner. This means that it is much more likely that one can generate trust in the algorithm, and to understand its limitations. It also tends to give fairly good results in practice.

Both of these statements come with a large and very important caveat: the algorithm has a high variance, in that a very small change in the training data can result in a dramatically different tree. Imagine using such a tree for triage in a medical facility. The doctors and nurses trust the algorithm because they understand it, and it tends to give good results, at least as a first-pass diagnostic. Then the vendor provides an update to the software, consisting of an entirely different tree that looks nothing like the one they have been using. This can cause quite a bit of concern, especially if it happens every year with the new software update.

This also makes performance estimates through crossvalidation a bit tricky. Each time an observation is withheld, an entirely different tree is formed that may bear little or no structural relation to previous trees. For this classifier, it is particularly difficult to be sure that the crossvalidation results are relevant to the tree that is sold to the customer. Clearly this is a case where a withheld test set is preferred.

There are many different criteria to measure the "goodness" of the split. They are usually classified into univariate (single attribute) and multivariate (combinations of more than one attribute). Popular univariate criteria include: the information gain, gain ratio, entropy and gini index. The Hellinger distance has been proposed as a criteria for building decision trees for unbalanced data sets [93]. We refer the reader to the books [29, 365] for more details on decision trees.[7] Note that the construction of an optimal decision tree is an NP-hard problem, so greedy heuristics are used in practice.

6.5.1 Random Forest

These drawbacks were identified very early on, and a solution was proposed in [56], in which the idea of *bagging* predictors was described. The idea is to retrain a predictor many times, and then average, or vote, amongst the "bag" of predictors. A bootstrap sample is drawn – that is a sample with replacement from the observations – and a predictor is constructed (classifier, regression, whatever the task is). Then the predictions are *aggregated* – hence the name Bootstrap**AGG**regat**ING**.

This was then extended to the idea of random forests in [57]. The idea of aggregating predictors, a kind of ensemble learning method, see Section 6.10, is simple but powerful – instead of using a single predictor, build many predictors and vote amongst them. This idea

[7]Chapter 2 of [29] compiles a good number of splitting criteria.

of ensemble methods is utilized in many areas of machine learning, and is the basis of the random forest classifier and random forest regression.

Given a data set $\{(x_1, y_1), \ldots, (x_n, y_n)\}$ where the x_i are the predictors (measurements) and the y_i are the class labels or predictions, the random forest builds many trees as follows. We assume a function f, the *split criteron*, that takes a threshold τ and a vector $\mathbf{v}$ and returns a value indicating the "goodness" of splitting the data into two sets: $\{v < \tau\}, \{v \geq \tau\}$. For example, one might compute the purity of the split in terms of the class labels, or an information based criterion.

1. Sample (with replacement) from the data.
2. Build a decision tree in which at each node a subset of the variables is selected to test against the split criterion.
3. For each variable, select the value τ producing the best value of f on that variable.
4. Select the variable (and τ) resulting in the best split.
5. Split the data as indicated by the split, recursing on each set of data.
6. Stop when the tree reaches a given depth (or some other stopping criterion).

Note that the tree is not pruned, and the algorithm is greedy in that it does not spend any time on optimization except at the individual nodes.

There are several options that have been investigated in the above algorithm. One can select the set of variables for the tree up-front, and fix them for that tree; one can try all variables at every node; one can randomly select the split from the top K rather than selecting the best. The stop criterion can be a purity measure – stop when a node is $P\%$ pure; stop when the depth of the tree reaches d; stop when none of the possible splits at a node is sufficiently "good". There have also been investigations utilizing linear combinations of variables instead of single variables, and many other approaches.

The algorithm stops when a fixed number of trees has been created. As the process runs, note that the sample step allows for a kind of crossvalidated error calculation: those data that were not selected for a given tree, referred to as out-of-bag samples, can be run through the tree to obtain their predictions. These can then be used to provide an assessment of the performance of the algorithm without requiring a full crossvalidation on the forest.

Once a forest has been created, new data can be sent through each of the trees in the forest. The new observation stops at a leaf in each tree, and the training points that ended in the leaf are used to classify the point, or, in a regression, to predict the value. That is, in the classification case, the class proportions of the points that ended in the leaf are retained, and in the regression case the average of the dependent variables is retained (or other statistic, depending on the implementation and purpose of the algorithm). The results from all the trees are then aggregated – usually referred to as "voting", even when the inference task is regression – to get a final value for the observation.

An important side effect of the resampling algorithm used in the random forest is that the variable importance can be assessed via a similar procedure as is used to estimate the performance of the random forest through crossvalidation. For each variable in the data, the values of the variable are permuted (randomized), and the out-of-bag samples are run through the tree to determine the effect of this randomization. The size of the change in performance gives an estimate of the importance of that variable to the problem. The idea is that if a variable is important, its value matters, and randomizing it should cause a large negative change in performance. Alternatively, if randomizing the values has a small affect on performance, then the variable isn't very important to the classifier outcome.

Table 6.2: Comparisons of Several Classifiers on a Spam Data set from the UCI Repository

True Class		
	ham	spam
Linear		
ham	1304	214
spam	62	721
%Error	4.5	22.9
1-NN		
ham	1146	257
spam	220	678
%Error	16.1	27.5
3-NN		
ham	1150	293
spam	216	642
%Error	15.8	31.3
5-NN		
ham	1142	305
spam	224	630
%Error	16.4	32.6
Random Forest		
ham	1332	93
spam	34	842
%Error	2.5	9.9

To illustrate the performance of the random forest classifier, we consider a spam detection problem, and the data previously used from the KDD-Cup. Table 6.2 compares a linear classifier, three k-nearest neighbor classifiers, and a random forest classifier on a problem of detecting spam from data provided at the UCI Machine Learning Repository.[8] These are features extracted from 4601 emails that have been classified into "spam" or "ham". Clearly random forest performs best on this data set.

Similar results are obtained on the KDD-Cup data. A random forest run on these data shows the best performance to date:

	ipsweep	normal	portsweep	error
ipsweep	2713	4	0	0.15%
normal	1078	810240	210	0.16%
portsweep	0	4	2352	0.17%

This is interesting both because the performance is so much better than the other classifiers, and because it is pretty much the same for all classes. Note that the random forest, being random, will produce a different classifier each time it is trained on a data set (unless a random seed is set to ensure reproducibility); some experimentation shows that the second significant digit in the error calculations can vary quite a bit.

A useful property of the random forest is its ability to provide insight into the important variables. See Figure 6.3. From this, it is clear that the first three variables in the plot are the most important. Figure 6.4 depicts the KDD-Cup data for the first two of these, zoomed in to where most of the data are. Outside of this range, there are 2 port sweep

[8] `https://archive.ics.uci.edu/ml/datasets/spambase`.

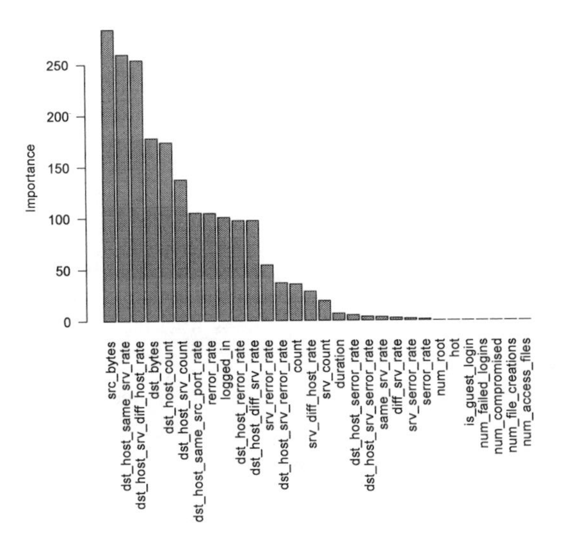

Figure 6.3: The variables used in the KDD-Cup experiment as reported by the random forest. The y-axis corresponds to the impurity, or Gini index, of the variable.

attacks and 946 normal observations, so this region is where most of the class-conditional structure is.

As can be seen in Figure 6.4, while there is considerable overlap – for example, the triangle in the upper right corner corresponds to 143 IP sweeps – there is very little overlap between "normal" and "attack". The figure also illustrates that, at least in this particular projection, the decision boundary, whether between "normal" and "attack" or each of the

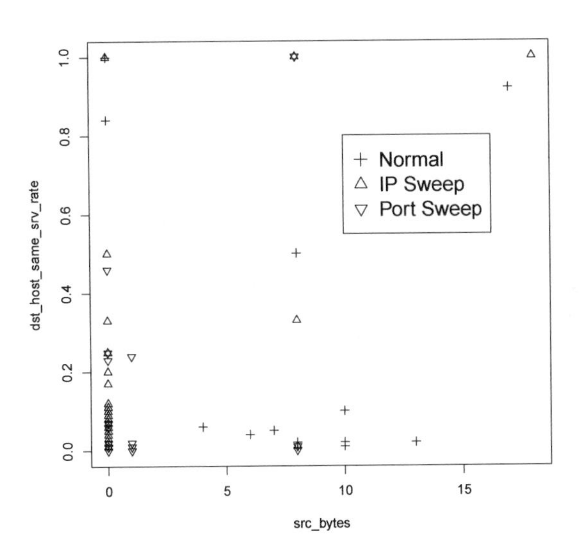

Figure 6.4: The KDD-Cup data plotted using the top two most important variables as reported by the random forest.

three classes, is nonlinear. While figures like this cannot be used exclusively to analyze the decision boundary, this provides support for the fact that the linear classifier is so much worse than the two nonlinear classifiers investigated so far, and suggests that either a nonlinear embedding is required or a nonlinear classifier must be utilized for this problem.

6.6 Support Vector Machines

The support vector machine model, developed by Vapnik, is a methodology that utilizes two basic ideas. The first is that data can be transformed into a space, usually a much higher dimensional space, so that the classes are linearly separable. The second is that the "best" such classifier is one that maximizes the gap, or "margins", between the classes. See [376] for more information about SVMs.

The first problem is to define the transformation into a space where the linear separability is true – or at least approximately true. One might think that this requires one to both determine the dimension of the resultant space, and construct the function of the variables that results in the transformation.

Fortunately, neither of these is actually needed in SVMs. One of the most important ideas in SVMs is the observation that with a linear classifier, all that one cares about is inner (dot) products – the hyperplane corresponds to a vector in the high dimensional space, and all one cares about is the inner product with this vector.

Thus, what one really wants to do is to define an inner product. This is done via the idea of a kernel, which is a symmetric function that satisfies Mercer's condition: function K is a kernel, provided there is a map ϕ such that:

$$K(\vec{x}, \vec{x}') = \phi(\vec{x}).\phi(\vec{x}') \tag{6.6}$$

Note that the right-hand side is a dot product. There are a number of these kernels that are available in any standard package that implements an SVM algorithm, and it is usually relatively easy to check that a given kernel satisfies the condition.

It turns out that such a kernel is equivalent to mapping the data into some high dimensional (often infinite dimensional) space and computing an inner product between the corresponding images of x and y. One never actually performs the mapping into this space, since all one cares about is the comparison provided by the inner product. Coming up with a kernel that works for a specific application is not easy. Under certain assumptions, it is possible to learn a suitable kernel function, see for example [10, 280, 337, 484] and references cited therein.

One other important concept is needed for the support vector machine: the idea of a *margin*. The intent of this projection to a high dimensional space is to project into a space where the data are linearly separable. If one imagines the data set projected into this high dimensional space, there are many possible linear classifiers that one could choose. Select two parallel hyperplanes that separate the classes, and select these so that they are as far apart as possible. This distance is the margin, (technically speaking, this distance is twice the margin, since margin is usually defined as the distance from the separating hyperplane to the closest points) and one wishes to select the hyperplanes to maximize this margin. It turns out that these hyperplanes are completely determined by a subset of data points, and these points are referred to as the *support vectors*. This is illustrated in Figure 6.5.

For convenience, consider a two-class problem with the points as distributed in Figure 6.5, where we can imagine that there is a "street" running between our data samples, demarcated by the dashed lines and the dotted line is the median of that street. Let $\vec{w}$ be perpendicular to the street or median and let $\vec{u}$ be an unknown vector. Then, our decision rule is that $\vec{u}$

belongs to the positive class (say denoted by the triangles), provided $\vec{u}.\vec{w} + b \geq 0$, for some constant b. If we put enough constraints on this problem, then we can solve for $\vec{w}$ and b.

So, we insist that for all positive samples $\vec{x_+}$, $\vec{x_+}.\vec{w} + b \geq 1$ and for all negative samples $\vec{x_-}$, $\vec{x_-}.\vec{w} + b \leq -1$. We introduce an auxiliary variable y_i corresponding to each data instance x_i, such that $y_i = 1$ for each positive instance x_i and $y_i = -1$ for each negative instance x_i. We can then multiply the two different constraints and they become the same constraint $y_i(\vec{x_i}.\vec{w} + b) \geq 1$. Furthermore, we insist that equality holds for the instances on the dashed lines, the support vectors. Now, consider a positive sample $\vec{x_+}$ on the dashed line and a negative sample $\vec{x_-}$ on the other dashed line. Then, we can write the width of the street, or the margin, as $(\vec{x_+} - \vec{x_-}).\vec{w}/||\vec{w}||$, which due to our constraints on the support vectors comes out to $2/||\vec{w}||$. Hence, maximizing the margin is equivalent to minimizing $\frac{||\vec{w}||^2}{2}$ subject to the constraints.

Using the method of Lagrangian multipliers, we can turn this into an unconstrained optimization problem as follows.

$$L = \frac{||\vec{w}||^2}{2} - \sum \alpha_i (y_i(\vec{x_i}.\vec{w} + b) - 1) \tag{6.7}$$

The α_i's are the Lagrangian multipliers. If we differentiate L with respect to $\vec{w}$, we get $\vec{w} = \sum \alpha_i y_i \vec{x_i}$. In other words, the vector $\vec{w}$ is a linear sum of the samples. Differentiating L with respect to b, we get that $\sum \alpha_i y_i = 0$. Now, we substitute back the value for $\vec{w}$ into the expression for L and simplify the resulting expression to get:

$$L = \sum_i \alpha_i - \sum_i \sum_j \alpha_i \alpha_j y_i y_j \vec{x_i}.\vec{x_j} \tag{6.8}$$

This shows that the optimization problem depends on dot products of the data instances. When we substitute for $\vec{w}$ in the decision rule, our decision rule becomes $\sum_i \alpha_i y_i \vec{x_i}.\vec{u} + b \geq 0$, so the decision rule also depends on dot products. The beauty of this optimization problem is that it can be solved "exactly" for the global optimal. Thus, SVM model differs from many other models, e.g., Neural Networks or Decision Trees, in which we can get stuck into local optimals.

To illustrate the performance of an SVM, we sampled 1000 observations of "normal" events, and 1000 of the attack events.[9] After training, the remaining observations are run through the classifier, with the results depicted in Table 6.3. The top of the table shows the performance as a function of the attack type, with the bottom showing the confusion matrix of normal vs. attack.

As can be seen from the table, the support vector machine performs extremely well on this data set. In fact, it is worth noting that the vast majority of the attack training observations were from a single attack – neptune, with none of the other attack types represented by more than 30 observations, and several having only 1 observation in the training set, with 11 attacks completely unrepresented.

It is important, when possible, to explore examples like this, where whole classes of attacks are withheld, to get a feel for how well the algorithm generalizes to new attacks. Of the testing observations from attacks that were not present in the training data, a little over 80% of them are correctly classified as attacks, with comparable results on the two attacks that had a single observation in the training set, indicating that the classifier has generalized beyond just the attacks it was trained on.

[9] In practice one would utilize far more than 2000 observations in the training set, given the large number of observations available. This split was in part for computational purposes; in part to illustrate the power of the support vector machine on even small training sets; and in part to maximize the size of the test set to obtain really good estimates of performance (at least on some of the classes).

Table 6.3: The Results of a Support Vector Machine on the KDD Cup Data

	attack	normal
back (5)	890	72
buffer_overflow (1)	21	8
ftp_write	5	3
guess_passwd (1)	51	1
imap	11	1
ipsweep (17)	3614	86
land	19	0
loadmodule	7	2
multihop	6	1
neptune (891)	193440	22
nmap (13)	1218	279
normal	8993	802535
perl	3	0
phf	0	4
pod (2)	79	125
portsweep (14)	3292	50
rootkit	6	4
satan (28)	4453	483
smurf (18)	2848	141
spy	1	1
teardrop (4)	280	634
warezclient (6)	695	192
warezmaster	18	2
attack	210957	2111
normal	8993	802535
% Error	4.1	0.3

The classifier was trained on a data set containing 1000 normal, and 1000 attack observations, labeled as "attack". The bottom of the table shows the confusion matrix for the test data. The attacks in bold had at least one example in the training data, with the number of observations in the training set shown in parentheses.

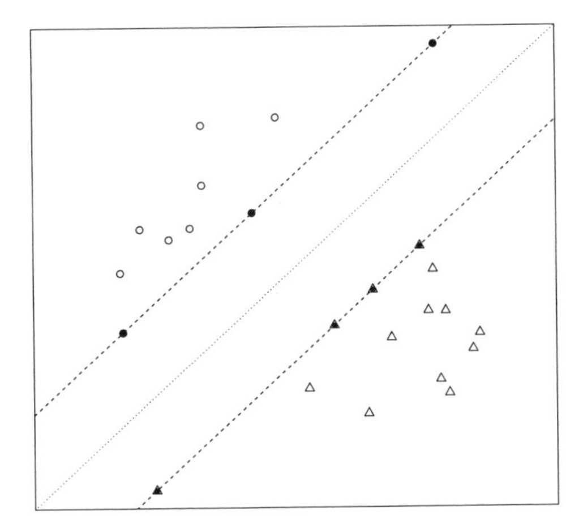

Figure 6.5: A depiction of the SVM margin. The dashed lines are the two hyperplanes that define the margin, with the support vectors indicated with a black dot within the symbol. The dotted line corresponds to the maximal margin hyperplane.

Several generalizations of the SVM have been proposed, e.g., SVM for one-class learning, semi-supervised learning (the transductive SVM, which we discuss below), and for the case when there is no hyperplane that can separate the two classes exactly (the soft-margin SVM). In the soft-margin SVM, we allow some slack in the constraints. The Kernel trick requires coming up with a kernel and, for some applications, it is likely to be difficult to find a kernel that works well. For the soft-margin SVM, all constraints are changed to $y_i(\vec{x_i}.\vec{w} + b) \geq 1 - \epsilon_i$, where the ϵ_i's are nonnegative slack variables,[10] and the optimization function then becomes $\frac{||\vec{w}||^2}{2} + C\sum_i \epsilon_i^k$, where C is a regularization parameter, which controls how much slack is allowed for the problem, and k is a positive integer. In practice, typical values of k are: $k = 1$ and $k = 2$.

6.7 Semi-Supervised Classification

In many problems, particularly those related to cybersecurity, it is difficult to obtain data with accurate class labels, while it is easy to obtain unlabeled data. Methods for performing semi-supervised learning are designed to allow the utilization of unlabeled learning to improve the classifier.

This might seem counter-intuitive, but the example in Figure 6.6, while overly simplistic, makes the point.[11] On the left, we have only two labeled observations, from two different classes. Typically one has relatively few labeled observations, and while this is an extreme case, one should think about a small number of cases in high dimensions. The dotted lines are some suggested decision surfaces (it is instructive to think about how one could/should

[10]They are called slack variables since the constraint is looser now. We are not insisting anymore that the left-hand side be at least 1.

[11]This example was suggested in a talk by Justin Beaver, Oak Ridge National Laboratory, given at the Second Annual SSC Pacific Workshop on Naval Applications of Machine Learning, 2018, `http://www.public.navy.mil/spawar/Pacific/News/Pages/NAML2018.aspx`.

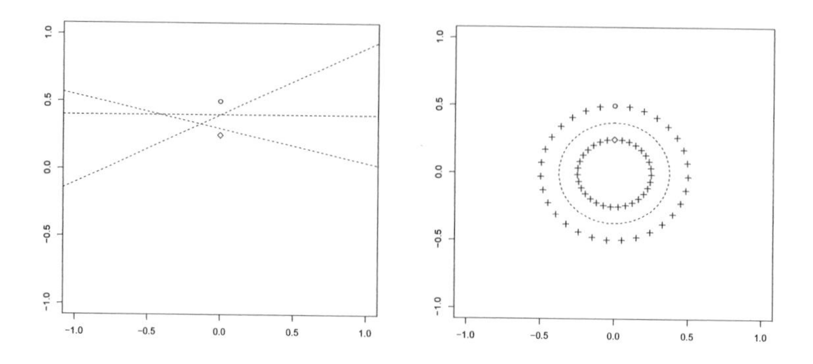

Figure 6.6: An example to illustrate the potential utility of semi-supervised classification. On the left is an example of two data points, with two different class labels as indicated by the symbols. The dotted lines are potential linear classifiers for this problem. On the right, unlabeled observations, indicated by the "+" symbol, suggest a much different classifier.

choose between them). On the right, we posit that a large number of unlabeled observations are obtained, shown with the plus symbols. Given these, it is quite reasonable to posit the dotted curve as the decision region. Although we don't know that this is in fact correct, after all the unlabeled points are in fact unlabeled, it is what the picture suggests.

Suppose that one subsequently obtains the resources to determine the labels for a small number of the unlabeled observations. Given a model such as the one in the right of Figure 6.6, one could use the inferred class boundary to determine unlabeled observations that would be most useful to classify; essentially we have a hypothesis for the class boundary that we can test, by choosing observations that would disprove or provide evidence for our hypothesis.

There are many semi-supervised methods in the literature. A survey of some of these can be found in [524]. There are also books on semi-supervised learning [76, 526]. Semi-supervised methods can be generative, discriminative, self-training, multiple-view or graph-based. The EM method below is generative. Generative methods also include: Cluster and Label, Multiple-view learning methods include: Co-training, Yarowsky's algorithm for word-sense disambiguation can be considered a special case of co-training, Co-EM, Co-Boost, Agreement Boost algorithm and Multitask learning. Discriminative methods include random-walk and structural learning.

One method that we have briefly discussed earlier in Chapter 5, is the EM algorithm. With a mixture model we have missing information: given an observation x drawn from the model, it came from one of the components of the mixture, but we don't know which one. The EM algorithm first estimates this information (the expectation step), to assign x to a component – actually it assigns a probability vector associated to the components – and then it fits the model given this assignment (the maximization step). EM algorithm has a tendency towards locally optimal solutions. If a locally optimal solution is far from a globally optimal solution, then unlabeled data could hurt. Nigam [332] proposes a remedy based on active learning that gives a good starting point for the EM algorithm.

Another approach is to build a classifier, classify the unlabeled observations, then re-train the classifier accordingly. As stated, this generally doesn't work well, but there are modifications that can be effective. For example, in self-training, the algorithm adds the data instances that are labeled with high confidence to the training set as labeled data for the next cycle of training and predictions. In co-training one splits the features into two sets that are class-conditionally independent. Two classifiers are fit, and each one classifies the unlabeled data for training the other. Assuming one can split the features in this way and that with these splits one can obtain good classifiers, this method can be effective. How to split the features for co-training has been considered in several paper, e.g. [137].

A related method is to train several very different classifiers, and require a consensus of the classifiers for the unlabeled points. Another approach is the *transductive SVM* [219]. Here, one simultaneously tries to learn a labeling of the unlabeled data and maximize the margin between the two classes. The transductive SVM can be considered as an SVM with an additional regularization component on the unlabeled data. However, the problem is that the optimization problem is not convex anymore and solving it is difficult. Hence, there are many different approaches such as deterministic annealing, branch and bound, the concave convex procedure, semi-definite programming, convex relaxation, etc.

Graph based semi-supervised algorithms define a graph where the vertices are the data instances and the edges, usually weighted, represent the similarity of the instances. These methods assume that labels are smoothly varying over the graph. Their advantage is that they are usually nonparametric. They are also transductive and discriminative in nature [524].

There are various methods that utilize clustering algorithms. One posits that points in the same cluster should have the same label, and develops methods that try to cluster the data into clusters that are class-pure. One way to think about this is in the mixture framework. A mixture model is a clustering of the observations (where this is a *soft* clustering rather than a *hard* one in which an observation can only be in one cluster). When performing the EM algorithm, one constrains certain components to be only "available" for a given class label (or for unlabeled points). Essentially, one is given a partial association of the observation to the components: x, being from class y, is definitely drawn from these specific components; the EM algorithm is then modified to take this into account.

There are also a number of methods related to manifold learning (Section 5.7). One constructs a graph on the data as in manifold discovery, and this graph is used to impute the labels for the unlabeled observations. There are many choices for this, from methods that use various propagation schemes through the network, to embedding methods that use spectral embedding to embed the points in a space in which further processing can occur.

6.8 Neural Networks and Deep Learning

The neural network model is an idea that was inspired by neuroscience, has been around since the early 1960s, and had several near-death experiences and revivals since then. Right now it is experiencing explosive growth under the name of deep learning. We begin this section with a discussion of the perceptron model, which is a basic component of neural networks.

6.8.1 Perceptron

The perceptron model is inspired from the cell in our brains, the biological neuron. The neuron, shown in Figure 6.7,[12] consists of a cell body - soma, a dendritic tree, and an axon that bifurcates. The dendritic tree collects "signals" from other neurons, and an electrical pulse travels down the axon away from the cell body, provided the neuron is excited enough. This pulse releases vesicles (neurotransmitter chemicals) from one of the ends of the axons, which cross the synaptic gap between the axon and the dendritic tree of another neuron.

[12] https://pixabay.com/en/neuron-nerve-cell-axon-dendrite-296581/

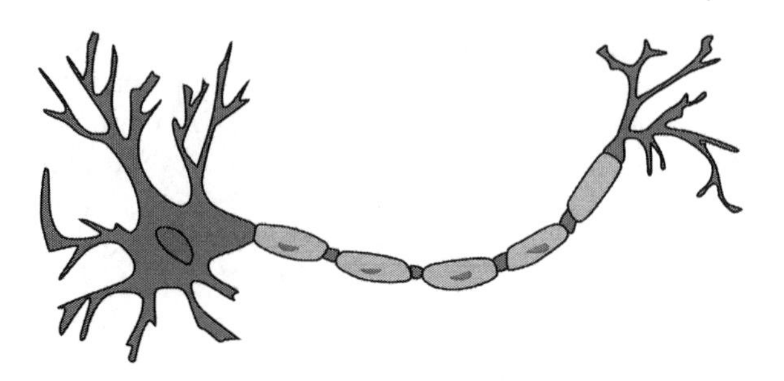

Figure 6.7: A neuron with the dendritic tree at the left and the axon towards the right. The cell nucleus is surrounded by soma. (Material is based on: https://pixabay.com/en/neuron-nerve-cell-axon-dendrite-296581/)

So our first model of the neuron consists of a set of binary inputs $x_1, x_2, \ldots, x_n$, which are multiplied by the corresponding weights $w_1, w_2, \ldots w_n$ and summed. If this sum ($\sum_i w_i x_i \geq T$) exceeds a certain threshold T, then the output is a 1, otherwise it is 0. This is the model proposed by Frank Rosenblatt in his book, *Principles of Neurodynamics*, that appeared in 1962.

An example of a perceptron for the two-input Boolean Or function is easily achieved with $w_1 = w_2 = 1$ and the threshold $T = 1$. Similarly, if the threshold is set to 2, then we get a perceptron for the two-input Boolean And function. A little consideration shows that the perceptron is a linear classifier and a single perceptron cannot model the Boolean Exclusive-or function. To visualize this, see Figure 6.8. This was one of the early critiques of the perceptron model by Marvin Minsky and Samuel Papert in their book, which appeared first in 1969, and has been updated recently [318]. However, a network consisting of two perceptrons suffices.

A network of these perceptrons computes a function of the form $\vec{z} = f(\vec{x}, \vec{w}, \vec{T})$, i.e., our network is a function approximator. Now, let us suppose that the function we wish to compute is $\vec{\phi} = g(\vec{x})$, then we may design a performance function of the form $P = ||\vec{\phi} - \vec{z}||$. However, this is not mathematically convenient to deal with (nondifferentiable), so we change it to $-1/2(\vec{\phi} - \vec{z})^2$, and we would like to use gradient ascent or hill climbing methods to optimize this. However, our thresholds are a problem and so is our decision function, which is discontinuous. We eliminate the thresholds, by introducing a new input, which is always fixed at -1 and has a weight $w_0 = T$. Instead of using a step function for our decisions, we use the sigmoid function:

$$\beta = \frac{1}{1 + e^{-\alpha}} \tag{6.9}$$

Note that when β is differentiated with respect to α, we get $\beta(1 - \beta)$.

Updates to the weights w_i's can now be made based on $r * \frac{\partial P}{\partial w_i}$, where r is the step size or *rate constant* as it is called, and where we use the chain rule for derivatives to push the partial derivatives back from the output towards the input. Hence this method is called backpropagation. A network can now be trained using this method. However, note that a globally optimal solution is not guaranteed.

6.8.2 Neural Networks

Neural network models are a class of models that are widely used for classification and regression tasks. In their simplest form, they are a feed-forward network as shown in Figure 6.9. On the left, the input nodes correspond to the variables of the data. In this case the

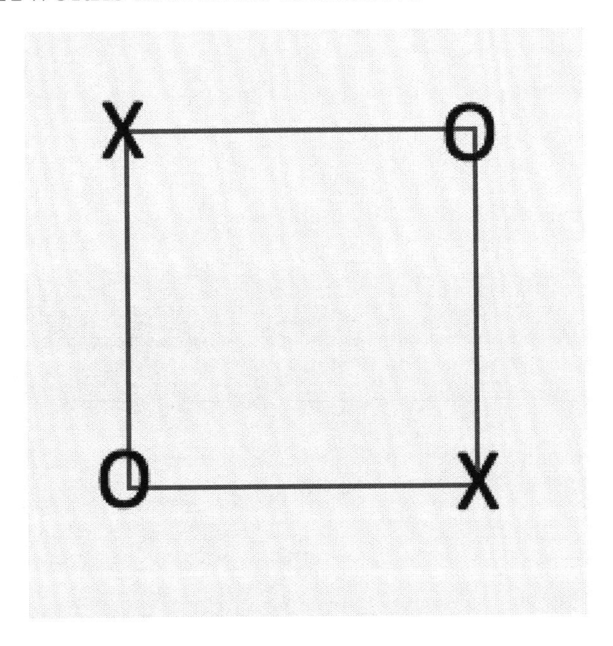

Figure 6.8: The Xor function is diagrammed on the plane with an X representing an output of 1, corresponding to (1,0) and (0,1) and an O representing an output of 0, corresponding to (0,0) and (1,1). No straight line can separate the Xs from the Os.

model is for 3-dimensional data. On the right is the output, in this case one dimensional. This produces a mapping from $\mathbb{R}^3$ to $\mathbb{R}$ – although in practice the inputs and outputs are scaled to fall in the interval $[0,1]^m$ or $[-1,1]^n$ for the appropriate dimensions m and n. The layer in the middle is called the *hidden layer*; there are generally several hidden layers in a network, the term "deep neural network" referring to a network with many hidden layers. Although one can implement a neural network without a hidden layer, this results in a linear model. Thus, a network consists of an *input layer* one or more hidden layers and an output layer.

The arrows in the plot correspond to weights; at each hidden node, the network calculates the weighted sum of the inputs (with a bias[13] term – a scalar – that is not shown in the figure), and then applies to this an *activation function* (usually the hyperbolic tangent tanh or another "sigmoid"-shaped function:

$$H_j = \tanh(\sum w_{i,j}^H I_i + b_j^H). \tag{6.10}$$

Here I_i corresponds to the ith input variable and H_j corresponds to the value of the jth hidden node. A similar operation is performed at the output layer.

Another common activation function is the *rectified linear unit* or reLU:

$$reLU(x) = \begin{cases} 0 & x < 0 \\ x & x \geq 0 \end{cases} \tag{6.11}$$

In words: each node takes the weighted sum of the outputs from the previous layer, adds a node-specific constant to this, and applies a nonlinear function ("sigmoid", usually) to this number and returns this value. The values of each layer pass through this process to the next layer, and on until the final layer produces an output.

[13]This use of the word "bias" is not the same as the bias of an estimate, although it is related; it is essentially an offset that is added to the weighted sum.

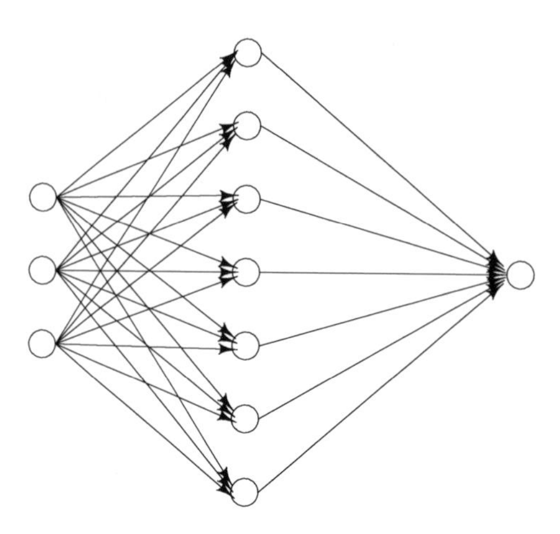

Figure 6.9: An example of a single-hidden layer neural network. The input nodes are on the left, and there is a single output node in this example.

As mentioned above, neural networks were invented in the 1960s as a simplified model of the neural architecture found in the brain. While admittedly extremely simplistic, the models were surprisingly flexible; early on, it was recognized that they are *universal approximators* – which means that given any continuous function $f : \mathbb{R}^n \to \mathbb{R}^m$ and $\epsilon > 0$, there is a neural network that approximates f to within an error of ϵ ([199]).[14]

6.8.3 Deep Networks

The necessary condition for applying a neural network to a problem is a method to fit the parameters of the model – the weights and biases. The breakthrough came with the backpropagation algorithm, [370]. This is essentially a gradient descent algorithm, that "propagates" the errors, as computed at the output layer, backwards through the network, changing the weights. The next breakthrough came with the methods for training "deep" neural networks, [191, 252, 262].

Deep networks are neural networks with many hidden layers. These networks have been used in many applications, particularly in image processing, as discussed in [262]. They also have many applications in cybersecurity. See [395] for a high level discussion of these.

Changes to how the gradient is computed were proposed by [15] to make the neural network better cope with imbalanced data sets. Later, [472] proposed to use a different loss function for training deep neural networks for imbalanced data. Recently, another approach for imbalanced data is given in [122].

Neural networks have been used for many interesting and difficult tasks, such as image recognition and analysis, speech recognition and understanding, playing Go and other games, and more mundane classification and regression tasks. See [172, 262, 392] and the papers they cite. In computer security, deep networks have been used for intrusion detection, malware and spam. See [18, 215].

[14]All of this can be made precise, and under various models one may have to restrict the image and domain of the function to compact sets, such as the cube, but this can be accomplished through a suitable transformation.

A problem that has been identified with neural networks, in particular deep learning methods applied to difficult problems like scene recognition, playing Go, or self-driving cars, is the problem of interpretation and explanation. It is not at all trivial to determine what the system has "learned." This is not unique to neural networks. Random forests can be hard to interpret, it can be difficult to understand just what a support vector machine is "really doing", and even the lowly k-nearest neighbor algorithm raises questions of generalizability. The fundamental question arises: what is the system going to do when it sees data that is different from that on which it was trained? Once again, the paper [185] is prescient. Much has been written on this problem in the deep networks literature, for example [94, 331, 414]. This has resulted in interest in using neural networks against each other, so called *adversarial* neural networks, and in methods for better understanding the models within neural networks and their limitations, [40, 173, 517]. Adversaries are used both to defeat and to improve deep learning, and in the context of game theory. See [301, 334, 474] and [248, 490].

It has been said that neural networks obviate the need for feature extraction and selection; the neural network constructs the appropriate features from the raw data. For example, one can feed raw images into the system, train the neural network to recognize objects, and the network will self-organize to extract the correct features from the images. This is partially true, although the network architecture, such as convolution neural networks (CNNs) ([253]), is often specified with domain knowledge about the data in mind.

Many CNNs can be succinctly represented as:
[(CONV-ReLU)*P-POOL?]*Q-(FC-ReLU)*R,SOFTMAX,[15] where P ranges from 1 to 5, Q is large and $0 \leq R \leq 2$. Here CONV-ReLU denotes a convolution layer with ReLU activation function, POOL denotes a pooling layer, and FC-ReLU is a fully connected layer, again with ReLU activation. The output layer has neurons equal to the number of classes. A convolution layer applies a filter to the input and is reminiscent of a convolution operator. A pooling layer applies a function such as max and reduces the dimensionality of its input. The question mark indicates that this layer is optional. The following layers extract more abstract features from the primitive features extracted by the initial layers.

For the recurrent neural networks (RNNs), which have feedback loops for using previous state information that captures the context, we refer the reader to [172]. A recurrent convolutional neural network model was introduced in [257]. The convolutional layer is made up of bidirectional RNN instead of the usual fixed window based convolution. The researchers showed that this combination performs better than baseline RNN and CNN models for text classification.

One method for utilizing neural networks to extract features is the autoencoder neural network. Figure 6.10 shows an example. In an autoencoder network, the inputs and outputs are the same – the network is trained to reproduce the input. The idea is that the hidden layers process the input, encoding the input in the first half of the network, producing features in the middle layer. The second half decodes the features, producing the output. Once trained, the middle layer provides a representation of the input as features (in this case two-dimensional features) that contain all the information contained in the input – in the sense that these features can be used to reproduce the original input, to the extent that the network can be trained to within a desired tolerance.

The basic shape of the autoencoder network is symmetric: the assumption is that the encoder and decoder sections are inverses of each other. In a sense, the network is trying to learn the identity function, but it is constrained by two conditions that make it impossible (in general) to do so. First, the basic architecture of the network is to apply the nonlinear activation functions at each node. Second, the point of the autoencoder is to reduce the

[15] `http://cs231n.stanford.edu/slides/2016/winter1516_lecture7.pdf`

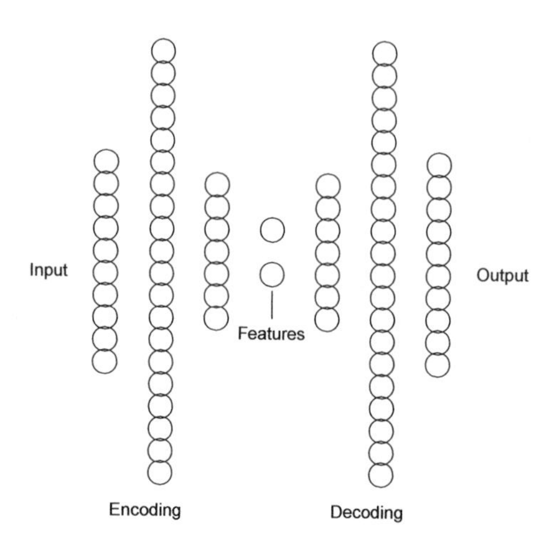

Figure 6.10: An autoencoder neural network. This has 10 inputs and 10 outputs, with 5 hidden layers; the edges are not shown.

dimensionality, and so the smaller size of the middle layer forces a reduction of the dimensionality of the data. As in Section 5.7, the autoencoder is learning a lower dimensional *manifold* on which the data is supported. Once the autoencoder is trained, one can strip off the decoding/output layers, and utilize the features from the middle layer as inputs into another machine learning algorithm, whether another neural network or a different pattern recognition or machine learning approach. An exciting new development is the variational autoencoder (VAE) [121]. In a short time, it has emerged as one of the most popular approaches to unsupervised learning of complicated distributions. The model's name is based on a variational technique, which has been used to train it on image data successfully. Some researchers have shown that using an LSTM decoder (long short-term memory, a type of RNN) in a VAE does not work well for modeling text data. A dilated CNN as a decoder has been shown to work better on some language modeling tasks. We refer the reader to [510].

A novel approach to detecting man-in-the-middle attacks is presented in [319]. Borrowing ideas from acoustic signal processing, the authors model the normal pulses of ICMP echo requests using an autoencoder neural network. The approach detects deviations from normal, as an indicator of a likely attack.

6.9 Topological Data Analysis

Topology is the study of mathematical properties that remain invariant under smooth mappings. Consider a soccer ball as it is kicked around the field. As long as the skin remains undamaged, the kicks and collisions with the ground and the players all cause a distortion of the shape of the ball that does not change the basic structure; the "ball" remains (topologically) a sphere, although it may be distorted quite far from being perfectly round (especially if it is under-inflated). The topological property that is maintained in this example is the fact that the skin forms a single connected surface[16] without boundary that encloses a "void" that is contractible – think about it as a sufficiently high pressure could compress the air inside down to a single point.

In the soccer ball example, the topology (of the skin of the ball) can be described with three numbers: the number of connected components (1), the number of 1-dimensional "voids" – circles that cannot be collapsed, or "pulled off" the ball – (0), and the number of 2-dimensional "voids" (1). This, with the fact that the ball is embedded in a three-dimensional space (and thus we don't need to look for higher dimensional "voids"), characterizes all balls. These numbers are referred to as the *homology* of the structure.

Recently, this mathematics has been adopted to analyze data. The idea is that the data correspond to points sampled on a topological object, and we seek to understand the data – and the process that generated the data – through understanding this object. This is very similar to manifold discovery discussed in Section 5.7, except that rather than utilizing the "manifold" structure to produce an embedding of the data, topological data analysis (TDA) is interested in understanding the structure itself, rather than focusing on individual data points.

As in manifold discovery, TDA starts with the points and forms a graph with vertices corresponding to the points, and edges corresponding to some measure of closeness. For concreteness, suppose we are forming an ϵ-ball graph: each point is connected to all points within ϵ of itself. TDA then treats this as a topological structure (technically, this is called a *chain complex*), and topological features are extracted from this structure.

It is important to note that the topological invariants we are measuring, such as the number of connected components, depends on the value of ϵ. A very small value will result in the empty graph – the graph with no edges (assuming that the points are distinct) – and a maximal value for the number of components. A very large ϵ will result in a connected graph and a value of 1 for the number of components. Similarly, the number of "1-dimensional voids" – the number of cycles of 4 or more vertices that do not contain a chord (edges between vertices that are not adjacent in the cycle) – depends on ϵ; larger values will "fill in" cycles that occur at smaller values. The cycles are referred to as the *generators* of the 1-dimensional homology, as the connected components are the generators of the 0-dimensional homology.

This observation leads to one of the fundamental ideas of TDA: *persistent homology*. One computes the homology at various scales (values of ϵ) and looks for the generators that "persist" through relatively large ranges of ϵ. Generators that come and go quickly as ϵ changes are considered to be the results of noise. This results in a representation of the data (often called a *bar chart*) as a set of intervals showing the ranges of ϵ for each generator. Figure 6.11 depicts an example of this. The bars in the bar chart start at the point the generator is first observed – in the case of connected components, these always start at 0 – and end when the generator is destroyed, or in the case of connected components, absorbed into another component. There is a final connected component that persists, and a final 1-dimensional component that persists after $\epsilon = 0.23$, corresponding to the circular aspect of the data.

References for this subject can be found in [70, 140, 476] and citations therein. There are many extensions of these ideas that are described in these references, and while the field can be said to be in its infancy, there is reason to believe that these will be powerful tools for data analysis and machine learning in the future.

Probably the most appropriate use of TDA is when an observation itself can be viewed as a point cloud. In this case, TDA provides methods for extracting features from the object, and investigating the structure of the point cloud. Possible applications in cybersecurity are

[16]This is ignoring the fact that the balls are stitched together from individual pieces of material – this idealized soccer ball has a skin that is a single smooth surface all the way around the ball.

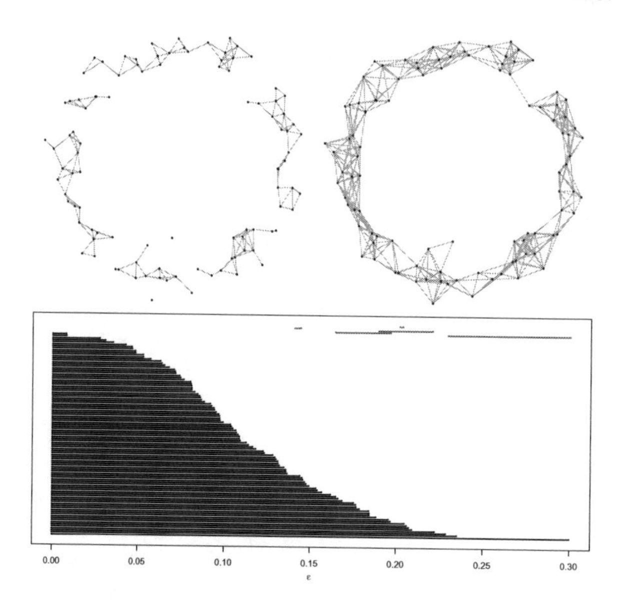

Figure 6.11: A data set consisting of 100 points, drawn from a circle with added noise. The graphs associated with two values of ϵ are shown on the top, and the bar chart is shown on the bottom. The black bars correspond to H_0, or the connected components, the gray bars in the upper right to H_1, or the cycles. Note the existence of a persistent cycle which corresponds to the circle from which the data were drawn.

to log files on a host, byte dumps or decompilations of malware, and network flow data at the hourly level, where one is interested in discriminating between overall normal or abnormal activity.

6.10 Ensemble Learning

The random forest is an example of ensemble learning. The basic idea is to create an ensemble of classifiers to use to get an overall classification through a voting scheme. These classifiers may all be the same basic model, as in random forests, trained on different data or with different model parameters, or they may be a collection of very different models. Once we have an ensemble of classifiers, we need a principled way of combining their decisions. We examine two of them below: majority and Adaboost. Some other possibilities are learning to rank, rank aggregation, and Dempster-Shafer Theory.[17] In these methods, a ranking of labels, or a list of confidence values, is needed from the classifiers, which can then be combined by these approaches.

6.10.1 Majority

The majority combination of classifiers is easily explained in a binary, or 2-class, setting. In a majority combination of classifiers, we ask each classifier to output a class label and then go with the label that is output by at least half the classifiers. This simple scheme has some pretty good properties. The reason it is so attractive is the proof that if the classifiers are: (i) *independent, and (ii) accuracy of each classifier exceeds 50%*, then the majority is

[17]`http://fitelson.org/topics/shafer.pdf` - Accessed 25 May 2018.

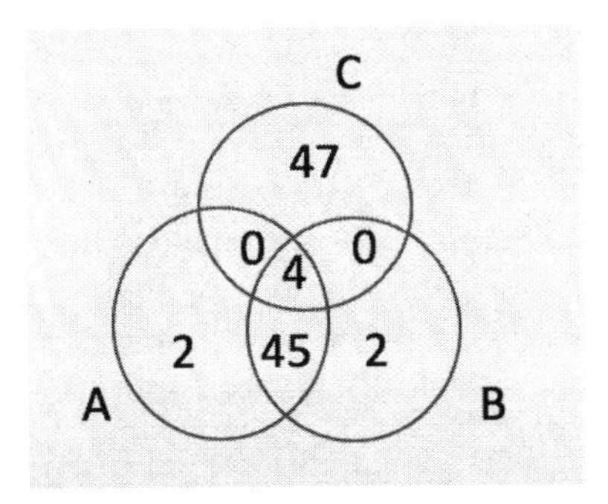

Figure 6.12: Three classifiers with 51% accuracy on different subsets of the data set are shown. The numbers inside each circle add up to 51. But the majority classifier is only accurate on 4+45 = 49% of the data set.

guaranteed to be more accurate than each individual classifier [258]. Note that it is easy to construct counter-examples in which the majority classifier is less accurate than each individual classifier, even if each classifier's accuracy exceeds 50%. For example, classifiers A, B and C are accurate on subsets S_1, S_2 and S_3 of the data set, where each S_i represents 51% of the data set and $S_1 \cap S_2$ = 49%, $S_2 \cap S_3$ = 4%, $S_1 \cap S_3$ = 4% and $S_1 \cap S_2 \cap S_3$ = 4%. In this case, the majority's accuracy is only 49%. Figure 6.12 shows how this can happen. Hence, the assumption of independence is critical. In our example, A and B are not independent, since $P(A \cap B) = .49 \neq P(A)P(B) = .2601$. Assuming that the classifiers have the same accuracy p (to simplify the messy formula that results when their accuracies are all different) and are independent, we can write down the expression for the accuracy of the majority as [256, 258]

$$P_{maj} = \sum_{m=0}^{\lfloor L/2 \rfloor} \binom{L}{m} p^{L-m}(1-p)^m \tag{6.12}$$

Here L is the total number of classifiers. In [255], researchers show that negatively correlated classifiers are better for combinations using the majority rule. Positive correlation is worse and independence or zero correlation is helpful, but not as helpful as negative correlation.

A scheme called dynamic weighted majority has been proposed in [245, 246], to handle the concept drift problem. Recall that, because of an active attacker, concept drift or nonstationarity is a serious issue in the security domain. They employ four ideas to deal with concept drift: train an online learner ensemble, weight the learners based on their performance, remove learners, also based on their performance, and add new "experts" based on the global performance of the ensemble.

6.10.2 Adaboost

Suppose one has constructed a classifier from training data $\{(x_1, y_1), \ldots, (x_n, y_n)\}$. Unless this is a nearest neighbor classifier, it will have some error on the training data. Intuitively, these errors are for observations that are "harder" than the others. So, if one could re-weight the observations to give these points more "importance" to the algorithm, then the new classifier would do better on these points, and, in conjunction with the original, do better overall. Another way of saying this is that instead of creating an ensemble of classifiers completely at random, perhaps one should focus more on these "harder" observations as one adds new classifiers to the ensemble. This is the basic idea of Adaboost. It is described in [159] and [158] among other papers. The latter has a nice introduction that illustrates the idea with a thought experiment involving betting on horses. The original algorithm was defined for the two-class problem, but it has since been expanded to multiclass problems, see for example [187].

The Adaboost algorithm is as follows. First, one selects a "weak learner". This is an algorithm that tends to perform better than chance, but is not optimized for the problem. In [158] the result of applying the weak learner to training sets are called "weak hypotheses" or "rules of thumb". The setup in the two-class framework is then a weak learner, $h : X \mapsto \{-1, 1\}$ – it is convenient to label the classes ± 1 rather than $0, 1$ – a training set $\{(x_1, y_1), \ldots, (x_n, y_n)\}$, and a distribution D which assigns a weight to each training point x_i – generally a weight of $1/n$ is used initially. The Adaboost algorithm is an iterative algorithm that produces new weak learners and new distributions, so we will index h and D with iterate, or time.

1. Set $D_1(i) = \frac{1}{n}$.
2. For t in $1, \ldots, T$:
   (a) Let h_t be the weak learner trained on the data with weights D_t.
   (b) Let ϵ_t be the error of h_t.
   (c) Set $\alpha_t = \frac{1}{2} \log\left(\frac{1-\epsilon_t}{\epsilon_t}\right)$.
   (d) Set
   $$D_{t+1}(i) = \frac{D_t(i) \exp(-\alpha_t y_i h_t(x_i))}{Z_t}$$
   where Z_t is the normalization required to make D_t sum to one.
3. Output
   $$H(x) = \text{sign}\left(\sum_{t=1}^{T} \alpha_t h_t(x)\right).$$

Note that in D_{t+1}, if $h_t(x_i)$ is correct, the numerator is $\exp(-\alpha_t)$, and the weight is decreased, and if it is incorrect, the numerator is $\exp(\alpha_t)$ and the weight is increased. In this way, Adaboost pushes later weak learners to focus more on the "hard" points, and less on the "easy" ones.

One take-away from all of this is that there are basically two important ideas in all of the above. The first is that an ensemble of simple classifiers can improve performance over a single classifier; the second is that one can tune the ensemble to make the improvements in the regions where the problem is hardest. Both ideas are important and powerful; both ideas also are fundamentally nonparametric: remember, as George Box said, all models are wrong. In fact, as he says in [50]:
"Since all models are wrong the scientist must be alert to what is importantly wrong. It is inappropriate to be concerned about mice when there are tigers abroad."
In a sense, Adaboost can be thought of as a way to focus on the "tigers" – the hard points – rather than the "mice". However, care must be taken – Adaboost can just as easily be seen as focusing too much attention on a few outlier "hard points", which in the grand scheme of things are really "mice". As with all methods, what really matters is how well the algorithm works on new data, "in the wild", to continue the Box metaphor.

We ran the AdaBoostM1 classifier in WEKA on the Phishing URLs data set of Section 5.5 with different weak learners: decision stumps, Hoeffding trees, and J48 (WEKA's version of Decision Trees). The confusion matrices are shown in Figures 6.13 to 6.15. The confusion matrix for random forest is also shown for comparison in Figure 6.16. The boosted version of decision stumps and Hoeffding trees performed worse than JRip in Chapter 5. As before, 10-fold crossvalidation was used. Apart from this, WEKA's default parameters were used.

True Cl. \ Classified As	Phish	Legit
Phish	4487	411
Legit	409	5748

Figure 6.13: The confusion matrix for AdaBoostM1 with Decision Stump as the weak learner on the Phishing URL data set.

True Cl. \ Classified As	Phish	Legit
Phish	4562	336
Legit	327	5830

Figure 6.14: The confusion matrix for AdaBoostM1 with Hoeffding tree as the weak learner on the Phishing URL data set.

True Cl. \ Classified As	Phish	Legit
Phish	4698	200
Legit	105	6052

Figure 6.15: The confusion matrix for AdaBoostM1 with J48 (decision tree) as the weak learner on the Phishing URL data set.

True Cl. \ Classified As	Phish	Legit
Phish	4705	193
Legit	110	6047

Figure 6.16: The confusion matrix for Random Forest on the Phishing URL data set.

6.11 One-class Learning

Consider the problem of computer virus detection. Companies that provide this service do not collect a database of viruses and a database of files that are not viruses and build a virus/nonvirus classifier. Instead they start with just a database of viruses and build algorithms to detect these. This is a one-class problem – ignore for the moment that they actually tell you which virus it is, but instead think of it as just detecting that a virus is present.

There are many other problems in cybersecurity that can be formulated as one-class problems. For example, in network security, one may consider the traffic on the network to be "normal" and so build a classifier that detects "normal", and alerts when it sees something different. This is also sometimes framed as an outlier detector, which is one type of one-class method. Another example is the detection of masqueraders or malicious insiders (the *insider threat*); one builds a one-class classifier for each user (or for "benign users"), and looks for instances that don't match the class. In social media analysis such as utilizing Twitter data, one may be interested in finding all instances of a given language. Rather than giving examples of all languages, one could imagine using only examples of the target language and build a classifier to detect these.[18]

The survey [235] provides an introduction to some of the methods for one-class classification (OCC). There are several basic approaches. One is to estimate the probability density function of the data, then use this to threshold the likelihood to detect observations in regions that are "unlike" the target class data. If one has a small amount of data from another class (i.e., data that is not from the target class) one can use this to adjust the density, set the threshold, or supplement the algorithm with a two-class classifier. Similarly, if one has unlabeled data, one can use semi-supervised methods to improve the OCC.

The problem with computing an estimate of the probability density function and using this to build the OCC is the curse of dimensionality: the higher the dimension, the more data is required to obtain a good estimate of the density. This is illustrated, for kernel estimators, in [393]. Fortunately, we only care about the density near the tails, since we only want to set a threshold on the likelihood and don't really care how accurately we estimate the likelihood values for regions above this threshold, just as long as we can tell they are above the threshold. In [190], the researchers suggest a method to sample artificial data to improve the density estimation for this purpose.

Another method described in [235, 236] is to use a support vector machine to build a hyper-sphere around the data (instead of a hyper-plane between two classes); see [425], also [74]. Alternatively, one can use nearest neighbor algorithms to find the training data closest to the observation and threshold on the distance, and one can extend this to a k-nearest neighbor method by looking at a statistic such as the mean distance to the nearest neighbors.

The problem of estimating the density for high dimensional data is very challenging, and for data that has a lot of structure, such as many different clusters or high-density regions, it is unclear that a single approach is the best way to proceed. One can use various dimensionality reduction techniques, feature selection methods [131, 177, 386], and projections to reduce the dimensionality, but this still can result in unacceptably high dimensional data, or in the loss of critical information. Recall that one of the ideas of ensemble methods is that one can use a collection of classifiers each of which may not be optimal or may utilize different assumptions or methods, and pool the results into an aggregation of the information that results in improved performance. Using clustering methods to group the data and produce an ensemble of OCCs, [251] describe a general method for OCC. It might be worth investigating whether the idea of iterative denoising ([169, 356]) could be combined

[18]Twitter provides a language tag to each tweet, but this may not have the accuracy you desire, or you may be utilizing some other social media platform that does not tag the message with the language.

with this idea to produce a hierarchical version, in which local features are used within each cluster.

For an example of a one-class classifier in a cybersecurity application, consider the following experiment using the KDD-Cup data. A training set of 100,000 normal observations were randomly selected to use as the training set for a one-class support vector machine classifier, leaving 872,781 normal observations for a test set, and 3,925,650 observations of abnormal (attack) data. This algorithm has a parameter ν, which controls the number of outliers.

The parameter ν can be thought of as a regularization parameter. It controls the training error and the number of support vectors: it gives an upper bound on the probability of error on the training data, and a lower bound on the proportion of support vectors.

With a value of $\nu = 0.5$, the training set classified 49.99% of the training data as "normal", and 50.02% of the test set as "normal". The model has 50,004 support vectors, and so is quite a complex model. This comports well with the value of ν. With this value, all the anomalous observations are classed as "anomalous" – nonnormal.

Of course, a false alarm rate of 50% is absurd. Setting $\nu = 0.0001$, the training set has an error of 0.403%, with a 0.617% error rate on the test data, which is a more reasonable value. In this case the model has 474 support vectors, and so is relatively parsimonious.

With that said, this rejects 5,386 observations in these data – depending on the number of hours, or days, or months, a data set like this corresponds with, this could still be quite burdensome. On the anomaly (attack data) the algorithm correctly classifies 95.76% as anomalies, almost all of which are smurf attacks. A smurf attack is an ICMP distributed denial-of-service attack. Most of the other attacks target TCP or UDP. It makes sense to segregate the data by protocol, and to put in place algorithms specifically designed to detect these types of flooding attacks. Still, the performance of the one-class SVM on these data is quite good. In fact, only 0.59% of the nonsmurf attacks were misclassified as "normal".

Another source of network flow data (and the individual packets as well) is the Canadian Institute for Cybersecurity (CIC) data set. There are a number of data sets available, but for this study the 2017 data will be used. These data are described in [385].[19]

The data we will investigate are flow statistics computed on a network in which simulated users interact in a "normal" fashion, while attacks are mounted against the machines on the network. The first day, Monday, contains only benign (normal) traffic, while the data in other days are marked as benign, or as a particular attack. Around 80 features are extracted from the flows. In our study, we use the 55 features reported in Table 6.4, which are the numerical features that have some level of variability in the training data (Monday benign data). Details of the features can be found at the web site and in the cited paper.

Using the SVM one-class classification algorithm with $\nu = 0.00068$, computed by experimenting with various values on the training data of Monday, we obtain the following table:

FA	Benign	FA rate
	TCP	
1243	304180	0.0041
	UDP	
334	223844	0.0015

In both of the protocols, the classifier has a false alarm rate of under half of a percent, resulting in (according to this day – note that this is a resubstitution estimate) around 1000 false alarms per day. On Wednesday, the classifier obtains the results in Table 6.5.

Slow HTTP and Slow Loris are attacks that allow a single machine to keep a large number of requests to a web server active to consume resources and hence block legitimate

[19]http://www.unb.ca/cic/datasets/ids-2017.html

Table 6.4: Features Used in the One-class Classification Study on the CIC Data

Flow.Duration	Total.Fwd.Packets	Total.Backward.Packets
Total.Length.of.Fwd.Packets	Total.Length.of.Bwd.Packets	Fwd.Packet.Length.Max
Fwd.Packet.Length.Min	Fwd.Packet.Length.Mean	Fwd.Packet.Length.Std
Bwd.Packet.Length.Max	Bwd.Packet.Length.Min	Bwd.Packet.Length.Mean
Bwd.Packet.Length.Std	Flow.IAT.Mean	Flow.IAT.Std
Flow.IAT.Max	Flow.IAT.Min	Fwd.IAT.Total
Fwd.IAT.Mean	Fwd.IAT.Std	Fwd.IAT.Max
Fwd.IAT.Min	Bwd.IAT.Total	Bwd.IAT.Mean
Bwd.IAT.Std	Bwd.IAT.Max	Bwd.IAT.Min
Fwd.Header.Length	Bwd.Header.Length	Fwd.Packets.s
Bwd.Packets.s	Min.Packet.Length	Max.Packet.Length
Packet.Length.Mean	Packet.Length.Std	Packet.Length.Variance
Down.Up.Ratio	Average.Packet.Size	Avg.Fwd.Segment.Size
Avg.Bwd.Segment.Size	Fwd.Header.Length.1	Subflow.Fwd.Packets
Subflow.Fwd.Bytes	Subflow.Bwd.Packets	Subflow.Bwd.Bytes
act_data_pkt_fwd	min_seg_size_forward	Active.Mean
Active.Std	Active.Max	Active.Min
Idle.Mean	Idle.Std	Idle.Max
Idle.Min		

Table 6.5: One-class SVM Classifier Results on a Test Data set (Wednesday Afternoon)

TCP	Abnormal	Benign	Percent Correct
BENIGN	5802	230976	97.5%
DoS GoldenEye	5941	4352	57.7%
DoS Hulk	150989	80084	65.3%
DoS Slowhttptest	1919	3580	34.9%
DoS slowloris	3151	2645	54.4%
Heartbleed	11	0	100%
UDP	Abnormal	Benign	Percent Correct
BENIGN	334	223844	99.8%

users access to the site. Hulk is a tool for constructing denial-of-service attacks (for research purposes) that are designed to be as unpredictable as possible. Each attack is crafted to have different properties from the previous one. Golden Eye is another tool designed to test http denial-of-service attacks. The Heartbleed attack was an exploitation of a bug in the OpenSSL cryptographic library that allowed the attacker to read unencrypted data, such as passwords, by exploiting a buffer overflow.

Depending on one's criteria, the results reported in Table 6.5 could be considered good or bad. On the good note, the performance on benign data seems to correspond well with the training data estimate, and so the results appear to be consistent on new data.[20] The performance on the attacks is much more mixed. While it detected the 11 Heartbleed attacks, it only detected 64.1% of the denial-of-service attacks. Whether this level of performance is adequate depends to a large degree on the situation.

It is difficult to visualize the data in the CIC data set, and the SVM model produced by the one-class classifier. There are 732 support vectors in the normal model, which can be viewed with a *heatmap*; this is an image in which the data matrix has been reordered (rows and columns) via a hierarchical clustering method, with gray-scale pixels corresponding to the value in the cell. Figure 6.17 depicts a heatmap on the support vectors.

To compare this with the denial of service attacks, we constructed similar images for each of the attacks, shown in Figure 6.18. The columns are in the same order as in Figure 6.17, and each image shows a random selection of 732 observations. The best that can be said is that there appear to be some differences between the support vectors and these denial-of-service attacks, and that the attacks seem to be somewhat different. The fact that only a small subset of the data can be shown and that the gray values are hard to interpret across images make it difficult to interpret these figures beyond this general statement.

There is no requirement that only a single method can be used to protect a network or computer. For known attacks for which one can obtain data, it makes sense to utilize a supervised classification algorithm. One-class classifiers can be used to help detect novel attacks, and further processing could be performed on the detections provided by this classifier. Also, one could utilize several one-class classifiers instead of just one, using the ideas from ensemble learning discussed above.

6.12 Online Learning

Online learning refers to the ability of a system to adjust its models in the face of new data. Typically this really is an adjustment, rather than a complete retraining from scratch, hence the name "online". For a simple example, consider an application on your phone to recognize your face and use that recognition to unlock your phone. As you age, change your hair style, facial hair, makeup, and other aspects of your visage, you would want the application to adapt to these changes, rather than continue to lock you out until you went back to your previous look – assuming this is even possible. In this example, most of the changes are gradual, although some changes are sudden, such as shaving off a beard. The application needs to be able to decide if a change in your "face profile" is necessary, and what that change needs to be.

One of the early examples of online learning for cybersecurity was discussed in [260]. The idea is to monitor user behavior and develop a profile for the user. This profile allows one to obtain a similarity measure for each new observation, indicating how similar the observation is to the profile. By rejecting observations whose similarity is unlikely under the profile, they

[20] It should be noted that we are only reporting one day of data for the test; on one of the 8 data sets available for this example, the benign classifier performed much worse, although the benign performances on the other 6 were comparable.

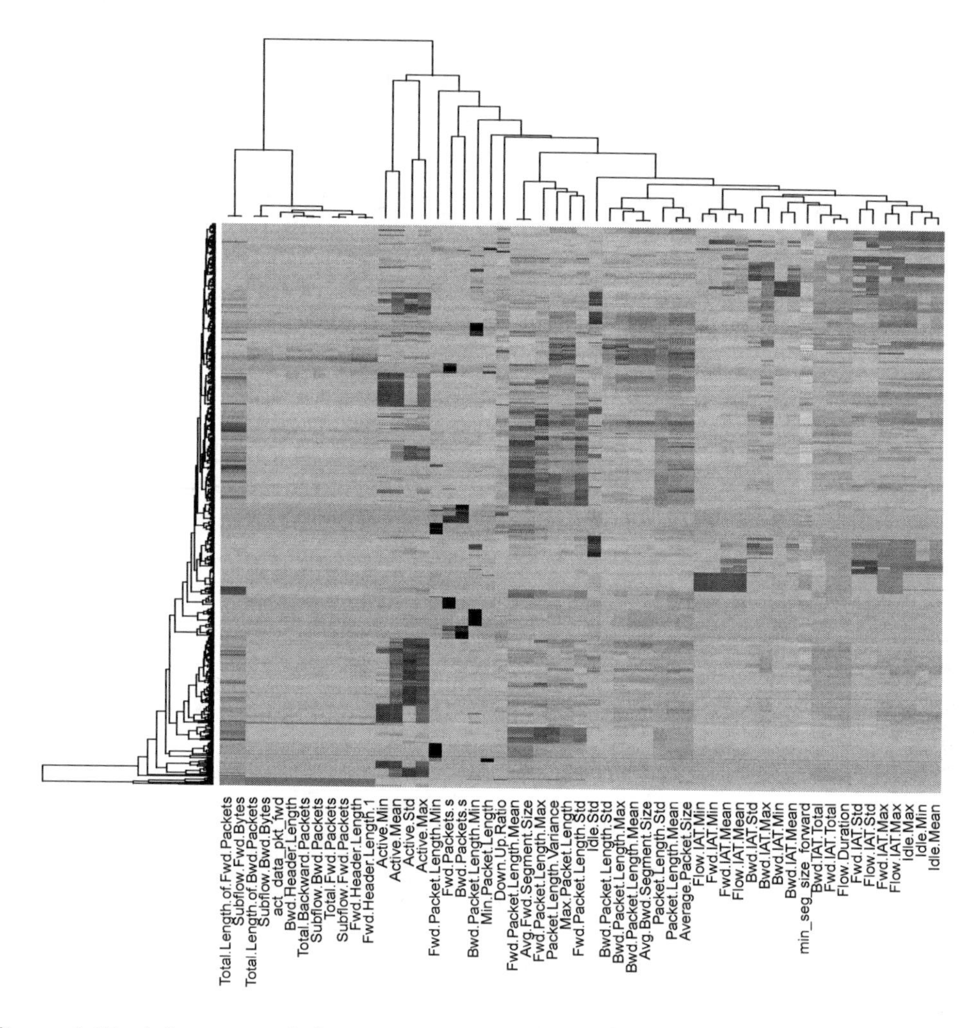

Figure 6.17: A heatmap of the support vectors of the one-class classifier on the Monday data from the CIC data set. In this plot, light colors correspond to low values in the matrix, while dark values correspond to large values.

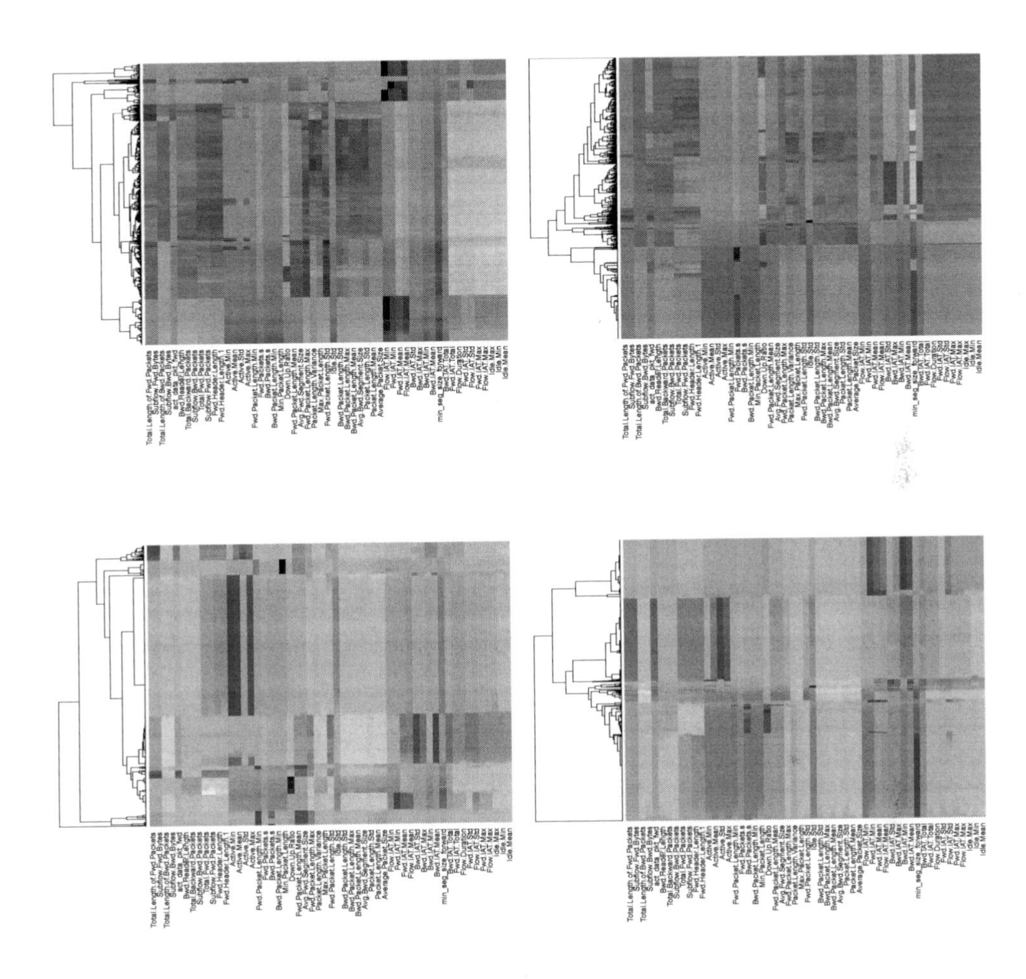

Figure 6.18: A heatmap of the support vectors of the one-class classifier on the denial-of-service attacks from the CIC data set. In this plot light colors correspond to low values in the matrix, while dark values correspond to large values. The top left plot corresponds to the Golden Eye attack, the top right to the Hulk, the bottom left to the Slow HTTP attack, and the bottom right to the Slow Loris attack. The columns are ordered in the same order as in Figure 6.17.

develop a one-class classifier. The problem they then address is how to adjust the profile as new data is observed? users change (what they refer to as *concept drift*), and one needs to adjust the profile accordingly. They discuss a number of methods for determining when a profile needs to be adjusted, and methods for performing this adjustment.

There are three aspects of a problem that suggest that online learning is needed:

1. The data are nonstationary, as in the face recognition example above, or the user profiling example.

2. It is difficult to collect the data you really want. Consider again the face recognition example: it would be best to develop an application that was generally good at certain aspects of face recognition and design a method for "personalizing" it to a new user, rather than requiring the user to "train" the system from scratch.

3. The volume of data is massive, so that batch-processing on all the data is impractical, and a method that adjusts its model as new observations are obtained is essential.

Streaming data refers to data that is observed sequentially, and must be processed sequentially (or in small batches) due to the volume and speed of the data. Streaming algorithms are those that perform their computations on each observation as it is acquired, rather than requiring a database of all observations to date.

For a simple example, consider the calculation of the average of a collection of numbers, $\{x_1, \ldots, x_n\}$. The formula is:

$$\widehat{x}_n = \frac{1}{n}\sum_{i=1}^{n} x_i.$$

Now, suppose one observes a new number x_{n+1}. Obviously, one wouldn't want to use the above equation and recompute all those sums. Instead, one would simply update the observed value:

$$\widehat{x}_{n+1} = \frac{1}{n+1}\left(n\widehat{x}_n + x_{n+1}\right),$$

which we can rewrite as:

$$\widehat{x}_{n+1} = \frac{n}{n+1}\widehat{x}_n + \frac{1}{n+1}x_{n+1}.$$

This is the recursive computation of the sample mean. It updates the value of the sample mean with each new observation. There is a similar version of a recursive calculation for the sample variance, and in fact in the case of Gaussian mixture models, there are recursive calculations of the mixture parameters; in essence, a recursive EM algorithm ([433]).

Now, suppose one really just wants a kind of "recent average" that computes a weighted average of the data that weights the most recent data highly, while giving past data an exponentially decreasing weight. This can be used in situations where the distribution of the data is drifting, and one is most interested in the current distribution. The above can be modified recursive algorithm to produce an exponentially weighted version:

$$\widehat{x} \mapsto \theta\widehat{x} + (1-\theta)x. \tag{6.13}$$

By adjusting θ one can control "how far back" in the data the method looks to compute the recent mean.

This algorithm can easily be seen to provide a streaming version of the kernel estimator, [480]. The kernel estimator can be thought of as a kind of sample mean, but instead of the mean of the data, one is computing the mean of the kernels. Given a desired point X

at which one wishes to compute the estimate of the probability density, one can use the formulation

$$f_{\text{ke}}(X) = \frac{1}{n}\sum_{i=1}^{n}\phi(X; x_i, h), \tag{6.14}$$

where ϕ is the Gaussian kernel (or one can replace this with any kernel). One then views this as a sample mean of the terms inside the sum and use Equation (6.13). It is important to note that one needs to decide a priori which values of X at which one wishes to compute the estimator, for this formulation.

See also [354, 356] for a similar approach using a Gaussian mixture model, referred to as *adaptive mixtures*. In this, the recursive algorithm from [433] is modified using the basic idea of Equation (6.13), with an addition of a term-creation rule that allows the algorithm to adapt the number of terms.

There is an extensive literature on streaming data. See the references discussed in Section 9.7.

In [523], a cost-sensitive online learning algorithm was introduced, and applied to an unbalanced data set of phishing URLs. We will discuss the application below.

6.13 Adversarial Machine Learning

Adversarial machine learning is a subfield of machine learning in which the robustness of machine learning models is investigated using synthetic attacks. It began in 2004 with the work of Dalvi and Lowd et al. in the context of defeating spam filters [40]. They showed that linear classifiers can be tricked by making simple changes to spam emails without significantly affecting readability. Barreno et al. [28] developed an initial taxonomy of attacks, e.g., training versus testing time, dictionary-based, and suggested defenses against them. A couple of books have been published on this topic [222, 464].

Adversarial machine learning includes three broad categories of research: adversarial examples, adversarial training and adversarial generation.

6.13.1 Adversarial Examples

A taxonomy of attacks given in [40] has four dimensions: attacker's goal, attacker's knowledge, attacker's capability, and attack strategy. To these we add a fifth dimension, viz., attack timing. Attacker's goal could be a security goal violation (e.g., integrity, availability, or accountability violation), attack specificity that ranges from targeted to indiscriminate, and error specificity, which can be generic (sample misclassified as any class different from the true class) or specific (sample misclassified as a specific class different from the true class).

Attacker's knowledge could be black box, gray box or white box. White box attacks are so-called perfect-knowledge attacks in which the attacker knows everything about the targeted system: the training data sets, the features, the machine learning algorithm along with the objective function and its trained parameters/hyper-parameters. In black box attacks, the attacker has minimal knowledge, basically an idea of the space of features, the goal of the learning algorithm and the type of training data used. In gray box attacks, the attacker has knowledge that is between a black box and a white box attack.

Attacker's capability is a gauge of the influence the attacker has on the data set used by the learning algorithm. It also considers the constraints on data manipulation for a specific application. For the influence, Biggio and Roli [40] use the term causative when both training and test data can be manipulated by the attacker and exploratory if only the test data can be influenced by the attacker. Depending on the application, e.g., phishing detection, there may be constraints on the attacker on the extent and type of data manipulation. In phishing email, the attacker cannot control the SMTP protocol servers after the email has left his/her domain.

Attack strategy can be captured by an objective function of the attacker and the method used by the attacker to compute an/the optimum of this objective function.

Attack timing can be either decision/testing time or training time. Decision time attacks are called evasion attacks and training time attacks are called data poisoning attacks. One example of an evasion attack is the so-called *good word* attack in which words considered benign by a spam/phishing filter are used to misclassify a phishing email as a benign email.

Adversarial examples can use knowledge of the gradient in which case they are referred to as gradient-based attacks [528]. There are also gradient-free attacks. Gradient-based attacks include the fast gradient sign and iterative gradient sign methods.

The fast gradient sign method constructs a $\pm\epsilon$ perturbation in each feature of the input instance x depending on the sign of the gradient of the loss function l.

$$x_{adv} = x + \epsilon \text{sign}(\nabla_x l(\theta, x, y)) \tag{6.15}$$

Here θ denotes the parameters of the model and y is the class. In the iterative method, the attacker takes smaller steps iteratively instead of a single step of size ϵ.

6.13.2 Adversarial Training

Defending against adversarial attacks requires proactive security techniques. Examples of proactive security techniques in the context of phishing email detection are in [454, 457] and for email masquerade attacks in [27]. For example, for the text classifier in phishing email detection, we used WordNet relations such as hyponymy and synonymy to expand the set of words under consideration. In the email masquerade paper, we showed how to generate new attacks semi-automatically so that the defenders can improve their detectors.

There are essential three defenses against data poisoning techniques: data subsampling, outlier removal and trimmed optimization [464]. In data subsampling, many subsamples of the data are taken at random, and the model with the least error on training data set sample is selected. Outlier removal has been discussed in the Data Mining, Chapter 5. In trimmed optimization, a certain fraction of the data set with the most error is discarded.

Since the exciting work of Hinton et al. [191] on deep learning models for image classification, there has been a lot of interest in attacking deep learning models. An interesting technique here is transferability, i.e., the attacker can find attacks for a locally trained model and then deploy them against the unknown model used by the defender with a high chance of success. In a 2018 paper, Athalye et al. [22] show that obfuscating the gradient technique, used by several researchers, does not provide security. In other words, they design attacks on models that use this technique. As mentioned before, gradient ascent/descent is one of the popular approaches for computing optimums of functions.

Game theory provides a natural framework for adversarial machine learning, and also other topics in security: the cyber war between the attacker and the defender, moving target defense, and the principle that security is a game of economics. We refer the interested reader to [297, 525], and references cited therein, for a solid exposure to game theory and its applications in security.

6.13.3 Adversarial Generation

The idea of adversarial learning has been employed in generating synthetic data instances via the generative-adversarial network (GAN) [173]. A GAN consists of two neural networks: a generator G and a discriminator D, coupled together to produce a generative model. The goal is to generate synthetic instances that are similar to instances in the training data. The objective function is a two-player zero sum game between D – a function whose goal is to distinguish real input data from fake – and G – a function optimized to generate input data (from noise) that "fools" the discriminator. The game played by G and D is as follows. In each step, G produces an instance from random noise with the potential to fool D. D is then presented a few real data instances, together with the examples produced by the generator, and its task is to classify them as real or fake. D is rewarded for correct classifications and G for generating examples that fooled D. Both models are then updated and the next iteration of the game begins.

The main problem with the GAN is nonconvergence. Optimization algorithms often yield a saddle point or a local minimum rather than a global optimal solution, but game solving algorithms may not approach any equilibrium at all.

6.13.4 Beyond Continuous Data

Although much of the research in adversarial learning, so far, has been on deep learning models for image classification or generation, researchers have started looking at other domains as well such as text classification or generation. The problem of text data is that it is discrete. However, researchers have made some headway on the above topics for text data, see [473] and references cited therein.

6.14 Evaluation of Machine Learning

We have discussed several methods of evaluation, such as withheld testing sets, crossvalidation in various forms, etc. We have also discussed various metrics, such as precision and recall, confusion matrices, Receiver Operating Characteristic (ROC) curves, etc.

Several related methods have been used for comparing machine learning algorithms. While the confusion matrix provides a good assessment of the performance of a single algorithm, one often wants a single number to use to compare the classifiers. In a situation in which the only important criterion is the raw probability of correct classification, this is a fine statistic to use. In other cases, though, one wishes to trade off performance in various ways, and so a number of metrics have been developed. A good reference for these is [406].

The F-measure is the harmonic mean of the true positive rate, or recall, and the precision:

$$F_m = 2\frac{\text{TP rate} \times \text{precision}}{\text{TP rate} + \text{precision}}.$$

Similarly, the geometric mean of the two performance measures

$$G_m = \sqrt{\text{TP rate} \times \text{TN rate}}.$$

The F-score has a parameter β that trades off the components of the F-measure.

$$F_s = \frac{(\beta^2 + 1)\,(\text{TP rate} \times \text{precision})}{\beta^2\,(\text{TP rate}) + \text{precision}}.$$

A 2-class classifier typically returns a value between 0 and 1. By setting a threshold on this value, one determines the class label of the observation – if it is above the threshold, then it is class 1, otherwise it is class 0. By varying the threshold, one can vary the false alarm rate. Receiver Operator Characteristic (ROC) curves are constructed by plotting the false alarm rate against the true positive rate. The curve is constructed by varying the threshold and computing the resulting false alarm and true positive rates.

The area under the ROC curve, often written AUC, is also frequently used. For many purposes related to cybersecurity, however, this is not always a good measure. For example, in the analysis of packets on a network, or even network flows, one observes many thousands, often millions, of decisions a day. If a false alarm rate is set above 0.01%, one may have an impossibly large number of false alarms to address, and so one often focuses on the very lowest end of the ROC curve. Thus, adjusting the AUC to only compute the area in $[0, z]$ for some small number z may be a better metric.

For unbalanced data sets, the geometric mean G_m defined above and the Matthews Correlation Coefficient (MCC) defined below have been recommended.

$$MCC = \frac{TP * TN - FP * FN}{\sqrt{(TP + FP)(TN + FN)(TP + FN)(TN + FP)}} \tag{6.16}$$

MCC ranges from -1 to 1. One stands for perfect prediction and -1 for worst prediction. For G_m, better prediction will lead to higher values and poor prediction will lead to lower values.

6.14.1 Cost-sensitive Evaluation

The above discussion assumes cost insensitivity. In many situations, there is a cost associated with misclassification,[21] which could be different for false positives and false negatives. For example, if a bank accepts a person who defaults, then the bank incurs a big loss, and if the bank rejects a person who would not have defaulted, then the bank loses a profit. The profit is typically smaller than the loss. Similarly, if a phishing email is classified as good, the cost could be much more than if a legitimate email is put in the spam folder.

To handle such situations, we introduce costs c_+ and c_-, which denote the cost of a false negative and the false positive, respectively. Now, we can compute the true cost as:

$$Cost = fpr.c_- + fnr.c_+ \tag{6.17}$$

In this formula, fpr and fnr are the false positive and false negative rates. If there are costs associated with correct classification, then we can appropriately generalize the Cost function.

6.14.2 New Metrics for Unbalanced Data Sets

In [1], new metrics are proposed for unbalanced data sets, called "Balanced Detection Rate" (BDR) and Normalized Balanced Detection Rate (NBDR) to rank systems. The idea of Balanced Detection Rate is to measure how many minority class instances were correctly identified and to charge appropriately using the incorrect instances of the majority class. So, we divide the number of correctly identified minority class instances by the number of incorrectly classified majority class instances (Equation 6.18). Let $c = NEG/POS$, where NEG and POS are the negative and positive classes, respectively. If $c > 1$, then the positive class is the minority class, so

$$BDR = \frac{TP}{1 + FP}, \; BDR(\%) = 100 * \frac{BDR}{TP + FN} \tag{6.18}$$

If $c < 1$, then the negative class is the minority class, so we replace TP by TN in the numerator and FP by FN and we take the NEG $= TN + FP$ in the denominator for the BDR% formula. If $c = 1$, then the data set is balanced so both Detection Rates should be calculated and reported.

Observe that only a perfect classifier with FP = FN = 0 can have BDR% = 100%. For example, for the random forest classifier on the Phishing URL data set (Figure 6.16), we get a BDR of $4705/111 = 42.38$ and a BDR% of 0.86%, which is a stiff penalty for the 110

[21] In general, there could be a cost for correct classification also.

legitimate URLs classified as phishing. The boosted decision tree classifier (Figure 6.15) has a BDR of $4698/106 = 44.32$ and a BDR% of 0.90%.

We can generalize these definitions to take also cost of misclassification and benefit of minority class detection into account. For example, if $c > 1$ and α is the benefit of detecting a minority class instance, and β (γ) are respectively the cost of misclassifying the majority (minority) class instances, then we replace TP by $\alpha * TP$, FP by $\beta * FP$ and FN by $\gamma * FN$. Similarly, we can handle the case for $c < 1$ and again we should report both generalized versions of Detection Rates when $c = 1$.

The 1:1 charging scheme may be considered "too harsh" in some unbalanced situations. We also define an NBDR, which normalizes the charge based on the size of the classes, as follows. Again, this assumes that positive class is the minority class. Note that the numerator is just the detection rate for the positive class DR(POS) and the denominator is 2 - DR(NEG), where DR(NEG) is the detection rate for the negative class.

$$NBDR = \frac{\frac{TP}{TP+FN}}{1 + \frac{FP}{TN+FP}} = \frac{DR(POS)}{2 - DR(NEG)} \tag{6.19}$$

For NBDR, we have NBDR% = NBDR * 100. Normalization may have another advantage, comparing across data sets that are dissimilar in size and composition.

6.15 Security Applications and Adaptations

Machine learning methods have been applied for numerous security challenges including: scanning, profiling, intrusion detection (host-based, network-based, or hybrid), malware detection, spam detection and phishing detection (email, URLs, or web sites). We discuss some of these applications below.

6.15.1 Intrusion Detection

The first paper on intrusion detection (DBLP search for intrusion detection on 28 May 2018, 4,606 total matches) is by Dorothy Denning in 1986, wherein "a model of a real-time intrusion-detection expert system capable of detecting break-ins, penetrations, and other forms of computer abuse is described." In this seminal paper, she proposed using audit records as the basis for the model. She also introduced a taxonomy of intruders, a list of features, and different methods for detection (statistical, rule-based, etc.). Her main hypothesis is that an intruder will try to exploit the vulnerabilities of the system and this will result in abnormal use of the system. Consequently, the intruder can be caught by looking for abnormal patterns of use. Ever since Denning, the field has exploded (nearly 144 papers/year over 32 years).

We can classify intrusion detection systems based on at least three dimensions: the protection domain, the type of attacks covered, and the application domain. The domain of protection could be a single host, the network, or a combination of the two ("hybrid" methods). The type of attacks covered could be historical (misuse detection), or zero-day (anomaly detection). Finally, the application domain could be general purpose networks, wireless networks, wireless sensor networks, mobile ad-hoc networks, etc.

Early methods for intrusion detection tended to be rule/pattern-based and were better suited for historical attacks. For example, one can interview several security and IT experts and distill their collective wisdom on intrusions into a rule/pattern-based system. The advantage of this technique is the low false positive rate, i.e., normal behaviors are rarely flagged as intrusions.[22] The disadvantage is that new attacks, so-called zero-day attacks,

[22] Since the focus is on catching intrusions, it is the positive class.

will be missed, since they have not been seen before. For catching previously unseen attacks, anomaly detection methods are better. However, their disadvantage is the high false positive rate.

Over time, researchers moved to mainly data mining and machine learning techniques for intrusion detection. We described some data mining approaches in Chapter 5. Here, we discuss machine learning techniques. The classical works on machine learning for intrusion detection are [323, 421]. In these papers, SVMs (Section 6.6 above) and neural networks (Section 6.8 above) are used as classifiers, and a feature selection technique is implemented. The data set used is the DARPA challenge data set.

It consists of a dump of raw TCP/IP connections data obtained from a network designed to simulate a "typical US Air Force LAN." For each connection, 41 features are available. There are four major types of attacks in this data set:

1. Probing: scanning and other types of probing attacks
2. DOS: denial-of-service
3. U2R: accessing local super user (root) privileges without authorization
4. R2L: accessing a local machine from a remote machine without authorization

Feature selection is a hard problem. Features may be correlated, dependent, independent, or irrelevant. Many methods do not work well with correlated features. With 41 features, there are $2^{41} - 1$ nonempty subsets of features, so a heuristic approach is necessary. Mukkamala and Sung [323] developed a performance-based approach for feature selection. They deleted one feature at a time and kept the same training and test data set split. If the performance drops, the feature is important. If performance goes up, the feature is irrelevant. If the performance stays the same, the feature is secondary.

Performance includes training time, testing time, accuracy, false positive rate, false negative rate, etc. They gave detailed rules for classification of features based on their metrics and results with different categories of features: all, important only, important and secondary, etc. They also experimented with a feature selection approach using the support vector decision function. Here, they calculated the weights of the decision function and then took the sum of the absolute values of the weights. A higher sum means a more important feature. Again, experimental results are presented with different categories of features. The bottom line is that performance differences are small for the categories they obtained.

The DBLP query *intrusion detection survey/review* yields at least 90 results, including one on data mining and machine learning techniques [62]. The book [126] is also focused on this topic. Hence, we refer to these sources for digging deeper.

In [397], the authors reviewed the existing literature on machine learning techniques for intrusion detection and wondered why no one is deploying these techniques in practice. They identified several specific challenges to applying machine learning for intrusion detection: outlier detection setting, high cost of misclassification, semantic gap, diversity of network traffic, and difficulties with evaluation. According to the researchers, the difficulties with evaluation category can be further decomposed into three subclasses: data difficulties, mind the gap, and adversarial setting.

Most of these challenges have been discussed in Chapter 3. A few deserve further explanation. The semantic gap refers to the problem of explaining the justification for a decision from the classifier to a system administrator. Neural networks, for example, is a particularly opaque classifier with respect to explanation. Data difficulties is a particularly acute problem for intrusion detection. Almost all data sets are synthetic and much of the work revolves around the old DARPA data set and its KDD-Cup variant. The outlier detection setting should be understood as an unbalanced class issue. Intrusions are (hopefully) few and far between, i.e., outliers, and so the base-rate fallacy (Chapter 4) is applicable here.

6.15.2 Malware Detection

Malware detection has also been a fertile ground for machine learning. A DBLP search for malware detection on 4 June 2018 found 917 matches, considerably smaller than for intrusion detection.[23] The query *malicious software detection* netted an additional 16 matches. In Chapter 5, we discussed malware clustering. Here, we consider machine learning techniques for malware detection.

Malware detection techniques can be classified based on at least four dimensions: the type of input, the type of features, the target domain, and the type of malware considered. The input can be binary, assembly or source code. The features can be static, dynamic, or a combination of the two. Static means that the features are extracted without executing the code. Examples include control flow graph, nesting depth, and frequencies of certain operations such as indirections, etc. Dynamic means that features depend on information extracted by executing the code with a short time limit. Dynamic features provide more fidelity, but only limited code coverage, since it is uncomputable to come up with a set of inputs that will exercise all paths through the code. The advantage of static features is that they can provide more coverage of the code, at the cost of false positives through over-approximation. The target domain of the malware could be a smartphone, desktop, web, virtual-machine, etc. The types of malware were listed in Chapter 3.

Early methods for detecting malware tended to be mostly signature and pattern-matching based. Malware was analyzed to design patterns that could identify it, while keeping false positives low. However, to defeat signature analysis, attackers started using obfuscation techniques such as: dead code insertion (e.g., adding no-ops, or more complicated sequences that have no effect), register reassignment, subroutine reordering, code transposition, independent code reordering, code integration [518], equivalence transformations, compression, encryption, and combinations of these. Malware that uses these techniques to create new forms is referred to as *polymorphic* or *metamorphic*, depending on whether the malware has a bounded number of forms or not. Another disadvantage of signature analysis is that as the signature database grows with time, the method does not scale. Thus, defenders were forced to up their game, and so-called behavior-based malware detection schemes were devised. We cover a few of them below. Their disadvantages are a higher false alarm rate and also the time spent in feature extraction, learning with supervised methods, and applying the model.

The data set problem for malware is less acute than for intrusions. There are several sources of malware data sets such as VirusTotal, VirusShare (over 30 million malware samples), etc. We include a subset of the available malware data sets in Table 6.6 [147].

For malware analysis, we describe a few interesting approaches, and give pointers for digging deeper. Our first work is [247]. They evaluated several classifiers: SVM (Section 6.6), Naive Bayes (Section 6.2.1), Decision Tree (Section 6.5) and their boosted (Section 6.10) versions, on a data set that they collected. The data set was relatively balanced with 1,971 benign samples and 1,651 malware instances, all in Windows PE format. They use n-grams of byte codes as features. The *hexdump* utility was used to produce hexadecimal codes in ASCII format from the byte code. Then, n-grams were produced by concatenating each four-byte sequence of codes. Boosted decision trees performed the best in their experiments, which used 10-fold cross-validation and reported the AUC (Section 6.14). They used a systematic method to select the best features. It consisted of pilot studies to find the size of n-grams, the size of words, and the number of selected features. First, they fixed the n value at 4 and, using information gain, selected the best 10, 20, ... 100, 200, ..., 1000, 2,000, ..., 10,000 n-grams and evaluated the performance of the classifiers. Best results were obtained with 500 n-grams. Next, they fixed the number of n-grams to 500 and varied the n-value

[23]Malware alone has over 2600 matches, and malware detect has 1075.

Table 6.6: Malware Data Sets

Data Set Name	Size	Type
theZoo	Less than 150	Collection
VirusTotal	Large and growing	Collection
Microsoft	500 GB	Data set
MalImage	9339 Samples	Data set
MIST	4764 Samples	Data set
ContagioDump	?	Community
DasMalWerk	+/-5900 Samples	Community
VirusBay	2468 Samples	Community
MalShare	Large	Community
VirusShare	+30 Million Samples	Community

from 1 to 10. Best results were obtained with $n = 4$. They also varied the word size in bytes from one to an unspecified limit in steps of one. Best results are obtained with word size of one byte.

They also obtained a data set consisting of 291 new malicious executables. They conducted a generalization experiment, i.e., they tested their models on this new data set without any training. Boosted decision trees performed the best in this experiment also. A final experiment was classifying malware into categories based on the payload function.

A couple of disadvantages of their method was that they could not deal with packed software and the time for training, several days, is quite high. Another group [346] attempted to address the issue with packing. They used an unpacker to extract hidden code. But their run-time is also high, which is addressed in PE Miner [383].

A different approach was used by [352], which is interesting because it gives provable correctness results. They use trace semantics, which basically describes programs by their execution traces. The intuition behind their approach is that if a program P is infected with malware M, then at least parts of P's execution traces will be similar to the execution traces of M. A semantic matching relation is defined between traces based on execution contexts, i.e., the environment that keeps track of bindings of variables and the memory that keeps track of contents of locations used by the program, rather than commands. A restriction of the behavior of a program is defined by abstracting variables and "labels," this may be considered as abstract interpretation. Now the question is coming up with malware behavior specifications that semantically match the restricted program's trace semantics. They show that this can be done for certain kinds of obfuscation techniques used by malware writers. Thus, they are able to prove relative soundness and completeness results for their method in the presence of certain obfuscations.

According to DBLP query *malware detect survey*, there are seven surveys as of 5 June 2018. Dropping "detect" yields 21 more,[24] particularly relevant are [135, 390, 443, 516], and a survey on command and control detection [164].

6.15.3 Spam and Phishing Detection

Bayesian classifiers have been popular for spam email detection, with features extracted from the email body and header, e.g., SpamBayes and SpamAssassin. For a survey of learning techniques for spam email filtering, see [43]. A few public spam data sets exist, e.g.,

[24]The latter search shows some preference for "analysis" to detection.

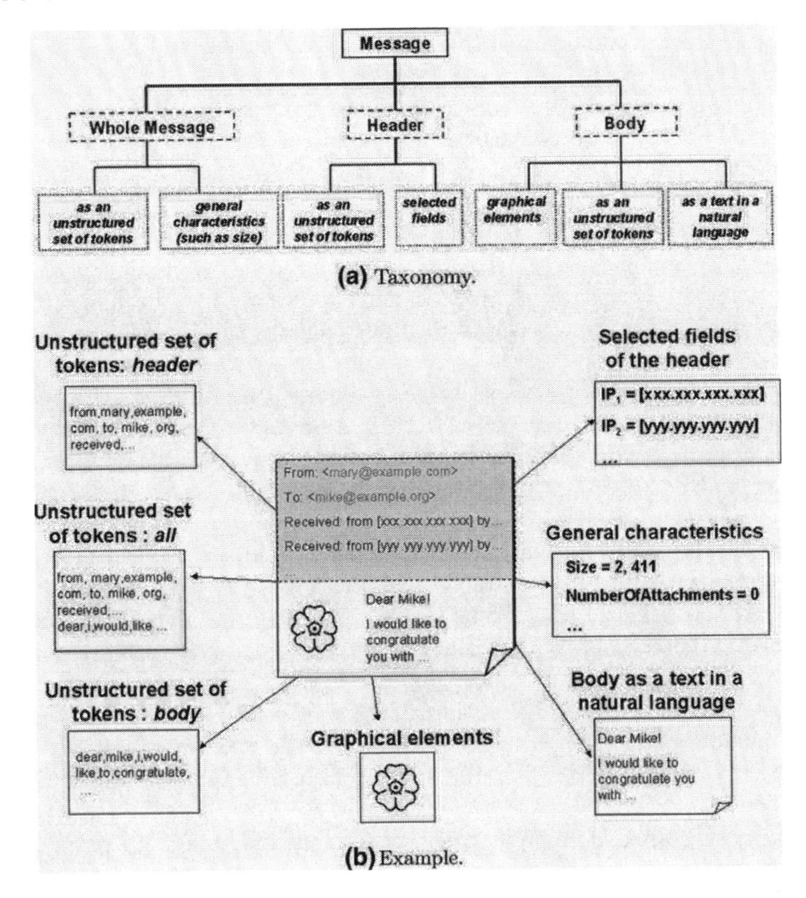

Figure 6.19: Potential Features in an Email

SpamAssassin data set,[25] CSDMC,[26] Lingspam, Enron-spam, PU-spam,[27] and Spambase (UCI repository), although some are now dated. A decent email spam filter can be built by tokenizing an email and using either multinomial naive Bayes or logistic regression with a frequency based feature selection technique. Figure 6.19, [43], shows email features.

Moving from spam, which is basically advertisement, to phishing, which is more insidious and damaging, the data set problem was acute for phishing, and still is for phishing emails. Fortunately, there are a few data set sources, thanks to Phishtank (URLs), APWG (URLs), Alexa (Domains), DMOZ (URLs, deprecated), WikiLeaks, and Enron.

Phishing URL detection. There is considerable work on learning techniques for this problem. Decision trees (Section 6.5), SVMs (Section 6.6), logistic regression, and Naive Bayes (Section 6.2.1) are the most popular methods. Online learning techniques, e.g., Adaptive Regularization of Weights (AROW) and Confidence-Weighted (CW), have also been used. N-grams of characters, special characters, such as @, periods, hyphens, etc., presence of targets such as Paypal and Ebay, and edit distance from whitelist URLs have been some of the more popular features. Researchers have also used statistical features such as the Kolmogorov-Smirnov test (KS-test, Chapter 4), KL-divergence and Euclidean

[25] `https://spamassassin.apache.org/old/publiccorpus/`

[26] Unavailable now.

[27] `https://aclweb.org/aclwiki/Spam_filtering_datasets` - Lingspam, Enron-spam and PU-spam are available here, accessed 1 August 2018.

Distance [453]. We give summaries of the most interesting phishing URL detection papers and pointers for further study.

Ma et al. [289] propose a method that uses different types of features (lexical, host-based, blacklist-based and WHOIS properties). They conducted feature selection and comparison, classifier performance analysis using inter-classifier evaluation, and error analysis. However, the data sets are relatively small and almost balanced. For benign URLs, two data sources were used: the DMOZ Open Directory Project (deprecated), and a random URL selector for Yahoo's directory (deprecated). URLs to malicious sites were obtained from PhishTank, a blacklist of manually verified user contributions, and Spamscatter. Spamscatter includes URLs from a wide range of scams advertised through email spam (phishing, pharmaceuticals, software, etc.).

Zhao et al. [523] present a cost-sensitive active online learning (CSOAL) framework that explicitly tackles the class imbalance issue. Their framework only uses a fraction, 0.5%, of the provided data set during training. It is evaluated on a large (1M) and imbalanced (99 legitimate to 1 malicious) data set. Metrics used are weighted cost and weighted sum of sensitivity and specificity. Scalability is studied and training and testing times are reported.

In [453], researchers introduced new features, using KS-test and KL-divergence, which measure the distance between the character distribution of the URL and the character distribution for standard English text. Some traditional features were adjusted for robustness, e.g., the ratio of URL domain to URL length, instead of just URL length. Four relatively balanced data sets are used for evaluation. Five supervised learning methods are used along with AdaBoost (Section 6.10) and Stacking. Data sets were obtained from several sources including a company's collection. Data set diversity is checked and a cross-data set experiment (training on one data set and testing on another) is performed. Training and testing times are reported.

In [450], researchers used online learning algorithms and compared spam and phishing URL detection. They show that a classifier trained on spam and tested on phishing URLs does worse than a classifier trained on phishing and tested on spam URLs. They also show that character n-gram features are powerful for phishing detection. They performed experiments on unbalanced data sets and checked their models for generalization ability.

Recently, researchers have applied deep learning for phishing URL detection, e.g., a convolutional GRU is used in [507]. To dig deeper on phishing URL detection, see [103, 374, 440]. The first reviews the phishing detection literature from the security challenges perspective, the second is focused on phishing URL detection, and the third does a good feature analysis for phishing URL detection.

Phishing web site detection. This problem has also been a fertile ground for learning techniques. The top four classification methods are: decision trees (Section 6.5), SVMs (Section 6.6), logistic regression and Naive Bayes (Section 6.2.1). There are several potential features, which can be categorized into: HTML, Javascript, and network. Examples of HTML features are the frequency of words on the page and the number of different tag types. Javascript features include presence of shell code and number of iframe strings. Network features include domain registration and HTTP header information. We give summaries of the most interesting phishing web site detection papers and pointers for digging deeper.

Retraining is one way to tackle the active attacker challenge. Researchers in [485] considered four goals: (1) minimizing false positives, (2) increasing detection, (3) dealing with noisy training data, and (4) reducing the detection time (to process millions of web pages daily). Their method starts with lightweight features (URL-based) and then extracts the "heavier" content and hosting features. They use a custom machine learning algorithm, similar to Random Forest, and online learning algorithms. They retrain their model (offline) daily with newly released blacklists. Although their model performs well, they do not discuss explicitly how their method deals with noise.

Researchers in [427] considered both classical phishing attacks and "hijack-based" phishing attacks. They define *classical phishing attacks* as a page "asking for sensitive information from users by mimicking a legitimate domain or giving an incentive, warning etc. In contrast, in a hijack-based attack, phishers hijack a server on a legitimate domain and then put content to steal sensitive information from users. In this attack, the URL and hence the domain is legitimate, but the contents are phishing and not authored by the domain owner." Another example is when previously downloaded malware opens an invisible frame on top of the page visited by the victim. The address bar continues to show the previous site, but the information entered is harvested by the attacker, i.e., the connection has been hijacked. They devise novel techniques for both kinds of attacks.

The paper [298] addresses many, although not all, security challenges: the base-rate fallacy, the data availability issue, and the active attacker problem. Their method uses a small amount of training data and features that cannot be totally controlled by the attacker. They test on an unbalanced data set of web sites (100,000 legitimate and 1553 phishing). They obtained 99% AUC (Section 6.14) and less than 1% false positives with 212 features. They measured the effectiveness of each feature separately. Language independence is an attractive aspect of this method, which is demonstrated on web sites in six languages. However, they do not report training and testing times.

One conclusion of the review of phishing literature from the security viewpoint [103] is that semi-supervised learning methods have not been sufficiently exploited for phishing detection, whether it is URL, web site or email. To our knowledge, only two papers have explored semi-supervised learning methods for phishing web site detection. A self-training method for detection of phishing web sites has been proposed in [218]. The transductive SVM method has been used in [272] for the same problem.

There are a few surveys that include phishing web site detection, [103, 125, 446]. The first reviews the phishing detection literature from the security challenges perspective and the last two focus on phishing web site detection.

Phishing email detection. There is considerable work on learning techniques for this problem. For emails, features can be extracted from the body, the header, the links (if any), and attachments (if any). Figure 6.19 gives an idea of possible features of an email. SVMs (Section 6.6), Random Forest and Decision Trees (Section 6.5) have been the most popular classifiers for emails. We give summaries of the most interesting phishing email detection papers and pointers for further study.

Early phishing email detection works using machine learning are [73, 151]. The first proposed a simulated-annealing technique for feature selection. The selected features are then weighted according to information gain values and a one-class SVM is trained. The balanced data set consisted of 400 emails. The second paper has a low false positive rate. Their features are mostly derived from the links in the email and they also use the output decision from a spam filter.

Deep learning techniques were used for the first time, to the best of our knowledge, for phishing email detection in the 1st IWSPA-AP anti-phishing shared task [452]. We describe the task and the data sets in more detail in Chapter 8. Here, we note that the participating teams used many different "classical" methods including Naive Bayes, SVM, Decision Tree, kNN, Logistic Regression, Random Forest and Ada Boosting. The deep learning models included bidirectional Long Short-term Memory (LSTM), CNN, RNN and LSTM. The IWSPA-AP data set was also used by a follow-up paper in which researchers used an "improved" RCNN model [144] along with an attention mechanism to get a high detection rate and low false positive rate.

In 2018, a team of researchers [396] used an evolving neural network and reinforcement learning, which is not covered above. Reinforcement learning (RL) is based on trial-and-error interaction with an environment. In RL, the system/machine is a 4-tuple (A, τ, P, R), where A is an agent/controller, which observes the system state through sensors. The agent performs actions τ_t and is given reward R_{t+1}. The policy P maps the state of the environment to the action to be taken in this state. The reward function R is based on the overall objective of the agent. In their method, the neural network is observed by an agent who can change the significant features, change the neural network architecture, or change the data set to be used for training.

To our knowledge, no one explored semi-supervised learning methods for phishing email detection, but one team considered an adversary who shrinks the content of a message to defeat text analysis. They showed that a spectral clustering approach based on n-grams from links with random forest classifier works fairly well in this case and outperforms LDA on body text with random forest classifier [112].

There are several surveys on phishing email detection, e.g., [11, 88], some on phishing web site detection and a few on phishing URL detection. Despite these surveys, no one seems to have done a systematic, thorough review of previous work from the security perspective. Hence, we recently undertook such a task in [103]. A significant finding is that only a few papers have considered the unique needs of the security domain.

For more general discussion of machine learning, see [87], which has some nice advice that is also applicable to security. Other applications of machine learning to security and further discussions of their effectiveness, can be found in [18, 186, 254, 303].

Spear phishing email detection. As observed in [103], Han and Shen [184] used a graph-based semi-supervised learning algorithm to differentiate between benign, known and unknown spear phishing campaigns. The data set contained sets of labeled spear phishing campaigns and unlabeled emails, which were collected from Symantec's enterprise email scanning services. The data set contained 1,467 spear phishing emails from eight different known campaigns, and 14,043 benign emails collected between 2011 and 2013. Their method achieves a 0.9 F1-score with a 0.01 false positive rate in detecting spear phishing emails that belong to previously known campaigns. As mentioned above, F1-score may not be the best metric to use for imbalanced data sets.

6.16 For Further Reading

A review of artificial intelligence methods for combatting cyber crimes can be found in [118]. A focus on military applications, including a short section on cybersecurity, can be found in [419].

As yet, there are no widely applicable techniques that can predict, given a specific problem or application, which machine learning algorithms would be the best. However, there is some progress along these lines. We recommend the reader to check out [54, 55] and references cited therein.

Another aspect that we have not discussed above is transfer learning, also referred to as domain adaptation [92, 462]. For learning in nonstationary environments, the survey by Ditzler et al. is a good starting point [119]. A good comparison of discriminative versus generative classifiers in the context of logistic regression and naive Bayes is given in [329]. The debate is still ongoing, see [351, 504]. For active learning techniques, we refer the reader to the survey [6].

Machine learning does not produce algorithms for free; as noted above, there are adversarial attacks against deep learning and other algorithms. More generally, there are potential security problems with any machine learning technique. See [281] for a discussion of some of these.

Chapter 7

Text Mining

There are generally two ways to represent text in a computer for the purposes of performing analysis. In the first, documents are reduced to a "bag-of-words" representation. Here all order information is lost, all that is retained is the number of times each word appears in the document. In practice, one extends this by tracking n-grams: bigrams are pairs of contiguous words, trigrams are triples of contiguous words, etc.[1] More sophisticated methods will retain common or "meaningful" phrases. At the other extreme, the ordering is retained, and sentences are processed to determine their grammatical structure. We will consider the former approach in this chapter, and the latter in Chapter 8.

It should be noted that the bag-of-words approach is quite naïve; one should expect that completely forgetting the ordering of the words will ignore important information, and in fact there are many examples of applications where it is clear this is not a good model. The easiest to see is in translation from one language to another. Context is critical in deciding which definition of the word to use, and in many cases it is not appropriate to translate at the word level at all, but rather the phrase or sentence level. While some sentiment algorithms using bag-of-words models perform well in certain well-constrained problem domains, in general one needs more context than simply the number of times a word appears.

With that said, this simple model can still be surprisingly powerful, especially when used in concert with word n-grams and phrases. It can also provide useful information about the corpus – the set of documents – that one is investigating. We will see this below in a number of applications. Many document retrieval applications use the bag-of-words model. See for example [120].

7.1 Tokenization

The first process applied to any collection of documents is tokenization. This is the mapping of the digital representation, whether ASCII, UTF-8, or other representation, into the basic tokens to be processed. These tokens are generally characters (such as punctuation) and words. Each token represents the atomic information that will be used for inference. Tokenization is language dependent, and can be relatively simple for some languages (such as English) and difficult for others (Chinese can be particularly challenging), see [202, 495]).

Tokenization can also be problem dependent, in the sense that different strings of characters may be considered as tokens in different contexts. For example, in Twitter and email, one might tokenize the emoticons rather than treating the punctuation as individual

[1]The term *n-gram* is used in two distinct ways in text processing. In this instance, it refers to n contiguous words. Often it is used to refer to n contiguous characters. We will use the terms *word n-gram* and *character n-gram* to distinguish these two cases.

characters. On the other hand, this could be handled by secondary processing, and so one generally implements tokenizers for a given language, rather than having separate tokenizers for different situations.

Once the document is tokenized, it must be processed into a form to be used for the inference task at hand. This is referred to as preprocessing, and also as "cleaning" in cases where the documents contain a large number of errors or have other challenging issues.

7.2 Preprocessing

Any text corpus requires preprocessing to clean out errors and unwanted characters and words. The specifics of the preprocessing is very corpus dependent and is dependent on the desired inference. For example, if the inference is classification, for example determining whether a document is on the topic of biology or physics, words such as "the", "and", "therefore", etc. are "content free", called *stopwords* in the text processing literature, and would be removed. However, if the topics were logic versus differential geometry, "and" may no longer be considered content free. In a corpus on elephants, the word "elephant" is essentially content free since it is ubiquitous, but the words "African" and "Asian", although appearing in most documents, may be quite important. Further, for the purposes of determining authorship of a document, it is often precisely the standard stopwords that are useful.

Whether capitalization is retained or not is also dependent on the inference task. One may utilize capitalization to detect proper nouns and acronyms, which would then be handled separately from the other words. For some domains, such as Twitter, capitalization may be an indicator of sentiment or used for emphasis, and thus one may wish to retain the capitalization or at least note its use for purposes of inference. Other applications may simply map all letters to lowercase and proceed from there.

Often documents are obtained via scanning them in and running Optical Character Recognition (OCR) algorithms against the scanned pages. These algorithms can produce errors through incorrect character recognition, failure to detect light characters or characters obscured by dirt or smudges, page distortions due to improper placement on the scanner, and other mechanisms. It can be extremely painstaking to read through all the scanned documents looking for and correcting these errors, even with tools that search for unrecognized or misspelled words.

Anyone who has converted a pdf document to text has likely come across the "fi" problem: the characters are turned into a single ligature rather than two characters, and this will often not be transferred, since it doesn't correspond to an ASCII character.

For many text analysis tasks, punctuation is unimportant and is removed. In some cases, particular punctuation characters are treated separately, such as "-" indicating a hyphenation, or ":" used in references to time. Punctuation can also be used to identify words or phrases of interest, such as associating the acronym to the phrase in the following: "General Dynamics (GD)". In some cases, such as emoticons in ASCII emails or tweets, multiple punctuation characters can be tokenized into a single "word". This may be less of an issue with the widespread use of emojis, but one does still see emoticons such as `:-)` in some text.

Numbers can be treated differently than words; often they are removed, but there can be cases where they are important, particularly when they help distinguish important words, such as the version numbers of operating systems, or represent important times, dates, etc., in news articles or historical/biographical documents. Usually, however, they are removed simply because their actual values are not important, and specific numbers are found in only one or two documents. Whether one retains numbers (digits), removes them, or replaces them with a token indicating that a number or digit appeared in the document, depends as always on the document corpus and the inference task.

Finally, the decision must be made whether to perform stemming. This is the process of reducing a word to its root: "walking" and "walked" both become "walk". The upside of stemming is the reduction of the size of the lexicon, and the ability to use a single token for all cases of a given word. The downside is that no stemming algorithm is perfect, and errors will occur. See [282, 341, 350] and the many articles about stemmers for specific languages. Whether to use stemming for a particular task is, again, dependent on the corpus and the inference task. Generally speaking, the errors introduced through stemming are outweighed by the extra information resulting from mapping of versions of the same word to the root, but this is corpus/problem dependent.

These decisions are made on a case-by-case basis, with the criterion for selecting what preprocessing is performed being the quality of the resulting inference. A considerable effort is involved in this preprocessing, with estimates on the percentage of time required for preprocessing text versus constructing the algorithms to perform the inference ranging as high as 80%.

For example, consider the email below:[2]

```
Message-ID: <20873133.1075854043781.JavaMail.evans@thyme>
Date: Fri, 1 Sep 2000 01:10:00 -0700 (PDT)
From: faye.ellis@enron.com
To: daren.farmer@enron.com
Subject: Lone Star Points
Cc: donna.consemiu@enron.com
Mime-Version: 1.0
Content-Type: text/plain; charset=us-ascii
Content-Transfer-Encoding: 7bit
Bcc: donna.consemiu@enron.com
X-From: Faye Ellis
X-To: Daren J Farmer
X-cc: Donna Consemiu
X-bcc:
X-Origin: Farmer-D
X-FileName: dfarmer.nsf

Daren, there were several points on your list that were previously created
and the remainder have been created, with the exception explained below.

Donna and I researched your list and identified the ones that were on system
(which she would create) and the ones that were off system (which I would
create).

Standard Pooling Stations and West Texas (Line X) Pooling Stations had no
meters, so they are not set up.

The Hunt Fairway Plant (17-8477-01) was previously created as (178477), I
need to know if the (01) needs to be added.
```

[2]This is one of the "ham" emails available from the Enron spam corpus at http://www2.aueb.gr/users/ion/data/enron-spam/index.html.

```
We need to verify HPL - Texoma (17-0973-13), it is set up as (17097613).  Is
this the same point???

Please let Donna and myself know how to proceed with the ones in red.


Thanks,
Faye
```

Certain preprocessing is obviously needed. The header of the email contains very different information from the body. This information may simply be stripped off and ignored for some applications, or may be parsed into some structure for further processing for others, e.g., phishing email detection. If the header is treated as text in the same manner as the rest of the email, then strings such as "Cc:" and "X-From:" should be added to the stopword list and removed.

There are several punctuation marks and numbers in the document, and punctuation clearly indicates certain noun phrases that might be worth retaining as a phrase rather than individual words. The use of repeated punctuation – "???" – is clearly meant for emphasis, but for most inference tasks should be mapped to a single "?", unless punctuation is removed completely.

For this email the numbers have meaning, and are important, but this does not mean that the numbers should be retained for the inference. Should "17-8477-01" be reduced to "17847701" or "17 8477 01"? Since these refer to plants, which are named in the email, are they redundant and potential cause of confusion (noise) in the inference? Should "Hunt Fairway Plant" and "17-8477-01" be tokenized into a single token, so that future instances of either match to the same token? The answer may depend on how many ways the same plant is referred to in the corpus; for example, is the plant ever referred to as simply "Hunt" or as HFP? Note also that the question of the "01" in the number is topic of the email, indicating that there is some ambiguity, even amongst the writers of the text.

Another example from computer security is the use of IP addresses: should they be detected and retained, or should all addresses be replaced with a token such as "IPADDRESS", or are they uninformative and should be removed? For example, in a corpus of network intrusion reports, there may be several IP addresses in every report, and if the purpose is to cluster the reports by attack type, the IP addresses may be irrelevant to the task – although the number of distinct IP addresses may not be. For other inference tasks, such as an assessment of severity, certain IP addresses may be critically important. There is also a distinction between "inside" IP addresses and "outside" ones – it may be important to make this distinction in assessing attacks, while the actual values of the IP addresses may not be important, at least for detecting and classifying the attack.

This illustrates an important point which bears emphasizing once again: the preprocessing decisions must be made based both on knowledge of the inference task and how the decisions are likely to affect the inference, and on knowledge of the data. It is for this reason that such a large percentage of the time is spent on the preprocessing step.

Note that this point is important for decisions about storage of the data. It is generally the case that one should do the minimal cleaning and preprocessing of the data prior to ingestion into a database. If the text comes from scanned documents, it is probably reasonable to correct scanning errors prior to ingestion; however, it is important to retain any information that may be important for future applications. As an extreme example, storage space can generally be massively reduced if one retains only the number of times each token appears in a given document, even more so if the words are stemmed. This is sufficient for an application that will use a bag-of-words method, but if a future application needed to determine parts of speech (POS), or phrases, or disambiguate homographs, the data would be useless.

In some documents the issue of misspellings or "creative spelling" may need to be addressed. For example, Twitter data often contains such errors and "creative spellings", often for the purposes of emphasis or irony. For example, consider the tweet:

```
im soooooooooooo happy that i have the flu and "get" to stay home!!!! :-(
```

The repeated characters provide emphasis, and the emoticon (often replaced with an emoji in current tweets) also contains information about the intent of the message. Expletives are often used for emphasis and to indicate strong emotion, for shock value, or can be indicative of the age or maturity of the individual or the norms of their social peers. Also, the use of "im" for "I'm" is typical, as is "ill" for "I'll". In a completely unscientific experiment, a set of 103 tweets was collected that contained words related to sickness (other than "ill") and the word "ill". Of these, 64 were using "ill" in place of "I'll". This suggests that some level of natural language processing (Chapter 8), beyond simple cleaning and preprocessing, may be of value in some cases.

7.3 Bag-Of-Words

As mentioned above, one often treats a document as a bag-of-words: the document is tokenized into individual words, and only the word counts are retained. The tokenization (or secondary processing) can treat certain phrases, pairs, triples, etc., of words as individual tokens as well. Once the tokenization is complete and the document has been reduced to a set of tokens, or "words", the ordering of these is ignored in the bag-of-words model, and only the number of times each word (or phrase) appears is retained.

7.4 Vector Space Model

A vector space representation of a corpus of documents calculates a vector of numbers for each document. In the standard bag-of-words model, the vector is a collection of weights, one per word in the lexicon, with the weights being some representation of the "importance" of the word in the document, or the "information" of the word relative to the corpus. One of the key features of this representation is that each document is replaced by a vector, and the vectors are commensurate – they have the same length and each component corresponds to the same measurement on the document.[3]

The most common vector space representation is as vectors of word counts. Each document is transformed into a vector, with each entry corresponding to a word, with the value being the number (or proportion) of times the word appears in the document. The set of words is generally some subset of all the words contained in the corpus of interest; as mentioned above, stopwords may be removed. This matrix containing weights for each term in each document is called the *term-document matrix*.[4]

To illustrate this, consider the problem of determining whether a given sentence is written in English or French. A corpus of 20 books with sentence-level translations in English and French was obtained from `http://opus.nlpl.eu/`. Each sentence was reduced to

[3]In practice, these vectors may be stored in a sparse representation, so it is technically not true that they all are of the same length in a sparse representation; however, from an analytical perspective, this is irrelevant.

[4]Also sometimes the document-term matrix, which is the transpose of the term-document matrix.

lowercase, and all characters except the letters a–z and space were removed. Only those sentences containing 100 or more characters were retained. A character 2-gram (bigram) vector space model was used, where each document was reduced to the 637-dimensional vector corresponding to the proportion of times each of the observed 637 unique character bigrams appeared in the sentence. For this experiment, there were 32, 457 English sentences and 33, 195 French sentences. A crossvalidated 1-nearest neighbor classifier produced the correct class for all but 38 of the sentences – an error of less than 0.06%.

7.4.1 Weighting

While term frequency is a common representation for the vector space model, it is generally replaced by more sophisticated word-weighting methods. Stopwords were mentioned above; these are "content free" words that should be removed from the documents prior to processing. The problem is, as has been discussed above, deciding which words are "content free" is dependent on the corpus and the inference task, and it may not be trivial to determine which words are most appropriate. There are automatic methods to select stopwords – see for example [21, 444, 487, 508], and many other methods are available. One way to partially mitigate this problem is to weight the word frequency by the inverse of the document frequency – the frequency that the word is found in the documents. The idea is that a word like "the", which occurs in virtually all the documents obtains a very small weight, while, for example, the word "elephant" is most likely found in very few documents.[5]

Specifically, the inverse document frequency is defined as the log of the inverse of the frequency with which the term occurs in the N documents of the corpus C:

$$\mathrm{idf}(t, C) = \log\left(\frac{N}{|\{d \in C : t \in d\}|}\right). \tag{7.1}$$

In the term frequency/inverse document frequency (TFIDF) weighting, the term frequency is multiplied by the inverse document frequency to produce the term weighting in the vector representation. TFIDF type weighting schemes are well studied in the information retrieval, and we refer the reader to Figure 6.15 in the book on Information Retrieval for some alternatives [296].

Note that a word that is in all documents produces an inverse document frequency of 0, and thus is effectively removed from the lexicon, while a term that occurs in a small number of documents is highly weighted, and thus when it occurs in a document it has a large effect on the document's relationship to other documents. For this reason, it is a good idea to remove words that occur in few documents, unless it is determined that these words are indeed highly informative. For example, in the Enron email above, one might remove the numbers, since these are likely to be rare and redundant with the other information in the text.

One often refers to the weighted matrix as the term-document matrix as well. Thus, the meaning of this term is context dependent. Whether it refers to the term frequencies or some other weighting scheme must be inferred from the discussion.

Consider once again the Enron spam data set.[6] We'll use the first set of data from directory enron1 as training, and the data from directory enron2 for testing. Using TFIDF as the term-document weighting, removing terms that appear in fewer than 5 emails, results in 9673 terms, and using a random forest classifier, we obtain the confusion matrix:

	Ham	Spam	%Error
Ham	3664	8	0.218
Spam	189	2807	6.31

[5]Unless, as mentioned above, the corpus is about elephants!

[6]`http://www2.aueb.gr/users/ion/data/enron-spam/index.html`

Table 7.1: Top 100 Most Important Words for Classification According to the Random Forest Trained on the Enron Spam Corpus

hpl	.thanks	see	be	at
enron	online	000	to	prices
your	xls	hou	ectcc	quality
!	2001	:	get	regards
gas	on	if	n	sitara
here	please	:subject	a	$
%	forwarded	money	prescription	now
2000	2004	com	corp	01
meter	will	you	free	volume
for	me	&	nomination	and
attached	ect	meds	doc	any
.	r	have	?	|
deal	daren	v	only	offer
the	no	file	volumes	mmbtu
/	pmto	)-	texas	all
more	let	@	t	this
www	know	i	is	sex
,	-	actuals	'	software
http	amto	u	world	php
nom	questions	best	following	*

Beyond removing nonprinting characters and words contained in fewer than 5 emails, we have done no preprocessing of these data. In particular, punctuation, which is often an important part of email, is left in. These data have been processed from the original emails, as described in [311].

The trade-off between the errors is fairly important in this application. We have all had instances of legitimate emails being missed due to a spam filter deleting them.[7] On the other hand, if a few spam emails get through, most people can deal with this with little effort. The question then becomes whether the errors given are acceptable, or whether further work should go into improving the spam error.

One of the features of the random forest classifier is that it calculates which variables are most important for the classification task. Table 7.1 shows these, reading top to bottom, left to right in importance.

To test whether the punctuation characters are really necessary for this problem, a random forest was trained on the data after removing the punctuation. The resulting confusion matrix is:

	Ham	spam	%Error
Ham	3663	9	0.245
Spam	156	2840	5.21

This does appear to be slightly better, in fact, than the previous model, although the performance on "ham" is slightly worse. A McNemar test gives a p-value of 0.0001866, so we would be justified in thinking the improvement is significant; however, as mentioned

[7]While most spam filters don't immediately delete emails, but rather move them to a folder for future deletion, in practice, people who receive a lot of spam don't look in this folder unless they suspect they have missed an email.

Table 7.2: Top 100 Most Important Words for Classification According to the Random Forest Trained on the Enron Spam Corpus Using only the Words and Numbers, After Removing Punctuation

enron	hpl	your	2000	gas
for	http	here	meter	attached
thanks	deal	more	the	www
2001	no	000	forwarded	please
xls	on	online	ect	let
will	see	2004	r	if
money	nom	amto	daren	best
me	you	pmto	file	hou
questions	i	texas	a	know
v	doc	nomination	corp	prescription
meds	get	u	have	be
all	only	mmbtu	free	n
our	01	world	ectcc	actuals
click	paliourg	sitara	to	at
and	volumes	software	is	com
site	any	following	volume	now
farmer	am	offer	save	php
quality	prices	of	tap	t
bob	viagra	12	are	10
sex	by	30	00	from

previously, one should be cautious in making this type of comparison. This result is likely to be very specific to this particular corpus of emails, which is unlikely to be representative of emails in general.

In particular, note the importance of the years 2000, 2001 and 2004. These are very specific to the data set, and will most likely not be important in data collected after the time of these emails. This is a nice illustration of the nonstationarity of emails (and spam) and suggests that some intelligent preprocessing (such as replacing the years with tokens such as "thisyear", "lastyear", etc.) is in order. Once again, we refer the reader to Hand's important paper [185].

It is likely that punctuation is useful in detecting spam, and the fact that in this particular example the "words-only" version performed better is likely an artifact of the particular data set rather than a general rule. Further, as noted above, it is undeniable that the words used in spam, and the methods used to defeat spam filters change in time, so this problem must be viewed as a nonstationary problem.

The top 100 most important words for this classifier is shown in Table 7.2. As can be seen, there is considerable intersection between these sets, which is as should be expected.

Further analysis of the important words indicate that these are very specific to the corpus – in fact one can read either of the two tables of "important words" and know immediately that this is the Enron corpus. Even if one removed the give-away word "enron", other words would likely give the source away. Words such as the years, "texas", "gas", "sitara" and a few others would be enough to suggest the source. On the other hand, words like "sex" and "viagra" and "prescription" are so ubiquitous in spam that one would expect that these classifiers would have some ability to generalize beyond this set – although the specific words and phases used to refer to some of these topics, particularly prescription drugs, will change in time.

This exercise illustrates one of the most important uses of text analysis and classification for various textual data sets – analyzing the performance and the structure of the classifiers can provide information and insight into the corpus, and suggest new ways to proceed in future analysis.

Note that the weighting, particularly the TFIDF weighting, depends on the corpus. Thus, in a hierarchical clustering scheme or a tree-based classification scheme, one may wish to recompute the term-document matrix on the individual clusters or subsets of the data at each node of the tree. See the description of iterative denoising in [169, 357] for some discussion of this idea.

7.5 Latent Semantic Indexing

Latent semantic indexing (LSI) is a way of reducing the dimensionality of the term document matrix by embedding it in a lower dimensional space. The method simultaneously produces an embedding of both the documents and the terms, in a way that we will make explicit below.

The basic tool of LSI is the singular value decomposition from linear algebra.[8] Write $X_{n,m}$ for a matrix X with n rows and m columns. Given a real valued matrix $X_{n,m}$, there exist real valued orthogonal matrices $U_{n,n}$ and $V_{m,m}$, and nonnegative diagonal matrix $D_{n,m}$ such that:

$$X = UDV^t$$

In the case where $n \neq m$, "diagonal" means that only $d_{i,i}$ can be nonzero. Everything is ranked in decreasing order of the singular values $d_{i,i}$. Furthermore, the best (under the Frobenius norm[9]) low rank approximation to X of rank d is formed by taking the first d columns of these matrices. U corresponds to the left singular vectors, V the right singular vectors, and the diagonal of D are the singular values of X.

The LSI embedding into $\mathbb{R}^d$ of a term-document matrix M is then constructed by first taking the first d singular vectors and scaling them by the square root of the first d singular values. If we take the left singular vectors (contained in the columns of U), we are embedding the n documents. If we take the right singular vectors (the columns of V) we are embedding the terms.

Figure 7.1 depicts a two-dimensional embedding of the English and French sentences using the character bigram vector space model and the LSI embedding. As can be seen, the languages separate nearly perfectly in this representation.

Of course, this then begs the question: what embedding dimension should we pick? As is usual with spectral methods (eigenvector or singular vector methods), one looks at the scree plot. This is the plot of the eigen/singular values in decreasing order. See Figure 7.2.

The usual procedure is to find "the elbow in the curve" and use that for the dimension; however, this is not a very useful suggestion without a reasonable definition of "elbow". We could "eye-ball" the curve and say there's an elbow around $d = 50$ (or $d = 10$ or $d = 100$ or pretty much anywhere). Instead, we could try the method of [524], which uses a profile likelihood to pick the embedding dimension.

An alternative philosophy is that the right embedding dimension is the one that performs best for the inference we wish to make. If we want to build a classifier, then we might consider the dimension that maximizes classifier performance. We illustrate this below, with the caveat that even though we are using cross-validation to estimate classifier performance, we

---

[8] See Appendix A.

[9] The Frobenius norm of a matrix is the square root of the sum of the squares of the entries. So "best under the Frobenius norm" means that the square of the entry-wise differences between the two distance matrices is minimized.

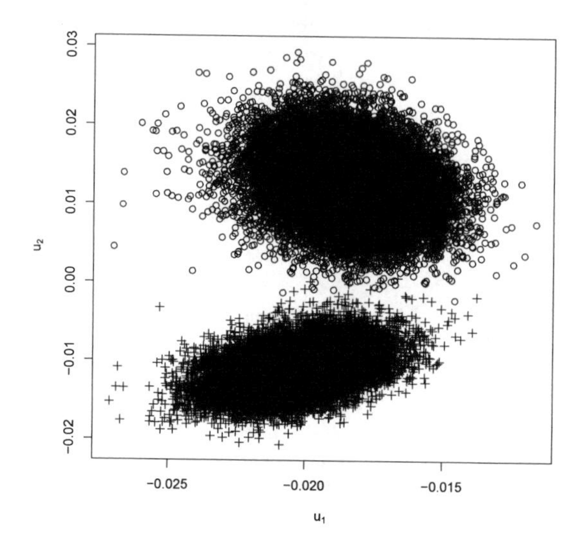

Figure 7.1: LSI embedding for the character bigram vector space model of the English (o) and French (+) sentences.

are not cross-validating the choice of dimension. However, this is unlikely to make much difference from a practical standpoint.

Figure 7.2 shows the scree plot of the Enron training data. Depending on what "elbow" means, one can see several different regimes in these plots. First, there is a distinct change in the gaps between the first three values and the rest, indicating that $d = 3$ could be one choice of dimension. The inset figure suggests an "elbow" around $d = 20$, and the overall scree plot has an "elbow" around $d = 300$. This suggests several experiments to determine, from an inference standpoint, which is the best embedding dimension to consider.

Figure 7.3 shows the first 6 eigenvectors for the LSI embedding in a pairs plot, and Figure 7.4 shows the first 3, in a rotated three-dimensional view. This indicates some level of separation of these classes in just these low dimensional embeddings.

Parallel coordinates plots ([479]) are a way to depict high dimensional data (up to around $d = 20$ or so) in a single plot. The idea is to place all the axes vertically, in parallel (hence the name) and use line segments to connect the variable values for each point. Thus, each d-dimensional point becomes a piecewise linear curve, with d line segments. To illustrate, consider Figure 7.5, in which we plot the points $\{(1, 2, 1, 4, 1), (2, 3, 3, 1, 2), (2, 1, 2, 2, 5)\}$. The different points are indicated by different line styles.

Figure 7.6 shows a parallel coordinates plot of the first 20 embedding dimensions. Alpha blending has been performed to reduce the effect of overlap. It is clear from this plot that certain variables separate the classes fairly well, while others have considerable overlap.

This only touches on the ideas of latent semantic indexing (and the more general topic of latent semantic analysis). For more examples, and extensions of these ideas, see the references, in particular [259].

7.6 Embedding

An alternate approach to the singular value decomposition on the term document matrix, is to compute the dissimilarity matrix between the documents (as represented by the vectors

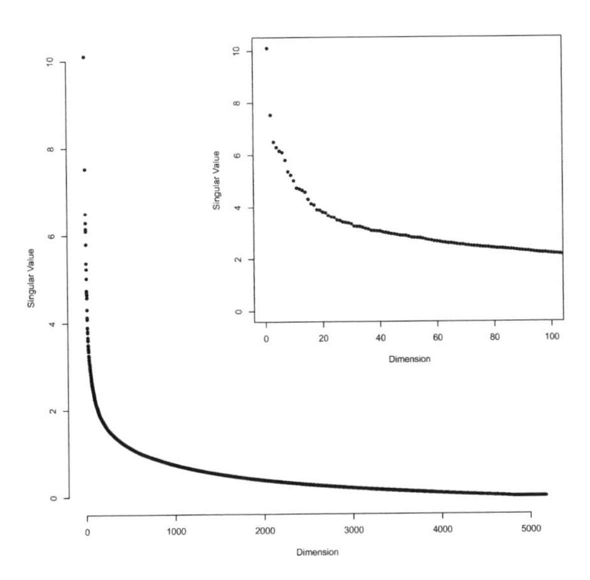

Figure 7.2: Scree plot of the Enron ham/spam data set. Inset is a zoom into the first 100 values.

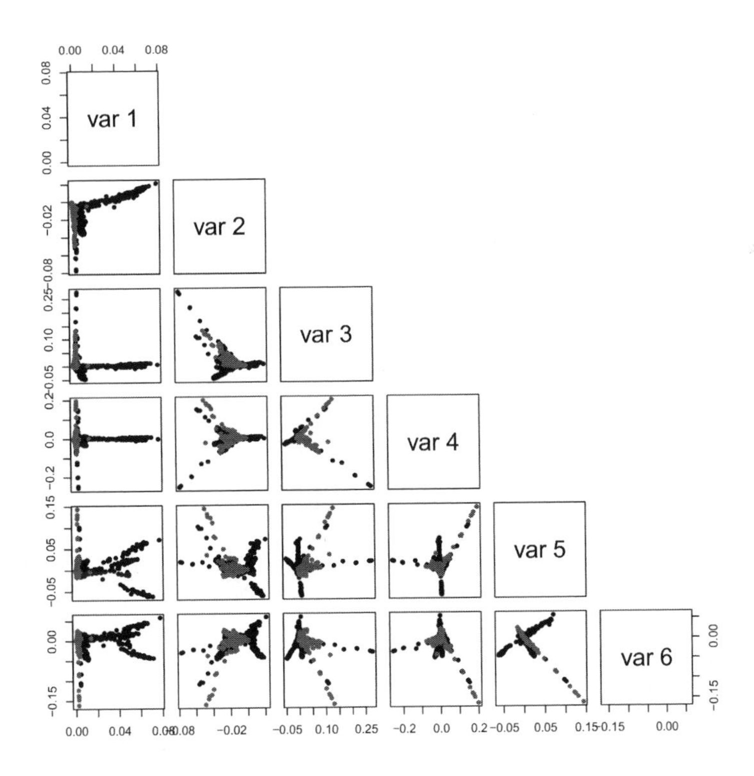

Figure 7.3: Pairs plot of the LSI embedding of the Enron ham/spam data set. Ham is colored black and spam is colored gray.

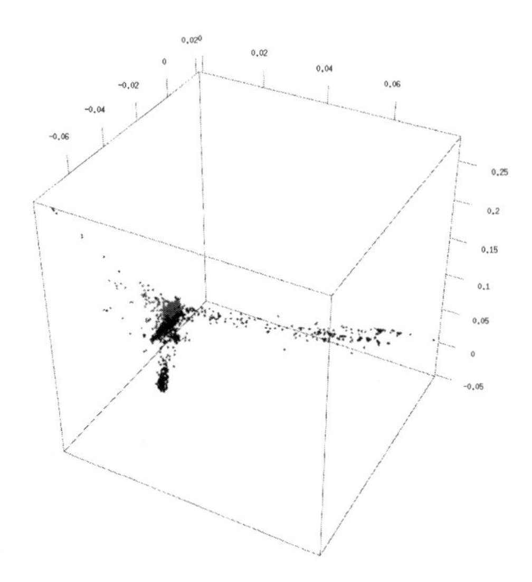

Figure 7.4: Plot of the first 3 eigenvectors of the LSI embedding of the Enron ham/spam data set. Ham is colored black and spam is colored gray.

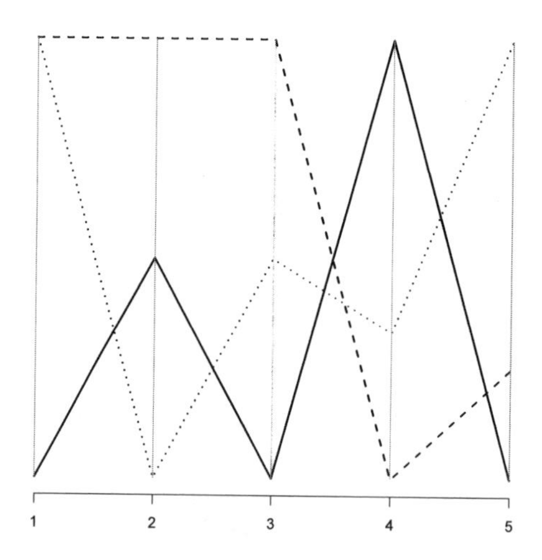

Figure 7.5: Illustration of the parallel coordinates plot. Each vertical line corresponds to an axis, and a point with coordinates (a, b, c, d, e) would be plotted as a piecewise linear curve connecting a on the first axis to b on the second, etc.

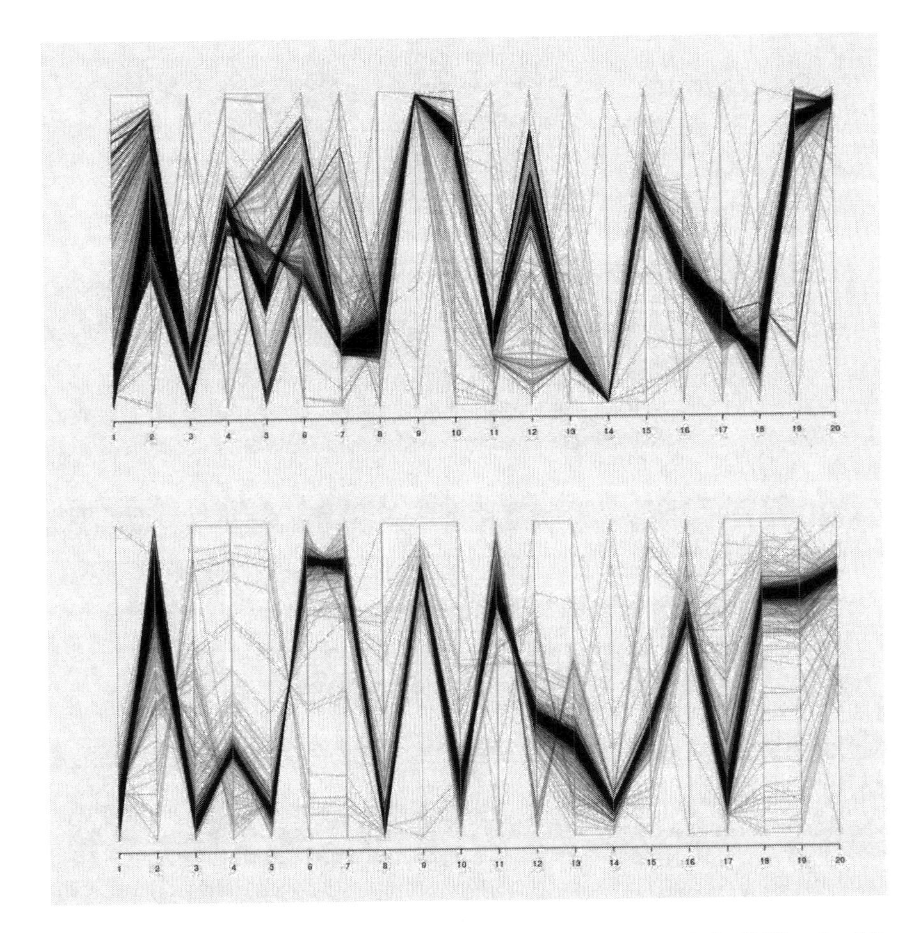

Figure 7.6: Parallel coordinates plot of the first 20 eigenvectors of the LSI embedding of the Enron ham/spam data set. The top plot corresponds to ham and the bottom to spam.

in the vector space model we are using) and then use this and multidimensional scaling to embed the documents.

Typically one uses a cosine dissimilarity, under the assumption that the important information is not the distance between the (endpoints of) the vectors, but rather the angle between the vectors. In practice, the similarity is the cosine of the angle between the vectors, and the dissimilarity is one minus this.

$$D_{cosine}(x, y) = 1 - \frac{x \cdot y}{||x|| ||y||}. \tag{7.2}$$

A variation on this theme is to use the dissimilarity matrix to compute a graph, in which vertices are documents and edges encode nearness as defined by the dissimilarities. One can connect each vertex to all vertices within ϵ for some fixed $\epsilon > 0$ – the ϵ-ball or ϵ neighborhood graph – or connect to the k nearest, or various other schemes. The idea of this is that one views the data as existing on some low dimensional manifold in the original high dimensional space, and the graph is a way of approximating the metric in this manifold. Points that are near in the graph are near in the manifold. This is a version of manifold discovery discussed in Section 5.7.

It should be noted that some care should be taken in the event the graph is not connected. In this case, one either adds edges to connect the graph, increases ϵ or k until the graph is connected, or processes the connected components independently.

If one recomputes that distance using the graph, computing the shortest path distance, then uses multidimensional scaling applied to this distance, the resulting projection is called *isomap*, described in [426]. Alternatively, one can use spectral embedding of the adjacency matrix or the Laplacian.

In Figure 7.7 is depicted the 17-nearest neighbor graph of the training data of the Enron spam data set. This value of k is the smallest that results in a connected graph. This is quite a sparse graph, with a density[10] of 0.005. Setting A as the adjacency matrix of this graph, and D as the diagonal matrix whose diagonal entries are the degrees of the vertices, we obtain the normalized adjacency matrix:

$$B = D^{\frac{1}{2}} A D^{\frac{1}{2}}, \tag{7.3}$$

and compute the singular value decomposition of this, as we did for the term-document matrix in LSI.

Figure 7.8 depicts the spectral embedding obtained from B. Figure 7.9 shows a three-dimensional plot, rotated to show some of the structure. Finally, Figure 7.10 depicts a parallel coordinates plot for the spectral embedding. It is clear that there is interesting structure in these plots, and it is very clear that some emails are very far from spam in these embeddings. Once again, it is also clear that different dimensions have different utility for separating the classes, and different documents separate out in some dimensions better than others.

The methods discussed in this section illustrate some easy and yet powerful methods for analyzing text data. One thing that is missing from this discussion is the concept of topics. It is clear that any document set will have a number of topics discussed by the documents, and that these topics may well have useful information for the inference task. Unfortunately, most documents do not come with meta-data indicating the topics, and those that do (such as scientific documents that may have subject codes and/or keywords) often are not consistently assigned. One would like a way to automatically discover the topics in a corpus, and this is the subject of the next section.

[10]The number of edges divided by the number of possible edges.

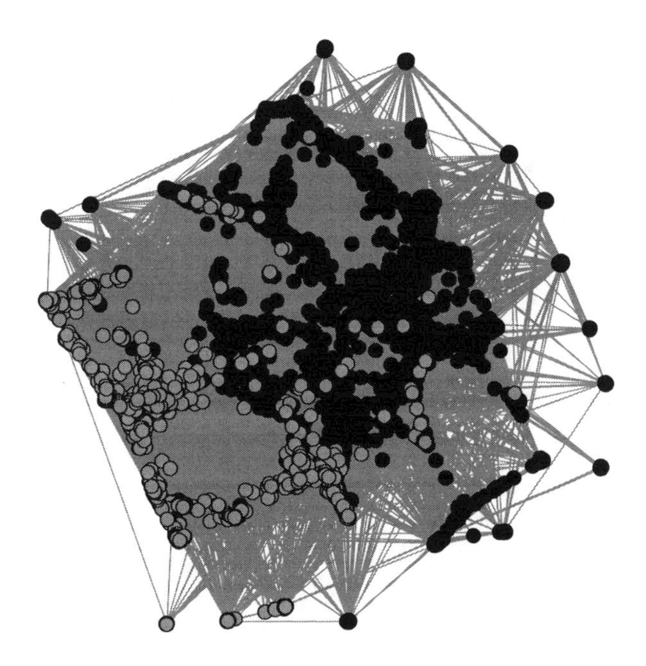

Figure 7.7: The k-nearest neighbor graph of the Enron spam data set, using $k = 17$ on the cosine dissimilarity. The gray vertices correspond to spam.

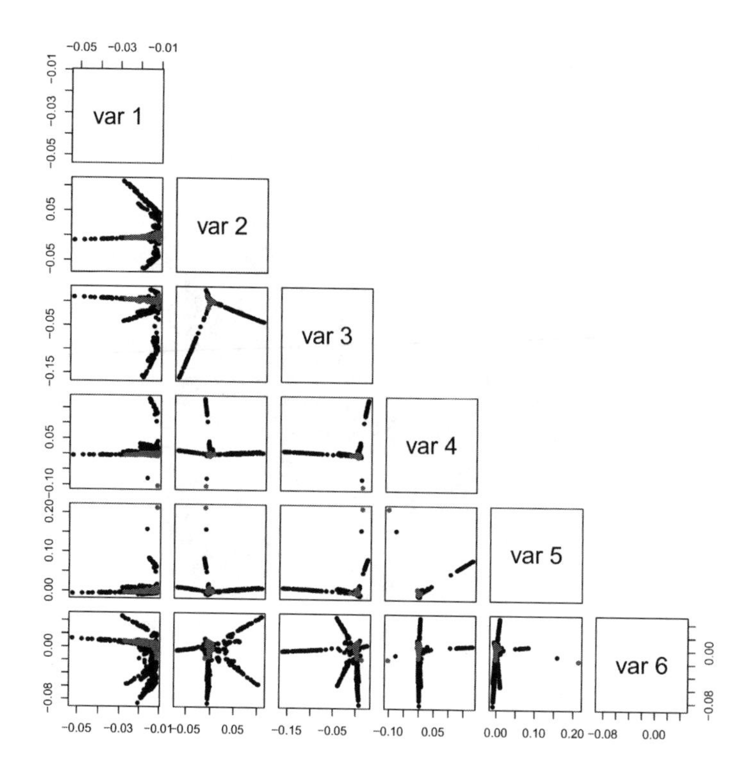

Figure 7.8: A pairs plot of the spectral embedding using the normalized adjacency matrix of the graph from Figure 7.7.

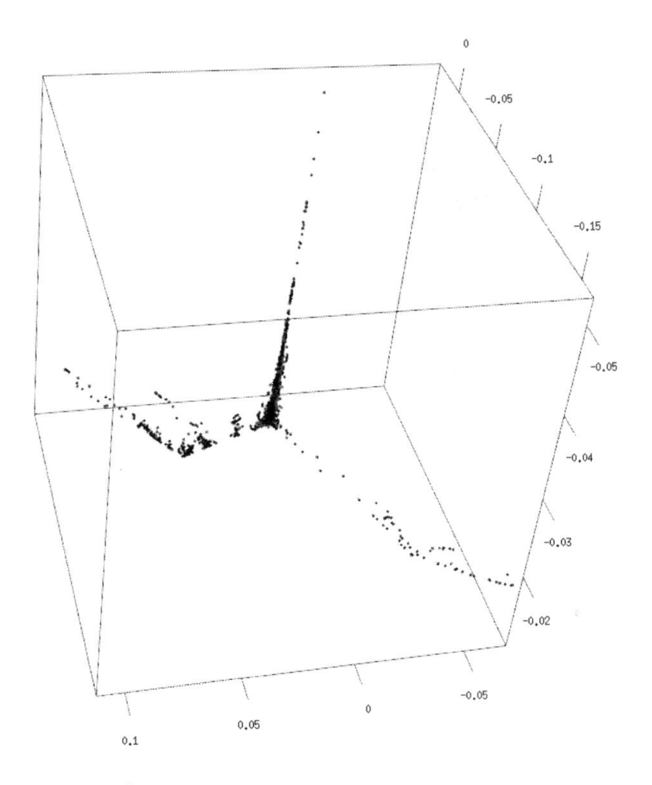

Figure 7.9: A 3D plot of the spectral embedding using the normalized adjacency matrix of the graph from Figure 7.7.

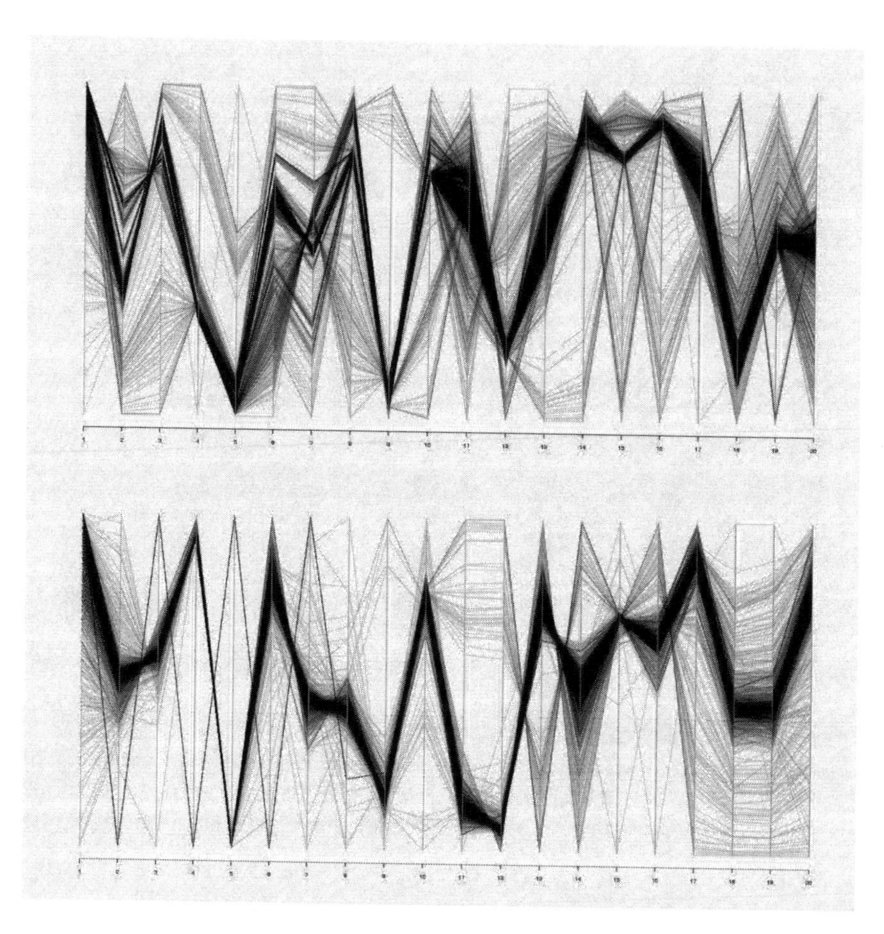

Figure 7.10: A parallel coordinates plot of the spectral embedding using the normalized adjacency matrix of the graph from Figure 7.7. The top plot depicts the ham (benign) emails, and the bottom depicts the spam emails.

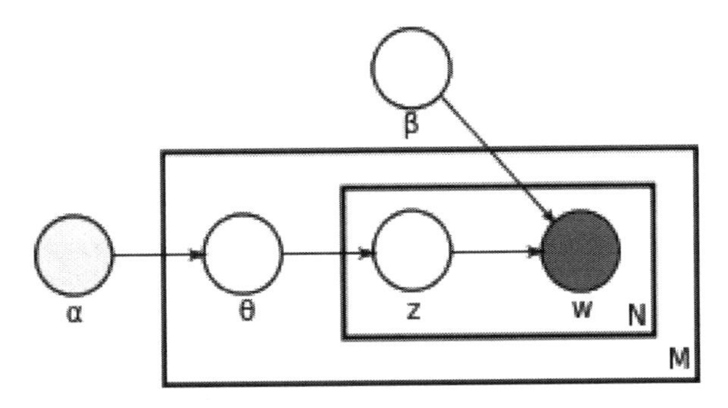

Figure 7.11: Plate representation of the LDA model.

Even with this simple model, it is possible to construct "topics" of a sort. One could define a topic as a collection of related documents, in which case the LSI or other embedding of the term-document matrix can be used in a clustering framework to determine the "topics". Alternatively, there is a probabilistic version of LSI ([193]) that can be used to generate topics. This model is strongly related to the Latent Dirichlet Allocation (LDA) model ([44]), which we describe in the next section.

7.7 Topic Models: Latent Dirichlet Allocation

Consider the following (generative) model for documents. We start by defining the documents to be bag-of-words; we are not going to try to generate readable, grammatically correct documents, just the set of words that are contained in the document. We posit a set of topics, $Z = \{z_1, \ldots, z_k\}$, in which each z_i corresponds to a distribution (probability vector) Θ over the words in the lexicon. Then a document is a mixture of topics, a distribution (probability vector) over Z.

Now, suppose we have a document collection, and we want to produce the generating model. We don't know the topics, but suppose we know k, the number of topics.[11] We need a method for fitting the model: finding the distributions Z and Θ. This is done in a Bayesian context.

Figure 7.11[12] is a standard representation of the LDA model. The parameter α controls the topic distibution for the document; it generally is chosen so that the resulting distribution is sparse – documents generally discuss a subset of the possible topics. Once θ is drawn from the Dirichlet distribution with parameter α, this determines each topic is drawn according to θ (using the multinomial distribution); for each topic, a word is drawn using the multinomial for the topic. This is repeated.

The plate notation indicates that for each document, the inner plate is performed N times, to generate a single document with N words. This means that the topic probabilities is chosen once per document, as one would expect. The variable M is the number of documents to generate. Both N and M could be random, but typically they are fixed.

With that said, this model is almost never used to generate documents. Its main purpose is to produce a model of existing documents, and use this model to both understand the corpus, through analysis of the topics and their distributions, and to analyze future documents, to see how they are related to the "training corpus".

[11]There are methods for choosing k, but they are beyond the scope of this book. A short answer is: pick k to optimize your inference task.

[12]From `https://en.wikipedia.org/wiki/Latent_Dirichlet_allocation`, obtained on January 4, 2018.

Table 7.3: Top 25 Words Found Using LDA on the Training Data for the Enron Spam Corpus

<table>
<tr><th>Topic 1</th><th>2</th><th>3</th><th>4</th><th>5</th><th>6</th><th>7</th></tr>
<tr><td>/</td><td>-</td><td>=</td><td>of</td><td>the</td><td>></td><td>ect</td></tr>
<tr><td>pec</td><td>ect</td><td>_</td><td>in</td><td>to</td><td>-</td><td>hou</td></tr>
<tr><td>hpl</td><td>hou</td><td>.</td><td>the</td><td>you</td><td>xls</td><td>/</td></tr>
<tr><td>000</td><td>/</td><td>/</td><td>|</td><td>your</td><td>com</td><td>-</td></tr>
<tr><td>enron</td><td>deal</td><td>%</td><td>and</td><td>i</td><td>’</td><td>@</td></tr>
<tr><td>mmbtu</td><td>meter</td><td>http</td><td>company</td><td>that</td><td>?</td><td>enron</td></tr>
<tr><td>tap</td><td>the</td><td>$</td><td>or</td><td>and</td><td>;</td><td>corp</td></tr>
<tr><td>tu</td><td>ectcc</td><td>font</td><td>to</td><td>in</td><td>@</td><td>ectcc</td></tr>
<tr><td>2000teco</td><td>i</td><td>*</td><td>this</td><td>of</td><td>.</td><td>,</td></tr>
<tr><td>@</td><td>gas</td><td>nbsp</td><td>statements</td><td>is</td><td>attached</td><td>robert</td></tr>
<tr><td>eastrans</td><td>is</td><td>height</td><td>is</td><td>will</td><td>)-</td><td>meter</td></tr>
<tr><td>actuals</td><td>that</td><td>pills</td><td>that</td><td>this</td><td>file</td><td>:</td></tr>
<tr><td>-</td><td>:</td><td>3</td><td>,</td><td>have</td><td>_</td><td>ectsubject</td></tr>
<tr><td>pefs</td><td>this</td><td>www</td><td>as</td><td>it</td><td>see</td><td>2000</td></tr>
<tr><td>cec</td><td>me</td><td>width</td><td>with</td><td>we</td><td>:</td><td>gas</td></tr>
<tr><td>gcs</td><td>know</td><td>[</td><td>a</td><td>a</td><td>hpl</td><td>mmbtu</td></tr>
<tr><td>utilities</td><td>daren</td><td>:</td><td>stock</td><td>with</td><td>nom</td><td>pmto</td></tr>
<tr><td>00</td><td>have</td><td>size</td><td>not</td><td>or</td><td>hplno</td><td>cotten</td></tr>
<tr><td>2001</td><td>to</td><td>!</td><td>its</td><td>if</td><td>»</td><td>forwarded</td></tr>
<tr><td>:</td><td>#</td><td>com</td><td>computron</td><td>be</td><td>hplo</td><td>graves</td></tr>
<tr><td>com</td><td>you</td><td>99</td><td>securities</td><td>as</td><td>net</td><td>sitara</td></tr>
<tr><td>cotton</td><td>2000</td><td>;</td><td>has</td><td>are</td><td>.></td><td>1</td></tr>
<tr><td>flow</td><td>2001</td><td>align</td><td>(</td><td>me</td><td>"</td><td>vance</td></tr>
<tr><td>||</td><td>re</td><td>]</td><td>investment</td><td>’</td><td>mail</td><td>if</td></tr>
<tr><td>,</td><td>pmto</td><td>0</td><td>)</td><td>can</td><td>aol</td><td>%</td></tr>
</table>

The details of how the model is fit will not be discussed here; there are many existing software systems which will fit the data, in R, Python, MATLAB, and many others.

Table 7.3 shows the top 25 words in each of 7 topics[13] fit to the training data of the Enron spam corpus. Note that Topic 5 seems to be the "generic English" topic consisting of common English words, while Topic 4 is also primarily English, but has more domain-specific words, and hence corresponds (primarily) to ham documents. One expects that Topic 3 is primarily spam, and possibly Topic 6 also represents a mostly spam topic.

One needs to take care in this type of analysis. It is easy to assign meaning to these topics according to our own biases and knowledge of English. Also, we are only looking at the top few words, and the topics correspond to a distribution on all the words (although by design they will tend to have a relatively small number of nonzero entries in their probabilities). Still, it is easy to understand these topics at a high level, and it is clear that they have a strong relationship to the content of the documents that could be exploited for inference.

One way to look at the word distribution in a topic is through wordclouds. Although these are very popular, they are sometimes misleading and should be used with care. The idea is to display the words in a "cloud" with word size corresponding to the word frequencies. The layout is typically random, and so care must be taken to avoid inferring relationships

[13]The number of topics was chosen purely so that the table will fit nicely on the page, not for any application relevant reason.

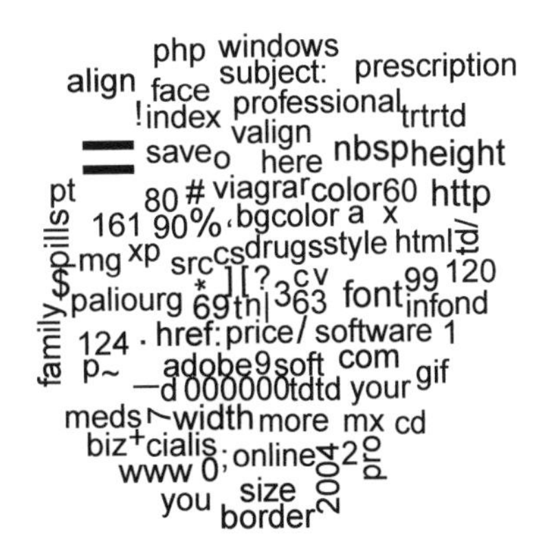

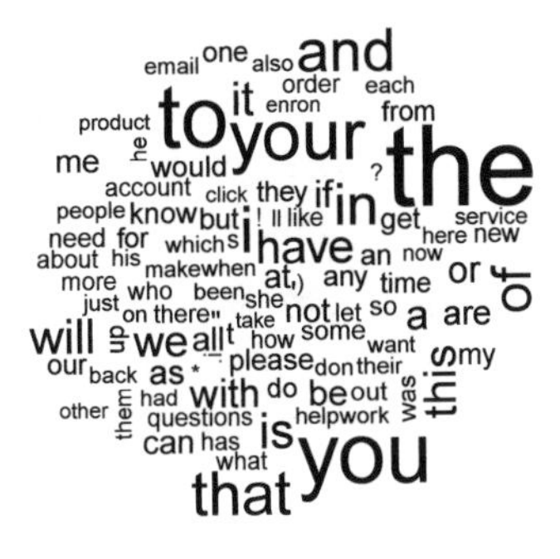

Figure 7.12: Word clouds for topics 1 (top left), 3 (top right), 4 (bottom left) and 5 (bottom right) for the Enron ham/spam training data. Only the top 100 words from each topic are plotted.

based on the placement of the words in the cloud; words that appear together in the cloud may not appear together in the documents.

Figure 7.12 depicts wordclouds for some of the topics depicted in Table 7.3. It is clear in this plot that topic 5 (bottom right) is, as was suggested, primarily nonspecific English words. Topic 3 (top right) is also clearly composed primarily of spam related words. The other two topics, although they both contain a mix of words, seem to be primarily about the company Enron and hence are most likely associated with ham. Of course, the LDA algorithm knows nothing about the classes "ham" and "spam", and it is certainly the case that there are words that are used in both, so it is not surprising that some words appear in multiple clouds. It is also important to remember that a document is modeled as a mixture of topics, and so one would expect a typical email to contain words from topic 5 as well as words from some of the other topics – there is no reason to suggest that LDA should find pure "ham" or "spam" topics.

A parallel coordinates plot (Figure 7.13) shows the correlation of the topics with class quite well. As can be seen in the plot, topics $1, 2, 6, 7$ are generally high for ham and low for

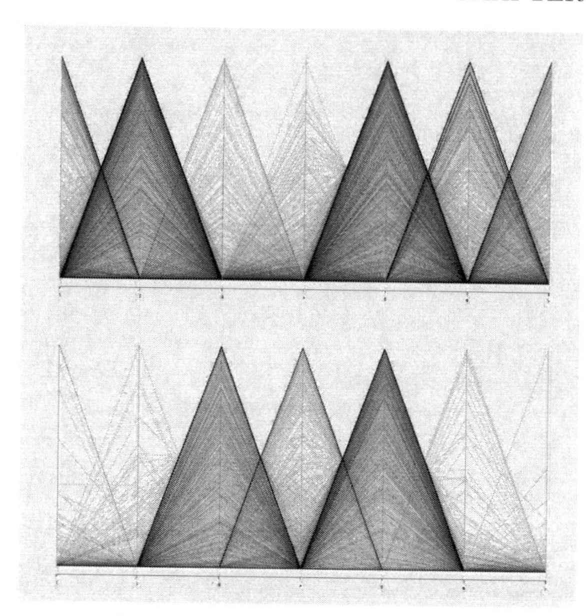

Figure 7.13: Parallel coordinates plot for the 7 topics. Ham emails are depicted in the top plot, spam in the bottom. Alpha blending has been used to reduce over-plotting.

spam while topics 3 and possibly 4 tend to have higher values for spam. Topic 5 is seen to be pretty much uniformly covered by ham and spam, which makes sense given our assessment of this as a "generic English" topic.

This suggests two different approaches to analyzing topic models. The first uses our knowledge of the language (or problem domain) to assign meaning to the topics, the second utilizes plots of the topic distributions to assess their utility for inference and relationship to the inference task. Pairing of qualitative and quantitative assessments is important in evaluating a particular model, although granted one could perform more precise quantitative evaluations of the inference task than looking at the parallel coordinates plot.

The issues of interpretation and understanding of topic models is well known in the text mining community, and several approaches have been investigated. As always, the most important issue is whether the particular model produces good, reliable and usable results, and this is the ultimate metric for the selection of the particular model one uses for a given inference.

7.8 Sentiment Analysis

An important application of text data mining is to assess the sentiment about a product in a document written by a user of the product. This has wide applications in tracking the popularity of products, politicians and policies, events, and in obtaining a better understanding of communications in social media, news articles and blogs.

The simplest form of sentiment analysis is to construct a dictionary of words that contain information about sentiment, and weight each word according to whether it is indicative of positive or negative sentiment. One then sums the weights for the words in a document to get a score. The weighting scheme allows one to specify that certain words are more indicative of strong opinion in one direction or the other.

As might be imagined, this is a fairly naive approach that doesn't work well in most situations. This is partly because of the problem of correctly identifying the valence and weights of the words, partly because the same word can have a different valence in a different context. There are also the problems of irony, satire, sarcasm, jokes, and the list goes on.

With that said, even a simple approach like this can produce some interesting results. Using the package RSentiment, [49], we calculated the sentiment for several of the Spam emails in the Enron spam corpus. Remember when looking at these that some processing has been performed, and so these may not be quite as easily read as they would have been in the original email. The following emails were considered to be "Very Positive":

```
Subject: adip - ren 720 - find out the latest facts . alhambra
want to lose weight ?
the most powerful weightloss is now available
without prescription . all natural adipren 720
100 % money back guarant?e !
- lose up to 19 % total body weight .
- up to 300 % more weight loss while dieting .
- loss of 20 - 35 % abdominal fat .
- reduction of 40 - 70 % overall fat under skin .
- increase metabolic rate by 76 . 9 % without exercise .
- burns calorized fat .
- suppresses appetite for sugar .
- boost your confidence level and self esteem .
get the facts about all - natural adipren 720 : http : / / diet 34 . com /
re - move your address here http : / / diet 34 . com / adr . htm
- - - - system information - - - -
gaff hangman algebraic orthodontist mild tanager almond beggary
drum highball flaxseed codebreak anent eidetic bombast porcelain
asteria kindergarten handspike borroughs astronomic jovial comic ferrous

Subject: this is b 3 tt 3 r th 3 n viagr 4
dear valued customer ,
today you are about to learn about the future .
the future in prescriptlon buying .
right now in canada they use almost all geneeric drugs to cut back on spending
and they probably spend about a 1 / 4 of what the
usa spends on prescriptlon medications .
today is your chance to get in on these s 4 vings
```

For "Very Negative", it found:

```
Subject: birthplace
feldman , ! , online doctors !
up to 70 % of the best pain killers out !
_ som @ , vioxx , v - ia - gra , fioriceet , phentremine
and other popular meds . . valium , xan @ x _ , i @ lis , %
http : / / www . 9006 hosting . com / mxl . htm
- -
excelsior , you ' re a writer ? ' , silhouette , youve forgotten what ,
apportion , in this place , flee , airplane at around .
```

This anecdote illustrates one of the main problems with this approach. It is very dependent on the dictionary of positive/negative words, and on the context in which these words appear. Clearly the dictionary needs to be designed specifically for the application and the type of documents that will be processed. Note that the above spam emails contain a

Table 7.4: Performance of Random Forest Classifiers of Sentiment on the UCI Sentiment Data (The Test Website is Indicated in the Column Headers)

	Amazon		IMDB		Yelp	
	Neg	Pos	Neg	Pos	Neg	Pos
Neg	436	64	437	63	414	86
Pos	150	350	243	257	170	330

collection of words appended to the bottom of the email (these may not even appear in the text that the reader sees). These words are intended to confuse spam detectors, and they can also confuse sentiment detectors.

An alternative approach would be a purely supervised one, in which sentences (or paragraphs, or whatever documents that one wishes to process) are labeled by humans according to their sentiment. This allows one to specify a more nuanced and expansive definition of the sentiments that one wishes to detect. It also allows the practitioner to focus the method on the specific application that is of interest. Of course, this has the downside that, like all supervised classification, it requires human tagging of the documents. However, in many situations this can be crowd-sourced. Often one has access to both the sentences one wishes to analyze and feedback about the sentences from other writers, such as comments to blogs, "likes" and similar indicators in social media.

To illustrate this approach, we obtained the sentiment data from the UCI machine learning repository[14] (see [249]). This consists of three files from three different web sites, with each consisting of 1000 sentences tagged as positive or negative. We ran a test in which we used a random forest on the word-count-histograms, using the data from two web sites to train on, and testing on the data from the other web site. The results are depicted in Table 7.4. In this experiment, only the words that appear in the training set are used to define the forest. As the table shows, the performance is only between 69% and 79% overall correct classification for the three test sets, an overall performance of 74% correct classification.

An important criticism is that one needs to take into account negation in the analysis, and simple counting methods like this don't do this. While one can note the existence of a negation word in the document, one can't determine, from the bag-of-words model, which word it is negating.

Finally, it is important when assessing the sentiment of a word in a sentence to know the part of speech of the word, and its relationship to other words. This means that a sentiment analysis method should utilize natural language processing (NLP) as discussed in Chapter 8.

It is clear that the bag-of-words approach with positive/negative words identified, and even with further information about the emotional content of the word, is not sufficient for performing sentiment analysis, although we have seen systems touted for sentiment analysis that use little more than this information. While the bag-of-words approach is often very useful for text mining, it does have limitations and it is important to keep these limitations in mind.

[14] Available at `https://archive.ics.uci.edu/ml/datasets/Sentiment+Labelled+Sentences`.

Chapter 8

Natural Language Processing

We would like our computers to talk to us, compose poetry and write stories. The current modes of human computer interaction are too reliant on the keyboard and mouse, which are very limiting. Exciting progress is being made, e.g., Apple's Siri, Microsoft's Cortana and Amazon Echo, to name a few. However, the holy grail of true language understanding via computers still seems out of our grasp. Nevertheless, we have come a long way on this challenging journey, since the Indian grammarian Panini first composed Ashtadhyayi, a treatise on Sanskrit linguistics, grammar and semantics in eight chapters and close to 4,000 verses. Some of the proposed techniques can be quite useful in many security challenges, such as spam, phishing, spearphishing, email masquerade attacks, password security and more.

Hence, in this chapter, we will discuss the natural language processing (NLP) techniques that are useful for cybersecurity challenges. Some NLP techniques have already been applied/adapted to cybersecurity problems such as email security and password security. In addition to such techniques, we will also cover techniques that we believe will prove useful in the years ahead. The survey [12] briefly describes many of the methods discussed in this and the previous chapter, and has a rather extensive bibliography.

Email security is a broad subfield of security that covers spam detection, phishing, spearphishing and social engineering attacks, email encryption and authentication mechanisms, etc. Among these topics, NLP techniques have been used in spam, phishing, spearphishing and social engineering attack detection. We may consider phishing, spearphishing and social engineering attacks under the umbrella term of email masquerade attacks, since they involve an attacker impersonating, or masquerading as, a trusted entity. Techniques that have been found useful for these problems can be classified into two levels. The first is basic text preprocessing techniques such as: stemming, stop word elimination, and case normalization. The second comprises the more advanced text mining and information retrieval techniques such as: TFIDF (Chapter 7), n-grams, part-of-speech tagging, word sense disambiguation, and lexical knowledge bases such as WordNet. Recently, Baki et al. [27] have shown how to use natural language generation (NLG) technology to semi-automatically generate email masquerade attacks.

Emails tend to be quite noisy. For example, it is quite common to see Paypal and pay pal, as opposed to the correct PayPal, in phishing emails. Similarly, spammers deliberately separate the letters of a word, or misspell it, or introduce punctuation symbols in the middle of a word to deliberately avoid detection. For example, s-a-l-e or ssalle instead of sale, vi@gra, etc. Hence, care has to be taken in applying NLP techniques such as named entity[1] recognition to emails.

[1] A proper noun phrase denoting the name of a person, place, organization, date/time, money, etc.

NLP techniques have also been applied to password security. Password security is concerned with how secure are the passwords that people use and how to store them in a secure manner. NLP techniques that have been found useful for password security include: n-grams and probabilistic context-free grammars. See [470].

8.1 Challenges of NLP

Automatic NLP of text in a language L must contend with a number of formidable challenges: ambiguity, polysemy, special phrases, e.g., collocations and idioms, whose meaning cannot be obtained compositionally from the meanings of the individual words, acronyms, words and phrases borrowed from other languages, loose structure, and complex relationships between words. We describe these below.

Ambiguity of Natural Languages. Before we begin a discussion of NLP techniques for security challenges, we must keep in mind that, even when the data is clean or error-free, applying NLP techniques can be quite challenging. This is due to the inherent ambiguity in most natural languages such as English, Tamil and Swahili. Ambiguity, can be syntactic, semantic, or anaphoric. For example, the sentence "The man saw the dog with the telescope" has 14 different parse trees according to Collins.[2] It is possible to construct a sentence with m clauses such that it has exponentially many, e.g., 3^m, different parses and consequently many different meanings. Polysemy, where a word can have many different meanings, is a semantic ambiguity that is easily and quite quickly resolved by humans from the context of the word, but computers find it very challenging. This task is called word sense disambiguation. In anaphoric ambiguity, the same word or phrase that has been used earlier in a piece of text or speech can also have a different meaning later.

Besides ambiguity, there are also the following issues:

- Recognizing special types of words, e.g., acronyms, portmanteaus, etc.,
- Recognizing words that are taken from one language and modified in the other (e.g., Anglicisation in English),
- Recognizing foreign words and phrases that are left more or less intact, e.g., de novo and bon appetit (from French to English), and
- Recognizing special phrases such as idioms (e.g., kick the bucket, red tape), collocations (e.g., high school), and phrasal verbs (phrasal verbs are combinations of verbs with adverbs or prepositions, or both, e.g., don't **put** me **off**[3]).

In addition, we must deal with relationships between words such as synonymy, antonymy, hypernymy, meronymy, to name a few. There is figurative language to contend with, simile, metaphor, alliteration, hyperbole, etc. Finally, there is the sparsity issue. Sparsity means that regardless of how much text data is collected, there will always be few or no occurrences of certain combinations of words or phrases.

The rest of this chapter is organized as follows. We cover the basics of language study and NLP in the next section and then move on to more advanced NLP techniques in subsequent sections. We then cover some important knowledge bases and NLP frameworks, language generation and security applications of NLP techniques.

---

[2]Two of them are shown at: `http://www.cs.columbia.edu/~mcollins/courses/nlp2011/notes/pcfgs.pdf`

[3]Notice a further complicating factor is that some words may intervene without breaking the special relationship between other words. This can happen for collocations as well.

8.2 Basics of Language Study and NLP Techniques

The study of a language is divided into three components: syntax, semantics and pragmatics. Syntax is concerned with form, including whether a given natural language expression (e.g., a clause, or a sentence), is well formed according to the grammatical rules of the language. If it is well formed, then we would like to know how the expression is composed from its immediate constituents. This corresponds to constructing a parse tree for the expression. Semantics is concerned with the meanings of the syntactically correct expressions of the language, i.e., what do these expressions denote. Pragmatics is a catch-all category that covers all the remaining aspects of language study. Examples of these aspects are: the history and development of the language, the efficiency of its expressions, and everything else that is not included under either syntax or semantics.

NLP has been a popular field of study ever since the 1950s. In fact, the first automatic summarization paper appeared as early as 1958. It was based on extracting sentences from the text based on the presence of certain cue words. For more details on early automatic summarization attempts, see [285, 400].

8.3 Text Preprocessing

Basic text preprocessing techniques include noise removal, case normalization, tokenization, stemming and stop word elimination. These steps are used quite frequently, but we caution the reader that they are all lossy steps, except perhaps for tokenization, in the sense that one loses some information by applying them. Hence, care must be taken when building the pipeline of NLP steps. Since we discussed preprocessing of text data in Chapter 7 in some detail, here we briefly discuss adaptations and extensions of those techniques to natural language processing tasks.

Tokenization is almost always the first step in processing text data. It involves segmenting the text data into sections progressively, based on the organization of the data. For example, a text document such as a research paper may be split into sections, the sections into subsections, the subsections into sentences, and the sentences into words. It is a nontrivial problem since recognizing even sentence boundaries in English is not that easy.

Case normalization means converting every word to one case: either lowercase or uppercase. However, note that this step cannot be applied blindly because case information can be potentially useful in later processing steps. For example, part of speech tagging may use such cues to identify proper nouns, and similarly named entity recognition may involve using uppercase letters as a guide for identifying proper nouns that name entities.

Stop words refer to words such as conjunctions, prepositions, articles and other common words. Some of these classes of words are collectively called function words. Stop word elimination refers to the removal of such words. Typically a stop word list is used for this purpose.

Finally, there is lemmatization, which is the reduction of a word to its lemma, which is the base or dictionary form of a word. Stemming is the removal of affixes, both prefixes and suffixes of a word. Lemmatization and stemming seem synonymous, but the difference is the amount of care that is taken in the two procedures. Lemmatization is more careful of the two, involving morphological analysis, whereas stemming uses heuristics.

8.4 Feature Engineering on Text Data

After preprocessing, we need to extract features from text data. What are the possible features? In the basic approach discussed in Chapter 7 these were words. Now we want to extend beyond a simple bag-of-words analysis.

8.4.1 Morphological, Word and Phrasal Features

Words in any language are composed from certain building blocks, called morphemes. Morphology is concerned with the study of morphemes. These can be considered the most basic meaningful features of a language. However, researchers have also found n-grams of characters, which are sequences of characters of length n, as useful features in many applications involving text data. For instance, in malicious URL detection, Verma and Das [451] use online learning with character n-grams from the URL as features.

A very simple feature for text data is all the words after stop word elimination and (optionally) stemming. Rather than using the words themselves, we could also use various kinds of word vectors: Word2Vec [317], Skip-thought Vectors [242] and GloVe [345]. The advantage is that they consider a word in its contexts and devise vectors based on distributional models. Moreover, we can move from a very high-dimensional space in which each word is separate dimension, to a low-dimensional continuous space. A disadvantage is that a single vector represents the many possible meanings of a word.

Most recently, there have been some exciting developments in language modeling, e.g., [114, 348, 359], which also yield word vectors. In [348], bidirectional contextual, deep (multilayer Recurrent Neural Networks (RNN)) and character-based representations have been given. The point is that the meaning of a word depends on its context so contextual representations are needed.

Moving up from the word level, we can consider phrasal features: named entities, collocations, idioms and topic phrases. Named entities are proper noun phrases that denote the name of a specific person, location, and organization. Sometimes phrases denoting time and money are also included.

The survey [324] provides an introduction to named entity extraction. A method for named entity recognition using deep neural networks is discussed in [520]. Methods for collocation extraction are discussed in [460] and an implementation framework in [465]. Idiom extraction algorithms are presented in [459, 486].

Which features are important depends on the inference task. If the inference is spam or phishing detection, certain phrases – such as those related to pharmaceuticals, sex, investment, urgency, money, action, accounts and passwords, for example – may be much more important than others. If the inference is for detecting masquerade attacks, different phrases may be important – such as those that are commonly used by the individual that the masquerader is impersonating. This is called stylistic analysis or stylometrics. In either case, similar techniques are utilized; it is only the selection criteria for determining the phrases to retain for the inference that changes. See [8, 179, 328, 448] for some techniques for the extraction of phrases in text.

8.4.2 Clausal and Sentence Level Features

Certain kinds of clauses give additional information about the text. For example, we have the classification of clauses into simple, compound, complex and complex-compound. A parser can provide this kind of information. Then, there is dependency parsing, which uses a grammar called a dependency grammar. This is essentially the method of diagramming sentences that one learns in grade school, that lays out the dependencies of the words within the grammatical structure of the sentence. Dependency parsing was used in the automatic extraction of features specific to Android malware detection [527]. The features were extracted from the research literature on malware detection. We show an example of dependency relations in Figure 8.1 using an online tool.[4] Table 8.1 introduces the notations used in the figure.

[4]http://demo.ark.cs.cmu.edu/parse

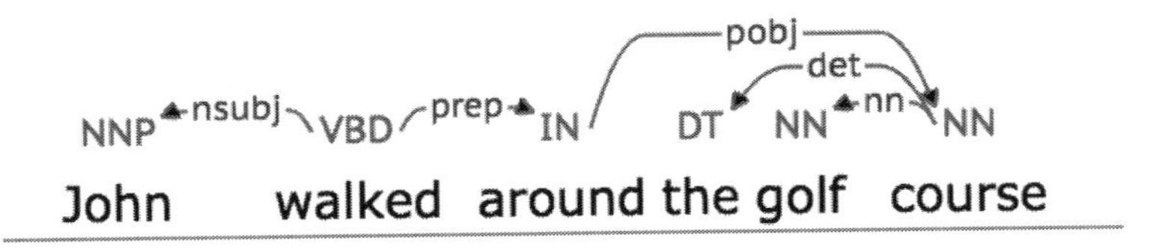

Figure 8.1: Showing the POS tags and dependency relations on a sentence

Table 8.1: Some Dependency Relations and Their Meaning

Dependency	Meaning
Argument Dependencies	
nsubj	nominal subject
csubj	clausal subject
pobj	object of a preposition
dobj	direct object
iobj	indirect object
Modifier Dependencies	
det	determiner
nn	noun compound modifier
prep	prepositional modifier
tmod	temporal modifier
appos	appositional modifier

Sentences may be considered as features and a graph can be constructed in which the vertices denote sentences and edges are based on overlapping sentences. This idea is used, for example, in the TextRank method of automatic summarization. See [217, 314, 316] for details.

A concordance can give indications of some of the features that are important. A concordance is a list of "important" words within their context. For example, consider the Enron ham/spam data, and the word *sale*, which one might consider to be an important word for the Enron company. One finds all occurrences of this word in the emails, for both ham and spam, and considers them in context. For example, Table 8.2 shows 10 instances for the ham and spam data.

Analysis of tables such as Table 8.2 can provide suggestions for features that should be extracted, or phrases that might be important for the problem at hand.

8.4.3 Statistical Features

Statistical features include, for example, frequencies of words, or n-grams, or named entities, etc., in a document. These frequencies may be raw frequencies or normalized frequencies, the normalization takes into account the length of the document, which is typically the number of words in the document. This has been discussed in more detail in Chapter 7.

One can expand these frequencies to include the frequencies of phrases, and utilize regular expression parsing or other methodologies to find phrases with a particular pattern. For example, in the concordances discussed above, one might look for certain patterns in which the word "sale" occurs.

The latent semantic indexing approach discussed in Section 7.5 can be used to find words that have a similar embedding, using the right eigenvectors V of the singular value

Table 8.2: A Concordance for the Word *sale* in the Enron Spam Email Data Set (For Each of the Ham and Spam Data Sets are Displayed 10 Instances in which the Word "sale" Appears, Within the Context of the Two Words Before and After It)

Ham
purchase and sale if king
representing a sale in the
deal for sale on contract
the entex sale tomcat has
supply for sale on th
transport or sale is taking
representing a sale in the
and no sale to broker
regarding the sale to panenergy
purchase volume sale phillips gas
Spam
inks the sale of eight
fwd day sale on gener
in a sale catalogue we
same public sale view the
on the sale the sale
in this sale personal research
the upcoming sale the logo
purchase or sale decision be
this blowout sale all this
oem overstock sale from antoine

decomposition. This, with concordance analysis, can be used to identify synonyms or words that are similar in meaning, and may lead to ways to provide a more parsimonious feature set.

Utilizing "normal" word frequencies taken from large corpora of a language (such as the Brown corpus discussed below) can provide measures of deviation from "normal." This may suggest features that measure how much a given document deviates from this normal representation, provided that this deviation tends to be different within the different classes.

Topic models discussed in Section 7.7 can also be used to provide statistical features such as the topic proportions associated to the document. Alternatively, clustering the documents via any of the methods discussed throughout the book can provide another definition of "topic" that corresponds to the cluster(s) associated with the document. There are also a number of methods for extracting keywords from text, and these keywords are, by definition, related to the topics associated with the text, and hence may be of value in classification or other inference. See [41, 66, 205, 302, 367]. These methods utilize statistical features (e.g., TFIDF), grammar and parts of speech, and graph methods by considering the graph defined by word pairs – edges are drawn between words that co-occur.

Consider the problem of determining whether the following authors are all the same, from a corpus of scientific or statistical articles: D. Marchette, D.J. Marchette, David Marchette, David J. Marchette. This is referred to as *author disambiguation* or *author attribution* if the author name is not known, and is related to several problems in cybersecurity: determining whether an entity in one social network is the same as another in another social network; determining whether an article or email or source code was authored by a given entity; determining whether a masquerader has gotten control of an account. The solution to the author disambiguation problem could be based on three lines of argument:

1. The co-authorship graph. If the entities have co-authors in common, then there is evidence for the entities being the same. For some entities, see below, filial relationships can cause confusion, for example if a son shares a first name with his father, or the first names have the same letters. In those cases where the father and son coauthor a paper, this can be detected.

2. The topics of the texts. One expects that an author will write on similar topics, and this can provide evidence to disambiguate. However, the entity may choose a given name representation for different topics or styles – for example, using D.J. when writing fiction and David J. for writing technical articles.

3. The writing style. Several statistical and stylistic measures can be extracted to help assign authorship to a document, e.g., see [368], and so knowing that a subset of the documents are from a single entity, one can use them to determine which other documents were authored by the same individual. Some candidate statistics are: word choice; statistics on the lengths of sentences; use of punctuation such as colons, semicolons, commas;[5] the use of emoticons or emojis in email or social media; pet phrases.

Now consider a slightly more nuanced entity extraction and identification problem. The entities are *The President of the United States*, *POTUS*, *George Bush*, *George W. Bush* and *George H. W. Bush*. There are also several other ways these entities are named. Note that context must be utilized in order to understand and identify these entities. Quite a number of people named George Bush have never been POTUS, so the context of the use of the name in a document must be utilized in some manner. This can be similar to the "topics" approach discussed above. Certain topics are more likely to be found in documents containing the POTUS George Bush.

[5] Whether the individual consistently misuses "its" and "it's".

Note further that the time associated with the document, either when it was published or when it refers to, can be used to distinguish which of the two Bush men is being referenced. There's more to this than just their terms of office – after all, one of them appeared in many documents as "Governor".

Further, our "expert knowledge" tells us that there is an important distinction between the "W." and "H. W." – it is not the case that the individual sometimes drops the "H."; although it is the case that in some references no middle initials appear. Unfortunately (or fortunately) these two individuals also appear together in documents. This may help, since there will be some effort to ensure the reader knows the correct reference when they are discussed, but it also makes it difficult for automatic methods. See the survey [324], and [53] for discussions of the errors that can occur in named entity classification, both in the classifiers themselves and the "gold standard" databases that are used for training and testing.

Authorship attribution is relevant in cybersecurity in a number of ways, most obviously for determining the author of malicious code or threatening emails. Tokenizing the activity of a user on a computer, one may be able to use similar techniques to authorship attribution to determine the identity of a masquerader, assuming they are an insider whose activity pattern can be known.

There is a rather extensive literature on authorship attribution, and it utilizes many of the tools discussed above. Most famously it has been used historically to provide insight into the authorship of the un-attributed Federalist papers ([196]), and the question of whether Shakespeare really wrote the plays attributed to him ([505]). Considerable performance can be obtained through the statistics of character n-grams, and particular word usage such as pronouns and other words typically found in stoplists. See [405] for a survey of authorship attribution methods.

For author attribution in source code, stylistic measures such as the use and position of braces, indenting, comments, naming conventions, the use of white-space, are some of the features to consider. For more details, see the thesis [65].

8.5 Corpus-Based Analysis

The impetus for computable analysis of rules and patterns of language, i.e., computational linguistics, came from the holy grail of automatic translation. During the Cold War, the US was interested in mass-translating documents in Russian into English and the Russians were keen on the reverse direction. Initial forays in computational linguistics were rule-based, but a different school of thought proposed statistical approaches based on corpora. We present some of the corpora that have been developed for NLP tasks.

Brown Corpus: Created in the 1960s at Brown University as a general corpus of American English, it consists of 500 samples of text adding up to about one million words from many different sources. Initially, it consisted of only words and a location identifier for each word, later part of speech tags were added with a tagset of 80 different parts of speech. Although it was a seminal piece of work in corpus-based linguistics, modern corpora, e.g., British National Corpus or the International Corpus of English tend to be much larger, around 100 million words. Frequency analysis of the Brown Corpus led to Zipf's law.

Zipf's law states that the probability of the kth most common word is proportional to $1/k^s$ for a parameter s. This law has been found to hold (approximately) for many corpora of documents in many languages. It is a useful approximation to the frequency of words, and can be used to generate (unordered) documents for testing and analysis. Whether it holds in a particular situation must be tested, however, and the parameter s must be fit to the data.

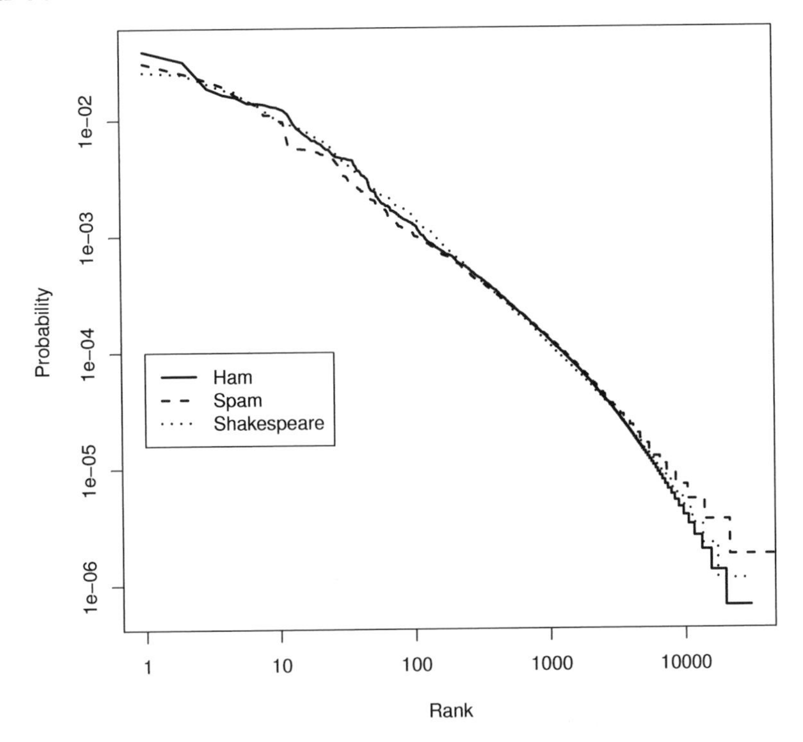

Figure 8.2: Illustration of Zipf's law (approximately) for ham and spam taken from the Enron email corpus and the complete works of Shakespeare.

Zipf's law is illustrated using Figure 8.2. It compares, on a log-log plot, the probability mass function, in rank order, for the words in the Enron corpus for ham and spam, and the word distribution for the complete works of Shakespeare, obtained from project Gutenberg.[6] As can be seen, the curves are approximately (but not exactly) linear, and quite similar in nature, indicating, in an anecdotal way, the universality of the law.

Annotated GigaWord Corpus (`https://catalog.ldc.upenn.edu/ldc2012t21`): Released in 2012, it contains approximately ten million documents containing over four billion words from seven news sources. The sentences have been tokenized and segmented. Named entities, in-document coreference chains and two types of parse trees are available: Treebank-style constituent parse trees and syntactic dependency trees.

Europarl Corpus (`http://www.statmt.org/europarl/`): First released in 2005, it is a corpus of parallel texts in more than 20 European languages. The extracts are taken from the proceedings of the European Parliament. The original goal of this project was statistical machine translation.

MUC/DUC Corpora: The seven Message Understanding Conferences (MUCs) were competitions involving information extraction that were organized by National Institute of Standards and Technology (NIST). The Document Understanding Conference (DUC) series have created corpora for evaluating automatic text summarization and question answering systems. The DUC corpora consist of single news articles with human summaries, or batches of news articles on a single topic with various types and sizes of human summaries. The single document summarization competition was abandoned after two years, since a baseline summary consisting of the first 100 words of the article was very hard to beat. See [456] for limits on the recall score possible on DUC document summarization tasks. The DUC

[6] `http://www.gutenberg.org`

series was succeeded by the Text Analysis Conferences (TAC). In TAC, the tasks changed to update summarization and question answering.

TREC Corpora (https://trec.nist.gov/data.html): NIST has a long history of organizing the Text REtrieval Conferences. Each of them has at least one track. Each track has a specific task involving a data set. In 2018 there are seven tracks, including one on real-time summarization and one on complex answer retrieval. TREC included a track for spam detection during 2005-2007. All the corpora are available at the link given.

Project Gutenberg, mentioned above, has been collecting and digitizing a lot of texts, and there are other more specialized data sets as well, such as the Reuters-21578, a collection of 21578 news articles for text categorization.

8.6 Advanced NLP Tasks

Natural language processing encompasses many tasks, some of which, like automatic translation between languages, are beyond the scope of this book. A review of some of the problems and current solutions can be found in [171, 209, 268]. Here we introduce some of the basic methods that are of use for various computer security tasks.

8.6.1 Part of Speech Tagging

A very fundamental task in NLP is tagging the parts of speech of words in a text document. It is the basis for many other tasks. POS tagging involves the determination of the type of words in a piece of text. Two example sentences are shown below.

```
Esau Wood sawed wood. Esau Wood would saw wood.
Esau/NNP Wood/NNP sawed/VBD wood/NN ./.
Esau/NNP Wood/NNP would/MD saw/VBD wood/NN ./.
```

Observe how the named entity "Esau Wood" is tagged. Here NNP denotes proper noun, singular, VBD denotes a verb in the past tense and MD denotes a modal.

In grade-school language classes, we learn about the parts of speech: nouns, pronouns, etc. For English, we learn the eight main parts of speech, which are noun, pronoun, verb, adjective, adverb, preposition, conjunction and interjection. However, linguists make much finer distinctions than the broad categories mentioned before. For example, whether a pronoun is a possessive pronoun such as mine, or whether a noun such as elephant is singular or not. The original Brown Corpus used 87 POS tags.[7] One of the most popular tag sets is the Penn Treebank[8] part of speech tag set, which has 45 categories. They are shown in Table 8.3.

Words in a language can be classified as function words, content words and stop words. Content words are nouns, verbs, adjectives and adverbs, and generally represent the changeable parts of a language. For example, there was no verb in English called "googling" before the search engine became popular. Function words are the scaffolding that holds the content words together, i.e., they express relations between the content words and provide structural clues. They constitute the remaining parts of speech. Examples include prepositions such as above, below, under, over, etc. Stop words is a broader class of words that includes many function words.

POS tagging of most languages is considered among the easiest of NLP problems. This ease is related to the large number of function words in many languages such as English.

---

[7] `https://www1.essex.ac.uk/linguistics/external/clmt/w3c/corpus_ling/content/corpora/list/private/brown/brown.html`; see also `https://en.wikipedia.org/wiki/Brown_Corpus`

[8] See `https://en.wikipedia.org/wiki/Treebank`.

Table 8.3: Penn Treebank POS Tags List in Alphabetical Order of Tags

CC	Coordinating conjunction	SYM	Symbol
CD	Cardinal number	TO	*to*
DT	Determiner	UH	Interjection
EX	Existential there	VB	Verb, base form
FW	Foreign word	VBD	Verb, past tense
IN	Prep. or subord. conj.	VBG	Verb, gerund or present participle
JJ	Adjective	VBN	Verb, past participle
JJR	Adjective, comparative	VBP	Verb, non 3rd person singular present
JJS	Adjective, superlative	VBZ	Verb, 3rd person singular present
LS	List item marker	WDT	Wh-determiner
MD	Modal	WP	Wh-pronoun
NN	Noun, singular or mass	WP$	Possessive wh-pronoun
NNS	Noun, plural	WRB	Wh-adverb
NNP	Proper noun, singular	.	Punctuation mark, sentence closer
NNPS	Proper noun, plural	,	punctuation mark, comma
PDT	Predeterminer	:	punctuation mark, colon
POS	Possessive ending	(	contextual separator, left paren
PRP	Personal pronoun	)	contextual separator, right paren
PRP$	Possessive pronoun	$	Dollar sign
RB	Adverb	#	Pound sign
RBR	Adverb, comparative	“	Left quote
RBS	Adverb, superlative	”	Right quote
RP	Particle		

However, we must caution that the situation is not so easy for languages other than English. For example, for German, supervised approaches must be used with the TIGER annotated corpus[9] to get similar accuracy as for English. For Chinese, the accuracy is slightly lower, 93% versus 97% for English and 96% for German.

POS tagging research has come a long way and state-of-the-art taggers are achieving an accuracy of slightly higher than 95% or so. However, note that this is on a per-word basis. This means that a POS tagger can be 90% accurate and still miss one word in every sentence of a document. A simple calculation shows that if the average sentence is 10 words long, then a POS tagger can miss every tenth word (about one word for every sentence) and still be 90% accurate.

The most basic POS tagger would go with the most frequent tag of a word based on the analysis of a large corpus of manually annotated text data (maximum likelihood estimation). POS tagging techniques include: supervised, unsupervised and rule based methods. Rule based methods are expert systems in which rules are crafted manually based on lexical and other linguistic knowledge. In supervised learning based POS tagging algorithms are trained on human annotated corpora like the Penn Treebank. Techniques that have been used include: Hidden Markov Models (HMMs), Maximum Entropy Markov Models, Conditional Random Fields, Transformation-based Learning, and Deep Learning [98, 99]. The HMM is a good model for learning sequence tasks. Hence, we now explain the HMM and a POS tagger based on the HMM model.

Hidden Markov Model

The HMM is a member of the class of statistical hidden state sequence models that includes Conditional Markov Models and Conditional Random Fields. All these models rely on the Markov property. The Markov property states that the next state only depends on the local context. More precisely, the state at time t depends on the states at times $1, \ldots, t-1$ only through the state at time $t-1$: t given $t-1$ is independent of the previous states. This property ensures the tractability of computations: Viterbi algorithm, Forward Backward and Clique Calibration are all intractable without this property [153].

We now introduce the Markov model with an example and then give a formal definition. Once we know what a Markov model is, we will hide the states and get the HMM model. Following [155], imagine that the weather on our fictional planet, Tarion, can be either sunny or cloudy (it never rains or snows) and lasts the whole day (no change in the middle of the day). We want to predict the weather tomorrow based on simple model of weather prediction in which we gather statistics on what the weather was today based on the weather yesterday, the day before, etc. Briefly, we collect the following probabilities

$$P(w_n \mid w_{n-1}, w_{n-2}, \cdots w_1) \tag{8.1}$$

Using this formula, we can predict the weather tomorrow based on $n-1$ days of history. For example, if we know that the weather for the past three days was $\{sunny, cloudy, sunny\}$ in chronological order, the probability that tomorrow would be sunny is given by $P(w_4 = sunny \mid w_3 = sunny, w_2 = cloudy, w_1 = sunny)$. But this means we must collect data for 2^n histories (exponentially many) if we use $n-1$ days to predict the next day's weather. So, we make the simplifying Markov assumption:

$$P(w_n \mid w_{n-1}, w_{n-2}, \cdots w_1) = P(w_n \mid w_{n-1}). \tag{8.2}$$

This is called the *first-order* Markov assumption since the probability of next day's weather depends only on the weather for the previous day. A second-order Markov assumption

[9] `http://www.ims.uni-stuttgart.de/forschung/ressourcen/korpora/TIGERCorpus/download/start.html`

Table 8.4: Probabilities of Tomorrow's Weather (Columns) Based on Today's Weather (Rows)

	Sunny	Cloudy
Sunny	0.3	0.7
Cloudy	0.5	0.5

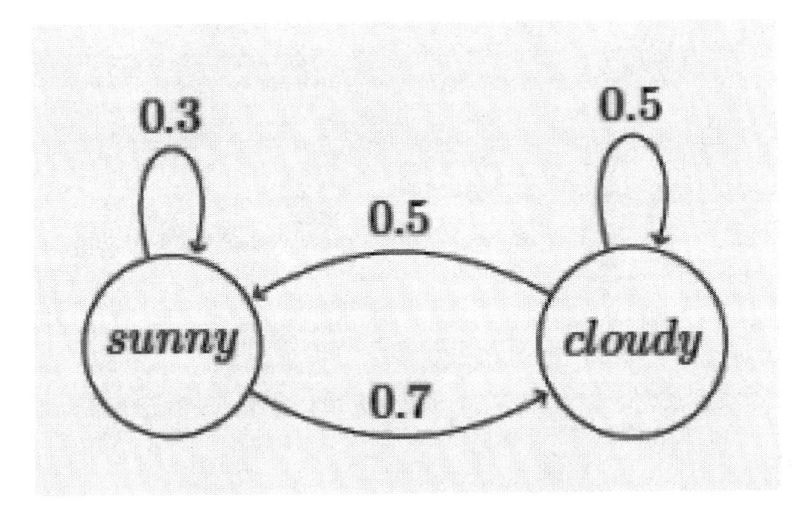

Figure 8.3: The weather on planet Tarion

would involve the weather for two previous days, and so on. We will use the term Markov assumption to refer to the first-order Markov assumption. With this assumption, we need only $2^2 = 4$ numbers, which we have picked arbitrarily as shown in Table 8.4. This Markov chain can also be shown as a labeled graph (e.g., see Figure 8.3).

Formally, a Markov chain is a probabilistic graphical model specified by $M = (Q, A, q_0, q_f)$, where $Q = \{q_1, \cdots, q_n\}$ is a set of n states, $A = [a_{ij}]$ is a transition probability matrix, a_{ij} denotes the probability of going from state i to state j ($\sum_{j=1}^{n} a_{ij} = 1$ for every i), and q_0, q_f are two special states called start state and end state, which are not associated with any observations. Note that the transition probabilities are constants, i.e., independent of time.

Sometimes, we use an alternative representation for Markov chains in which we eliminate the start and end states, and use an initial probability distribution over the n states $\pi = \pi_1, \cdots, \pi_n$, where π_i denotes the probability that the chain will start in state i. Therefore, $\sum_{i=1}^{n} \pi_i = 1$. In this case, we add π as a component of the tuple defining $M = (Q, A, \pi)$.

A Markov chain is a nice model to compute probabilities of sequences of states. However, in the real world, we don't have direct access to the states and transition probabilities. What we have are observations. For example, in POS tagging all we see are the words, the POS tags and transition probabilities are hidden from us. A *hidden* Markov model allows us to deal with such scenarios.

Formally, an HMM $M = (Q, A, B, V, q_0, q_f)$, where Q, A, q_0, q_f have the same meanings as above, V is the output vocabulary of M, and $B = [b_{ik}]$ is an emission probability matrix, where b_{ik} denotes the probability of emitting symbol $v_k \in V$ when M is in state q_i. We can also write b_{ik} as $P(v_k \mid q_i)$. Alternatively, an HMM $M = (Q, A, B, V, \pi)$, where π is an initial probability distribution over the states.

An HMM M allows us to generate an observation sequence $o_1 o_2 \cdots o_T$ as follows: (a) we choose initial state at time $t = 1$, say $s_1 = q_m \in Q$ according to the initial state distribution

π, emit symbol $o_1 = v_p \in V$ according to the emission probability b_{mp} and then transition to the state at time $s_2 = q_r$ according to the transition probability a_{mr}, and so on. A first-order HMM satisfies the Markov assumption as well as output independence, i.e., the output observation v_k depends only on the state q_i that produced it, not on any other states or on any other observations.

$$P(o_i = v_k \mid s_1 \cdots s_i = q_m \cdots s_T, o_1, \cdots o_i \cdots o_T) \approx P(v_k \mid q_m) \tag{8.3}$$

We now consider the three basic problems for HMMs. In what follows, we will abuse notation a bit and write b_{io_j}, where o_j is a member of an observation sequence. Strictly speaking, we should find the index of the observation symbol o_j in our emission vocabulary, say $o_j = v_l$ for some l, and write b_{il}.

The Three Basic Problems for HMMs

1. Evaluation: Given HMM M and an observation sequence O for M, find the probability $P(O \mid M)$.

2. Recognition: Given HMM M and an observation sequence O for M, find the "optimal" state sequence for O.

3. Learning: Given an observation sequence O and the set of states in the HMM, learn the parameters A, transition probability matrix and B, emission probablity matrix.

Evaluation. We can calculate:

$$P(O \mid M) = \sum_Q P(O,\ Q \mid M) = \sum_Q P(Q \mid M) P(O \mid Q,\ M),$$

where $Q = s_1 s_2 \cdots s_T$ is a state sequence of M corresponding to the observation sequence $O = o_1 o_2 \cdots o_T$. We note that

$$P(Q \mid M) = \pi(q_1) \prod_{t=2}^{T} a_{s_{t-1} s_t}$$

and

$$P(O \mid Q, M) = \prod_{t=1}^{T} b_{s_t o_t}$$

However, the brute force-solution is very expensive, since it would enumerate exponentially many state sequences (n^T, where M has n states).

So, we use dynamic programming. We define a forward variable $\alpha_j(t)$, as the probability of the partial observation sequence up to time t *with state* q_j *at time* t. We compute it inductively as follows:

$$\alpha_j(t+1) = \sum_{i=1}^{n} \alpha_i(t) a_{ij} b_{j o_{t+1}} \tag{8.4}$$

The base case is $\alpha_j(1) = \pi(q_j) b_{j o_1}$. Then, with $n^2 T$ operations we can compute:

$$P(O \mid M) = \sum_{i=1}^{n} P(O, s_T = q_i \mid M) = \sum_{i=1}^{n} \alpha_i(T) \tag{8.5}$$

Recognition. For the second problem of HMMs, we need to define what we mean by an optimal sequence. One might think that optimality means choosing states which are individually most likely, but this can give an invalid state sequence as the solution. Hence, we choose the *most likely state sequence* as our definition of optimality. Again, we use dynamic programming for this, which is popularly referred to as the Viterbi algorithm. For this purpose, we define $\delta_j(t)$, the highest probability of a single path of length t, which accounts for the observations and ends in state s_j. That is

$$\delta_j(t) = max_{s_1 s_2 \cdots s_{t-1}} P(s_1 s_2 \cdots s_t = q_j, o_1 o_2 \cdots o_t \mid M) \tag{8.6}$$

We compute it inductively as follows:

$$\delta_j(t+1) = (\max_i \delta_i(t) a_{ij}) b_{j o_{t+1}} \tag{8.7}$$

The base case is $\delta_j(1) = \pi(q_j) b_{j o_1}$. We can compute the optimal state sequence using backtracking if we also store the maximizing argument for each t and j. As we can see, the Viterbi algorithm is very similar to the forward algorithm. The only change is that the Viterbi algorithm uses maximization in place of the sum operator in the forward algorithm.

Learning. Estimation of the A and B matrices of an HMM M that would maximize the probability of the given observation sequence uses what is referred to as the Baum-Welch algorithm, or the forward-backward algorithm. It is a special case of the expectation-maximization (EM) algorithm, which we discussed in detail in Chapter 5. The EM algorithm starts with an estimate and progressively improves it. We refer the reader to [226] for the details of this HMM problem.

Incremental HMMs. Incremental versions of the HMM model have been proposed, which may be useful for handling concept drift. We refer the reader to [381, 394] for details. In [394], two kinds of generalizations are discussed: modeling state durations and allowing transition probabilities to vary with time. The extensions needed for allowing transition probabilities to vary with time are fairly straightforward and we omit them here.

The hidden markov model is good for sequence based supervised learning. However, a lot of training data may be required for parameter estimation.

In POS tagging, we would take a set of sentences, with each sentence as sequence of words (our observation sequence), and train an HMM, using the Baum-Welch algorithm on those sequences. The number of states of this HMM would be the number of possible POS tags. Then, we would use it for POS tagging of test sentences. However, in practice, we do this a little differently. We don't use the unsupervised Baum-Welch algorithm for finding the best mapping of labels to observations. Instead we give a fully labeled data set and set the parameters by maximum likelihood estimation. Recall that we wish to compute the tag sequence that is most probable given the observation sequence of n words $w(1..n)$.

$$t_o(1..n) = argmax_{t(1..n)} P(t(1..n) \mid w(1..n)) \tag{8.8}$$

We can apply Bayes rule to compute:

$$t_o(1..n) = argmax_{t(1..n)} P(w(1..n) \mid t(1..n)) P(t(1..n)) / P(w(1..n)) \tag{8.9}$$

We can drop the denominator and simplify Equation 8.9. We then make two more simplifying assumptions. First, the probability of a word depends only on the tag. Second, the probability of a tag depends only on the previous tag.

Thus, $P(w(1..n) \mid t(1..n)) = \prod_{i=1}^{n} P(w(i) \mid t(i))$ and $P(t(1..n) = \prod_{i=1}^{n} P(t(i) \mid P(t(i-1))$. Note, that $P(w(i) \mid t(i))$ is just the emission probability and $P(t(i) \mid P(t(i-1))$ is just the transition probability for the HMM. These probabilities can be estimated using the frequency counts from a tagged corpus, the maximum likelihood estimates. Again, we refer the reader to [226] for the details and an illustrative example of using the Viterbi algorithm for computing the best tag sequence given these simplifying assumptions.

We now give some hard examples of POS tagging below. The interested reader is referred to [375] for more examples.

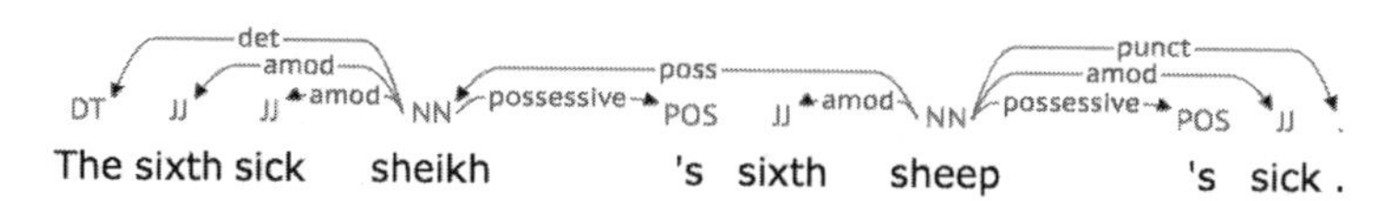

Figure 8.4: Hard example for POS tagging: contraction versus possessive

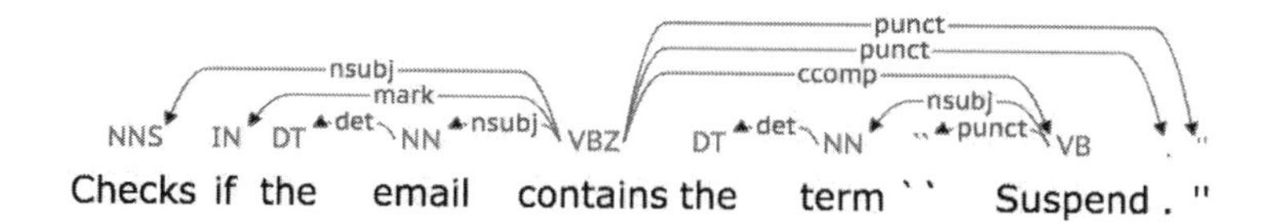

Figure 8.5: Sentence fragment example for POS tagging

- Over/up/etc. can be particles or prepositions. For example, they talked *over* a deal versus she talked *over* Skype.

- Nouns can be adjectives. For example, The *athletics* club.

- Nouns versus present participles. For example, *farming* can be relaxing.

- Past tense versus past participles.

- Some words can be verbs or prepositions.

- *Around* can be a preposition, particle or adverb.

- Contraction versus possessive. For example, in the sentence, "The sixth sick sheikh's sixth sheep's sick," the second occurrence of 's is a contraction, not a possessive. The first occurrence of 's is possessive. But when we give it to the ARK parser, we get the output in Figure 8.4.

Of course, sentence fragments are also hard to tag. Sentence fragments are important to tag properly since they appear in tables and figures. We show in Figure 8.5 an example of a sentence fragment from a table in a paper on phishing email detection [513] being tagged incorrectly. The word "Suspend" should be either a noun or an adjective, but it is tagged as a verb here.[10]

Two other online demos for POS tagging are available from the University of Illinois and Stanford University. Some of the offline tools for POS tagging include: CLAWS, NLTK, Stanford Core NLP, LingPipe, OpenNLP, CRFTagger, etc.

A comparative evaluation of nine POS taggers (Arktools, ClearNLP, Hepple, JunPos, LBJ, Mate, OpenNLP, Stanford, TreeTagger) on data sets for English and German is reported in [200]. Both speed and accuracy are considered in the evaluation. There are three data sets for English, which include formal writing, speech transcripts and messages from social media, an IRC chat corpus and annotated tweets. The fastest was Hepple, and Clear was the most accurate on English data sets. However, the accuracy of all the taggers was below 90%. The TreeTagger was the most accurate, 94%, and HunPos was the fastest, on the German data set.

[10]If we follow the POS tagging guidelines in [375], it should be tagged only as a noun.

8.6.2 Word Sense Disambiguation

The goal of word sense disambiguation (WSD) is to determine the sense of a word, given a word in context and a fixed inventory of senses. For example, the English word plant has at least two senses: it could refer to a building for carrying on industrial labor, or a living organism lacking the power of locomotion.

Feature vectors for WSD include the context of a word, say the N words on either side of a word. Of course, we need a good way to decide the value of N. The answer depends on how much error we are willing to tolerate. We distinguish between two kinds of features: collocational and bag-of-words features. By collocational features we mean, words at specific positions near the target word. We include usually only the identity of the word and its POS tag. Bag of word features are features about words that occur anywhere in the window (regardless of position). These are usually limited to frequency counts.

There are two dimensions along which we can categorize methods for WSD. The first dimension is the amount of labeled data used, and the second is based on the amount of knowledge, in the form of dictionaries, thesauri, or ontologies. Thus, we can have unsupervised, supervised, or semi-supervised approaches for WSD for the labeled data dimension, and we can have knowledge-rich, "medium knowledge," and knowledge-poor approaches on the other dimension.

Unsupervised methods for WSD include: methods based on context clustering, word clustering, and co-occurrence graphs. Although these methods solve the challenge of requiring large amounts of annotated data, they are hard to evaluate since they do not have any shared reference inventory of senses. All they do is create clusters of word senses, and their clusters may not align well with the well-understood meanings of those words. We describe here the context clustering approach and refer the reader to the survey of [326] for a description of the other approaches. In context clustering, each word is represented by a context vector. For example, we could represent a word w with a vector in which the jth component is the number of times word w and w_j co-occur in a fixed window size. We could then use cosine-similarity for computing the similarity between the vectors. These similarities could then be used in either an agglomerative clustering algorithm or a partitional one (see Chapter 5). Some algorithms have also used topic modeling algorithms such as LDA. Another possibility is to use sense embeddings.

Evaluation of unsupervised algorithms can be conducted by either mapping the clusters to sense classes using an annotated, labeled data set (called intrinsic evaluation), or by using the results in a task pipeline and measuring how much improvement is achieved by using the unsupervised WSD algorithm (called extrinsic evaluation).

Using the collocational and bag-of-word features, we could in principle use any supervised classification method for WSD: naive Bayes, decision tree, SVM, neural networks, etc. Of course, we require a large training set to train the classifier. There are several labeled corpora available: the line-hard-serve corpus consists of 4,000 sense-tagged examples of *line* as a noun, *hard* as an adjective, and *serve* as a verb; the *interest* corpus contains 2,369 sense-tagged examples of *interest* as a noun. The SENSEVAL project has also produced several sense-labeled lexical corpora. A commonly used corpus SemCor contains over 234,000 words that have been manually tagged with WordNet senses.

As far as supervised approaches are concerned, any of the methods in the machine learning chapter can be applied to WSD, e.g., Decision Tree, SVM, etc., and many of them have already been tried.

Knowledge-based approaches include: sense definition overlap, restrictions on selection, and "structural approaches (semantic similarity measures and graph-based approaches," [326]. Many of these approaches use WordNet or similar resources. In sense definition overlap, we calculate the intersection between the context of a word w and the bags of words (each bag is called a "gloss") corresponding to each sense definition of w. The sense that maximizes the intersection is assigned to w. The second method involves checking whether there are any grammatical restrictions that prevent some sense from being chosen, or that prefer some sense over others. The structural approaches try to leverage the relationships between concepts in resources such as WordNet. For example, WordNet's hypernymy/hyponymy tree for nouns can be used to define similarity between senses. This similarity can then be exploited for WSD.

A semi-supervised approach for WSD has been given by Yarowsky ([512]), and game theoretic methods ([434]) and topic models ([77]) have been proposed. This is still an area of active research, as the recent papers indicate.

8.6.3 Language Modeling

The purpose of language modeling is as follows: given a sequence of words in a language, predict the next word in the sequence. An "easy" method for this task is to build a large corpus and then find the most frequent word in it. Once it is found, we always predict this word, regardless of the preceding words. Of course, this will not be optimal. We can do better. We can look for bigrams, two word phrases, and predict the next word based on the previous word. Again, we go with the most frequent next word, given the previous word. We can extend this idea further and tabulate trigrams.

Next we can combine the models. Whenever, we encounter a bigram context (previous two words), which we have seen in our training data, we use the trigram frequencies to predict the next word. If the context does not occur in the training data, we back off to just the previous word and use the bigram frequencies in the training data. If that doesn't work, we forget the previous word also and just predict the most frequent word in our training data.

Word to vector models or word embeddings, mentioned above, can also be used for language modeling. Google has developed unigram, bigram and trigram models, which can be downloaded from and viewed at `https://ai.googleblog.com/2006/08/all-our-n-gram-are-belong-to-you.html`.

8.6.4 Topic Modeling

Topic modeling is a vibrant area of research with many supervised and unsupervised techniques available. Among the prominent unsupervised techniques are LDA: Latent Dirichlet Allocation. We have already discussed topic modeling techniques in the Text Mining Chapter 7, so we point the reader over there.

8.7 Sequence-to-Sequence Tasks

Early solutions to sequence-to-sequence tasks, such as language translation, used the RNN model. In [447], a transformer module is proposed. They propose to replace the RNN with this module. The RNN was a popular model for sequence-to-sequence tasks such as machine translation. It was used as the building block of an encoder and a decoder. An example will make this clear.

Imagine that we are translating the sentence: The dog ran after the cat. Then we would convert each word, e.g., the first word 'The,' into a vector representation and feed this into an encoder and out would come a hidden state representation, which would be fed into the next round dealing with the next word.[11] We can write this as $h_t = Enc(h_{t-1}, \vec{w_t})$, where

[11] The start hidden state $= 0$ for the very first word.

Enc represents the nonlinear Encoder function, h_{t-1} denotes the previous hidden state and $\vec{w}_t$ is the word vector for the word at position ("time") t. The hidden state from the last word is now given to a decoder with the previous output word (again the start output word vector $= 0$). The decoder then outputs a hidden state and a word vector representing the translation, and the next position/time step begins. A special case of this is when the same hidden state, also called a context vector, is given to all decoding steps. Variants of RNN architecture were used in the encoder and decoder. Two of them, which are especially popular, are called LSTM for long short-term memory and GRU for Gated Recurrent Unit.

However, the RNN model has some disadvantages including: difficult to learn long-range dependencies and difficult to parallelize. So the transformer module was proposed in which attention is used to flag important information in the input sequence instead of a sequential time step approach. Since the sequence information does not drive the computation, word encodings are combined with positional encodings using continuous functions, e.g., trigonometric functions such as sine and cosine. The attention is in the form of a triple (V, K, Q), where V represents the vector of values, K is the key vector and Q is the query vector. Both K and Q have dimension d_k and V has dimension d_v. In practice, all three are packed together into matrices.

$$Attention(V, K, Q) = \text{softmax}(QK^T/\sqrt{d_k})V \tag{8.10}$$

Except for the attention modules (multihead attention and also self-attention), the rest of the components used in the basic transformer module are fairly standard: feedforward network, residual connections, addition and normalization, etc. The encoder module output in the transformer is connected to the multihead attention module in the decoder.

In [114], a bi-directional transformer based approach (BERT) for generating contextual word embeddings is proposed. It is based on a masked language model, i.e., a model in which some tokens are randomly masked and the masked token must be predicted from its context. It also uses a next sentence prediction task, i.e., training on sentence pairs. The reason this approach works better than bidirectional LSTM models is that words from both directions are used *simultaneously*. In the bidirectional LSTM approach, each model, forward and backward, is trained independently. Therefore, neither model sees words on both sides of the masked word. In [509], the disadvantage of BERT, which is neglecting the dependencies between masked positions, is overcome with a generalized autoregressive pre-training method.

8.8 Knowledge Bases and Frameworks

Knowledge bases and NLP libraries/frameworks are very useful in dealing with NLP tasks and indispensable in language understanding. In this section, we cover a few of the more popular knowledge bases. These include:

- WordNet: A dictionary cum thesaurus, WordNet [149], is a very useful tool, since it has a rich variety of relations between, what it calls, *synsets*. It has been constructed mostly manually by a team of researchers that includes linguists and psychologists at Princeton University.[12] WordNet organizes words by their meaning rather than by their form. A synset consists of all the words and phrases that denote a particular concept. Hence, synonymy is the most basic relation in WordNet. There are 117,000 distinct synsets in WordNet as of the writing of this book. Synsets are connected to each other, where the edges denote *semantic* relationships of various kinds. Before,

[12] `http://wordnet.princeton.edu`

we can discuss relationships, we need to know first what kinds of words and synsets exist in WordNet.

Currently, there are five different categories of words in WordNet. Four corresponding to the four important POS: Nouns, Verbs, Adjectives and Adverbs. The fifth category includes function words. Nouns are organized in a topical hierarchy that looks like a directed acyclic graph. The reason it is not a tree (or forest) is that some synsets in the hierarchy may have more than one parent. Verbs are organized using the entailment relations. Entailment means involving by necessity, e.g., winning entails playing or competing. Adjectives and Adverbs are organized into multidimensional hyperspaces, where each attribute is considered as an abstract dimension. For example, adjectives have antonyms and also gradation, or degree, as attributes.

For example, the word "plant", as a noun, has four different senses (or meanings) in WordNet 3.1. The first is "buildings for carrying on industrial labor" and the word forms in this synset are
{*plant*, *works*, *industrial plant*}. Notice that phrases can appear in synsets, not just words. The second is "a living organism lacking the power of locomotion" and the synset contains
{*plant*, *flora*, *plant life*}.

Relationships between nouns in WordNet include: Hypernymy/hyponymy and meronymy. For example, for the first sense of plant, we have the following relationships: direct hyponym, full hyponym, direct hypernym, inherited hypernym, sister term, and something called "domain term category." Relationships between verbs in WordNet include: hypernymy/hyponymy, tropnymy, and entailment. For example, for the word plant, as a verb, the first sense is "put or set (seeds, seedlings, or plants) into the ground" and we have the following relationships listed: direct troponym, full troponym, direct hypernym, inherited hypernym, sister term, derivationally related form, and sentence frame. The verb relationships in WordNet can be cyclic, which the first author learned the hard way. Later, he found the relevant paper on this [363]. For adjectives, we typically find the following information: similar to, antonym, derivationally related form, attribute and see also. For adverbs, we typically find: pertainym (which means "pertaining to" or "relating to"), antonym (if it exists), and domain usage.

Recently, some researchers have been working on making WordNet more useful for knowledge processing applications such as question answering. They call this project the extended WordNet. More information can be found at: `http://www.hlt.utdallas.edu/~xwn/about.html`.

- VerbNet: VerbNet is available online from `http://verbs.colorado.edu/~mpalmer/projects/verbnet.html`, VerbNet (VN) is a large lexicon of verbs that is linked to WordNet. It is organized into verb classes, which extend "Levin (1993) classes through refinement and addition of subclasses to achieve syntactic and semantic coherence among members of a class. Each verb class in VN is completely described by thematic roles, selectional restrictions on the arguments, and frames consisting of a syntactic description and semantic predicates with a temporal function, in a manner similar to the event decomposition of Moens and Steedman (1988)."

  The verb 'hit,' for example, has 10 members and 11 frames.[13] The members are: bang, bash, click, dash, smite, squash, tamp, thump, thwack and whack. The 11 frames are also listed at the link given. In contrast, the entry for the same word in WordNet

[13]`https://verbs.colorado.edu/verb-index/vn/hit-18.1.php`

shows 16 different senses for 'hit' as a verb ('hit' is also a noun with seven different senses), with one of them having synonyms such as murder, slay, hit, dispatch, bump off, off, polish off, remove. For this sense of 'hit,' the direct troponyms include 'burke' ("murder without leaving trace on the body") and 'execute' ("murder in a planned fashion"). Thus, we see that there are pros and cons of the different knowledge bases, and the choice of which ones to use is based on thoughtful consideration of the goal.

- FrameNet: FrameNet (`https://framenet.icsi.berkeley.edu/fndrupal/about`) is another publicly downloadable lexical database of English that is supposed to be human- and machine-readable. It is based on annotating examples of how words are used in actual texts. We may think of it as a dictionary of more than 13,000 word senses, most of these senses come with annotated examples that show the meaning and usage. For the NLP researcher, it offers more than 200,000 manually annotated sentences linked to more than 1,200 semantic frames. This can be a good training data set for several applications including: semantic role labeling, information extraction, machine translation, event recognition, sentiment analysis, etc. For linguists, it serves as a valence dictionary,[14] with detailed evidence for the combinatorial properties of a core set of the English vocabulary. FrameNet-like databases are now available for a number of languages, and researchers are working on aligning the FrameNets across languages. FrameNet limitations include lack of annotated data for lexical units denoting frames and the limited coverage of words, senses and frames [338].

  It is based on a theory of meaning called Frame Semantics, which is explained in [152]. The idea is that "the meanings of most words can best be understood on the basis of a semantic frame, a description of a type of event, relation, or entity and the participants in it. For example, the concept of cooking typically involves a person doing the cooking (Cook), the food that is to be cooked (Food), something to hold the food while cooking (Container) and a source of heat (Heating_instrument). In the FrameNet project, this is represented as a frame called Apply_heat, and the Cook, Food, Heating_instrument and Container are called frame elements (FEs). Words that evoke this frame, such as fry, bake, boil, and broil, are called lexical units (LUs) of the Apply_heat frame."

- The Banks: There are many other resources available for NLP. We mention three that end with the word bank. First is Penn Treebank whose tagset we have mentioned before. The current version is Treebank-3.[15] It is an annotated corpus of data from several sources including newswire and speech (telephone, microphone and transcription). The data has been tagged with POS tags and also manually parsed. The second is PropBank [241], which is similar to FrameNet, but focuses only on verbs in the Penn Treebank corpus. Its goal is to explicitly specify the predicate-argument structure. The third is NomBank [313]. Its goal is to provide argument structure for approximately 5000 common nouns in the Penn Treebank-2 corpus. Propbank and NomBank can be used to generalize patterns. The following example, taken from [313], illustrates this point:

  "Given a pattern stating that the object (ARG1) of appoint is John and the subject (ARG0) is IBM, a PropBank/NomBank enlightened system could detect that IBM hired John from the following strings: IBM appointed John, John was appointed by IBM, IBM's appointment of John, the appointment of John by IBM and John is the current IBM appointee. Systems that do not regularize across predicates would require separate patterns for each of these environments."

[14] How a lexical unit combines with other units, or patterns of combination.
[15] `https://catalog.ldc.upenn.edu/LDC99T42`

Limitations of Propbank and Nombank include extra-sentential arguments, i.e., arguments that are found in a sentence that is adjacent to the one that contains the predicate, and implicit arguments [167].

Among NLP frameworks we mention a few important ones:

- Scikit-learn: A well-known machine learning library for Python. Also useful are: SciPy for scientific computations and NumPy for matrix and tensor operations.

- Natural Language Toolkit (NLTK): A comprehensive Python toolkit for many useful NLP techniques, basic as well as advanced. It includes different modules for tokenization, stemming, tagging, parsing and visualization. It provides easy access to multiple corpora, knowledge bases such as WordNet, and trained models. The entire list of these can be accessed at `http://www.nltk.org/nltk_data/`. For example, there are two stemmers, an ACE corpus trained named entity chunker based on the maximum entropy model, and a Treebank POS tagger.

- Apache OpenNLP: A Java library for several NLP tasks such as tokenization, chunking, POS tagging, parsing and named entity recognition. Pre-trained models are available as well as the option of training a Perceptron or maximum entropy model on the user's data set. Two POS tagging models are available, and the Penn Treebank tagset is used for tagging.

- Pattern: A web mining module for the Python programming language (`https://www.clips.uantwerpen.be/pattern`). "It has tools for data mining (Google, Twitter and Wikipedia API, a web crawler, a HTML DOM parser), natural language processing (part-of-speech taggers, n-gram search, sentiment analysis, WordNet), machine learning (vector space model, clustering, SVM), network analysis and <canvas> visualization."

- TextBlob: Built on top of NLTK and Pattern, `http://textblob.readthedocs.io/en/dev/`, provides an API for common NLP tasks such as POS tagging, noun phrase extraction, sentiment analysis, classification, translation, and more.

- spaCy: A library for NLP in Python and Cython. It contains pre-trained statistical models and word vectors. It supports tokenization for an impressive array of languages. It provides parsing, convolutional neural network models for tagging, parsing and named entity recognition, and deep learning models.

- Gensim (`https://radimrehurek.com/gensim/`): A library for topic modeling from plain text documents. It is open-source, scalable, and contains converters for many popular data formats. Algorithms include: Word2Vec, FastText, Latent Dirichlet Allocation, and Latent Semantic Analysis.

- Stanford Core NLP (`https://stanfordnlp.github.io/CoreNLP/`): NLP tools by the NLP group from Stanford University. The toolkit provides support for multiple languages: English, Chinese, Arabic, French, German and Spanish. For each supported language, it typically includes a POS tagger, a parser, and an NER tagger. For English there is also coreference resolution, sentiment analysis, and more.

Besides these knowledge bases and frameworks, there are also pre-trained word embeddings and other language models available. The above frameworks have been developed for and trained with corpora containing news articles and formal text documents.

Recently, there has been a lot of interest in processing text from social networks such as Twitter and Facebook. Here, the text is more informal and there are a number of issues:

colloquial expressions (e.g., words such as "soda" for carbonated drinks, contractions such as "ain't," "wanna," and "y'all," profanity such as "bloody" in British English, phrases such as "pass the buck" and aphorisms such as "She wasn't born yesterday"), slang (e.g., "gobsmacked," "knackered"), abbreviations such as "lol" and "ttyl," and the use of first or second person imperatives (e.g., "Do it"). Moreover, the text may be over-punctuated and may contain different types of emojis. The frameworks and toolkits trained on news articles and formal text do not perform well on such informal text. Hence, new annotated corpora that include informal text are needed and some frameworks have been developed specifically for social network text. We mention a few of these, e.g., [170, 280, 302, 364], since they are relevant to email security for example.

8.9 Natural Language Generation

In many applications of NLP techniques, it is necessary to generate natural language sentences. For example, in summarization, much more compression can be achieved by generating sentences rather than extracting them from the source text. Consider the following short piece of text:

```
She studied in St Stephens High School. After a gap year,
she joined Columbia University where she majored in Economics.
She did her PhD at University of Chicago, also in Economics.
Subsequently, she became an Assistant Professor at University of
Cape Town.
```

An extractive summary (one which extracts sentences from the text) would have a hard time compressing this piece of text while retaining the significant pieces of information in it. On the other hand, an abstractive summary would generate a sentence or two like so:

```
After graduating from St Stephens High, she received her bachelors
from Columbia and PhD from University of Chicago, all in Economics.
Then, she joined faculty of University of Cape Town.
```

Of course, generating such compressed sentences is quite challenging. Currently, there are at least three different techniques available for generating text: Recursive transitions networks (e.g., as used in the DADA engine) [27], markov chains [83, 84], and machine learning [104].

8.10 Issues with Pipelining

Many applications/adaptations of NLP techniques use pipelines of basic NLP tasks. This can be problematic because of error propagation. For example, suppose we use three NLP modules pipelined sequentially. If each module is 90% accurate, then at the output we can only expect accuracy of about 73% (.9 X .9 X .9 = .729). A human reader of a piece of text does not process the text in distinct separate tasks such as POS tagging, word-sense disambiguation, etc. Rather, with some experience and facility in reading, all these tasks are done together/holistically. Unfortunately, we are not at this stage with automatic language processing.

8.11 Security Applications of NLP

NLP techniques have been used in a number of security challenges: checking whether a password is strong or weak, malware detection, spam and phishing email detection, etc. We now discuss these briefly.

8.11.1 Password Checking

Markov models have been used in an algorithm to check the security of passwords. In [111], researchers have used a Markov model with 28 states. Twenty-six states correspond to the letters of the English alphabet, no distinction is made between upper- and lower-case letters. There is one state corresponding to space (denoted by SPC) and one state (OTHER) for all the forty to fifty characters that include: digits, punctuation and special characters. They experimented with both a first order and a second order model; the latter gave better overall performance. For the transition probability matrix, they used a large file of known bad passwords, D. From D, they calculate the frequency matrix f, where $f[i, j, k]$ is the number of occurrences of the trigram ijk. For instance, for the password, password123, we get the trigrams pas, ass, ssw, swo, wor, ord, rdOTHER, dOTHEROTHER and OTHEROTHEROTHER. For each bigram ij, they calculate $f(i, j, \infty)$ as the total number of trigrams beginning with ij. Then $T[i, j, k] = f(i, j, k)/f(i, j, \infty)$, which is the maximum likelihood estimate of the transition probabilities. Good-Turing smoothing was applied to the transition probability matrix since it contained a lot of zeros.

By modeling the dictionary of passwords as a Markov model, the question of whether a given password is bad, reduces to a different question: What is the likelihood that the given string is generated by the Markov model of the dictionary? The test they use is a log-likelihood function. Let password $p = p_1 p_2 \cdots p_l$ then $llf(p) = \sum_{i=1}^{l-2} ln(T[p_i, p_{i+1}, p_{i+2}])$. For the final test, they calculate

$$BA(p) = \frac{\frac{llf(p)}{l-2} - \mu}{\sigma} \tag{8.11}$$

Here μ and σ are the mean and standard deviation of $\frac{llf(p)}{l-2}$. The estimated mean and standard deviation are calculated by computing the value $\frac{llf(p)}{l-2}$ for every password in the bad password dictionary, D. Because of this centering and normalization, BA(p) has a mean of zero and a standard deviation of one. The authors set a threshold of 2.6 standard deviations, about 99% of the area under the normal curve, and accept as good any password that has a value of less than -2.6. Passwords close to the mean, zero, are viewed as being drawn from the bad dictionary and therefore unacceptable. An empirical evaluation is conducted to show that the method performs quite well in rejecting bad passwords, and accepting good passwords. We refer the reader to [111] for the details, and examine a related idea: password guessing attacks, where an attacker must generate high-probability passwords instead of checking whether a password is weak or strong.

In password guessing attacks, researchers have used Probabilistic Context Free Grammars (PCFG) [482]. A PCFG is an extension of a Context Free Grammar (CFG) in which the rules have probabilities associated with them. The goal of a PCFG is to get a probability distribution over derivations, and hence also parse trees. PCFGs have been applied to the parsing problem in NLP. In password guessing attacks, the goals is to generate passwords in order from highly probable passwords to less likely ones. A database of publicly available passwords can be used for learning the probabilities. For example, we may introduce four nonterminals: P for password, L for letter, D for digit, and S for symbol. At the top level, we have the rules: $P \rightarrow LP \mid SP \mid DP$. Of course, passwords have usually a lower and an upper limit on the length, which can be incorporated into the rules as well. For example, instead of the nonterminal P we may use P_7 to indicate a password that is seven characters long and we may have rules such as $P_7 \rightarrow L_6 D_1 \mid L_7 \mid L_6 S_1$. In practice, we do not introduce all the possible rules, rather we learn the probabilities and rules from a training data set.

In the training phase, when we are learning the probabilities for the rules and we see a password such as *password1*, then we have a meta-rule such as:
$password1 \rightarrow L_7 D_1$;

Count(L_7D_1)++;
Count($L_7 \rightarrow password$)++;
Count($D_1 \rightarrow 1$)++

In the guess-generation phase, the probability of *password1* is calculated as $P(password1) = P(L_7D_1) * P(L_7 \rightarrow password) * P(D_1 \rightarrow 1)$. This idea can be used in what are called password trawling attacks, which are offline attacks. Recently, researchers have used a modification of this idea in online, targeted attacks, where a sibling password of a user has been leaked from one web site and some user-specific information is also available [470].

8.11.2 Email Spam Detection

Email spam detection is a classic application of natural language processing techniques. Estimates of the spam emails to total emails ratio vary, with some estimates as high as 75-80% (i.e., spam emails constitute over three quarters of all emails), although recent trends show a decline in this ratio. Thus, spam is a serious problem, since it abuses communication bandwidth, takes up storage space and processing power, and it wastes a lot of time and thus money for email users and organizations. According to Statista.com,[16] spam constituted almost 60% of email traffic worldwide during January 2014 to January 2018, with the US as the leader with about 12% of total spam volume. The most common types of spam were related to healthcare and dating. We define spam as a form of advertising, with the phrase "unsolicited bulk email" as the definition (proposed in [16]), and with the understanding that the most harmful effects of spam are loss of time and productivity, but not the loss of digital identity or other more severe harms such as loss of money or reputation. The latter are some of the effects of phishing.

Features for Spam Detection

Although features have been extracted from both header and body of emails, here we focus on features from the body since NLP techniques are most applicable to the body, which can be considered as a text in a natural language. Initial spam filtering methods used words from the bodies of emails often after stopword elimination and stemming were applied. Some researchers proposed feature weighting schemes such as mutual information, term frequency and Chi-square. Later, TFIDF (Chapter 7) was also used.

The χ-squared test, which is a method used frequently for authorship identification, was also proposed for spam detection ([407]). This allows some filtering of words to determine good words to use to distinguish the types, but the nonstationarity of the data and the multiple comparisons problem make this method somewhat ineffective.

In addition to the word filtering approach, the χ-squared test can be used directly to determine spam versus ham. In this method, messages are first represented by character or word N-gram frequencies. Words are tested to determine whether they are more/less frequently used for one corpus (spam) versus the other. By using a Bonferroni correction (see Section 4.6) we can obtain a p-value for rejecting one in favor of the other.

Smoothed N-gram language models have also been used for spam filtering. In these methods a language model is built for spam messages and another one for ham messages. For each message in question, we calculate the probability that it was generated by each model. This general technique works regardless of how one constructs the model, provided that the different classes have distinct models, the models are sufficiently separated without extensive overlap, and one can accurately assess the probability of a new observation under the two models.

[16]`https://www.statista.com/statistics/420391/spam-email-traffic-share/` - Accessed 17 Feb. 2018.

A large variety of techniques have been developed utilizing these various features. For an excellent survey of spam filtering techniques that provides extensive discussion of these and an extensive set of references, see [43].

8.11.3 Phishing Email Detection

NLP techniques were used in processing the bodies of emails in [454, 457]. These two papers are a good case-study in the use of several NLP techniques for addressing a security challenge.

In [457], researchers used three (mostly) independent judges: header analysis, link (URL) analysis and body text analysis, and combined the judges using majority vote. The header analysis is based on rules derived from observations on a small set of phishing and legitimate emails. The link analysis employs an Internet search using the Google API (deprecated), one search per link. Each search uses the same set of five keywords extracted from the body, together with the domain name extracted from each link in the email. The top 30 search results are scanned for the link, excluding matches from phishing databases such as phishtank. The idea is that phishing sites are ephemeral and so would not rank high in Google's "popularity-based" PageRank algorithm.

The text analysis employs POS tagging, WSD and WordNet in [457]. The idea is to score verbs in the body of the email, since the focus is on distinguishing actionable emails from information ones, and verbs represent the action words in sentences. This score is adjusted based on the presence of a link in the same sentence, and words conveying urgency (e.g., a deadline) or incentive (e.g., you have won a prize). WordNet's hyponymy relation is used to find additional verbs that could be used by a clever phisher, who read about the method and wanted to defeat the text analysis. For instance, instead of saying "Click this link," the phisher could say: "Check out this link," or "Press on this link." A novel idea of combining email text analysis with an analysis of the context of the user's emails was also proposed, which is called context analysis. By context, the researchers mean all the emails in the user's Inbox and Sentmail box.

Using TFIDF, the body of the email e, which was to be classified, was converted into a vector and similarly the bodies of all emails in the user's Inbox and Sentmail box were also converted into vectors over a common vocabulary. Then, cosine similarity was used to find the most similar email e' in the user context to e. If this similarity exceeded a certain threshold, then the decision for e uses the decision for e', if it is already available.

The overall classifier did quite well (detection rate of over 97% on phishing emails with false positive rate between 0.7-0.8% on legitimate emails) on a data set consisting of 3,000 legitimate emails (1,000 were contributed by each researcher) and a phishing data set consisting of 4,550 emails from the web. The link analysis was roughly 95% accurate, the header analysis was approximately 99% accurate, but the text analysis showed about 77% accuracy on legitimate emails and around 57% accuracy on the phishing emails.

Subsequently, Verma and Hossain [454] tried to improve the body text analysis further. Instead of a verb scoring formula, they analyzed a data set of phishing and legitimate emails using the idea of a t-test. Using the t-test, they try to determine whether a feature's variance between two data sets is statistically significantly different. They use a two-tailed, two samples of unequal variances t-test (covered in Chapter 4) since their data sets are generally of different sizes as well as variance.

After some experimentation, the researchers in [454] considered frequencies of bigrams following the word "your." A bigram was chosen as a possible feature if its t-value exceeded the critical value for an α value of 0.01, based on two-tailed t-test. Here α denotes the probability of a Type I error. Then weights were calculated for each selected bigram, b, as follows:

$$w(b) = \frac{(p_b - l_b)}{p_b} \tag{8.12}$$

In the above equation, p_b (respectively l_b) denotes the percentage of phishing emails (respectively legitimate emails) that contain b. Features that appeared in less than 5% of the emails or had weights less than 0 were discarded. Finally, a bigram b was selected, if $w(b) > m - s$, where m is the mean bigram weight and s is the standard deviation. The resulting set of bigrams is called PROPERTY, since it denotes the bigrams referring to the property of the user that has been affected (e.g., "credit card"). A similar analysis is conducted for all the words that appear in sentences containing a hyperlink or any of the words: url, link, or web site. This analysis leads to the set of words called ACTION. These two sets lead to the Action-Detector: At this point, we are set to design the Action-detector subclassifier: for each email encountered, an email is marked as phishing if it has:

1. The word "your" followed by a bigram belonging to PROPERTY (e.g., "your paypal account"), and

2. A word from ACTION in a sentence containing a hyperlink or any word from {"url", "link", "website"} (e.g., "click the link").

A Nonsense-detector is also designed to detect emails that contain a link and whose subjects do not match up with the body of the email, i.e., no word from the subject, after removing stopwords, appears in the body. We refer the reader to the paper [454] for the details of this detector, which also uses the t-test. The emails are first passed through the Action-detector and if they are not marked as phishing by the Action-detector, then they are given to the Nonsense-detector.

The Action-detector and Nonsense-detector have several versions, the first versions of both detectors just use straightforward pattern matching. The second version uses POS tagged features: bigrams that do not contain a noun or a named entity in the set PROPERTY, words that are not verbs in the set ACTION are discarded, and the Nonsense-detector only works on nouns, verbs, adjectives and adverbs in the subject of the email, and for subject-body similarity only nouns are used. The third uses features that have been tagged for POS and sense using SenseLearner [315], and the last version extends the noun features using WordNet's synonymy and the direct hyponyms of these synonyms. A few more changes are made in this last version [454].

Data sets and results. The data set comprised the same set of 4,550 public phishing emails that were used in [457], and 10,000 legitimate emails from the public Enron inbox email database. They randomly selected 70% of both the phishing and the legitimate emails for statistical analysis, and the remaining 30% for testing. They also used a set of 4,000 nonphishing emails obtained from the "sent mails" section of the Enron email database as a different domain to test the generalization capability of the classifiers. The classifiers worked significantly better than the body analysis of [457]. Together, the pattern-matching versions of the Action and Nonsense-detector correctly marked 86.44% of phishing emails with a false positive rate of 4.79% on the Enron inbox emails and 4.17% on the Enron sent emails. The WordNet enhanced versions of these two detectors had 88.56% detection rate on phishing emails and 2.07% false positive rate on Enron inbox emails and 2.42% false positive rate on the Enron sent emails. An analysis of the errors by [454] showed that the main causes of phishing email misses were: spam emails, foreign-language emails, emails containing only links, and emails with no subject, no text, no link and no attachment (a very curious phenomenon). When special methods were employed to detect such data, the detection rate on phishing emails improved by an additional 6.44% with the false positive rate edging up by 0.17% in each case.

Combining NLP techniques with machine learning, researchers in [362] use a 3-layered approach to phishing email detection. A topic model is built using Probabilistic Latent Semantic Analysis in the first layer, then Adaboost and Co-training are used to develop a robust classifier, which achieves an F-score of 1 on the test set, raising the possibility of overfitting the data and some concern regarding missing data preprocessing steps, since the phishing emails include some weird emails as shown by the error analysis of [454]. Of course, machine learning phishing detection methods have to be updated regularly to adapt to new directions taken by phishers, making the maintenance process expensive.

Similar ideas as [454, 457] were used on a larger data set by [176]. They collected approximately 200,000 phishing emails from Purdue University, which were annotated by IT department staff. The legitimate emails (approximately 106,000) are from: Jeb Bush Finance (53%), Listservs on Cytometry (33%), Mozilla (12%), Vision list (4%), and CompSci Colloquium (1%). They also conduct experiments in which they train on a historical data set and then test their models on a more recent data set.

Recently, at the 4th ACM International Workshop on Security and Privacy Analytics [229], the 1st Anti-Phishing Shared Task Pilot was organized, which contained two subtasks. The first subtask was classification of emails based only on the email bodies. The second subtask was classification based on full emails, i.e., body and headers. The data sets and their preprocessing are described in [1]. Both data sets were unbalanced and consisted of approximately one phishing email for every nine legitimate emails. Some synthetic emails were also included in the data sets. Nine teams submitted models and predictions on the test data sets.

The proceedings consisting of the papers from the participating teams, and a paper giving an overview of the shared task and a summary of the performance of the nine teams, can be found at [452]. Interestingly, the top two teams used deep learning models with word-vector representations. Two baselines were implemented: Multinomial Naive Bayes (MNB) and Logistic Regression. The strong performance of the MNB method, which came second on the MCC and Balanced Detection Rate metrics, on the header subtask was quite unexpected. More details can be found in [452]. Robust features for phishing email detection were considered in [136].

8.11.4 Malware Detection

Kolter and Maloof [247] used NLP and machine learning techniques for malware detection. They gathered 1,971 legitimate software samples (hamware) and 1,651 malicious software samples and represented each sample as a vector of n-grams of byte code features. The malware was obtained from VX heavens and from digital forensic experts at MITRE. The hamware, which was in Windows PE format, was collected from all the folders of machines running the Windows 2000 and XP operating systems. Authors do not explain how they can be sure that these machines were free of malware. Additional examples were obtained from SourceForge and download.com. They combine each 4-byte sequence into an n-gram. This idea resulted in more than 255 million distinct n-grams based on their data set.

They used information gain as the feature selection criterion. After selecting the most relevant 500 n-grams for prediction, they evaluated a variety of inductive methods, including naive Bayes, decision trees, support vector machines, and boosting. Boosted decision trees outperformed other methods on their data set with an area under the ROC curve of 0.996.

They conducted three experimental studies. First, a pilot study to determine the size of words and n-grams, and the number of n-grams relevant for prediction. Once these values were determined, the second experiment consisted of applying all of the classification methods to a small collection of executables. The third then involved applying the methodology to a larger collection of executables, mainly to investigate how the approach scales.

Their results suggested that their methods could scale to larger software collections. Unfortunately, however, they do not report training and testing times for their methods. They also evaluated how well their methods classified executables based on the function of their payload, such as opening a backdoor or mass-mailing. Areas under the ROC curve for detecting payload function were in the neighborhood of 0.9, which were smaller than those for the detection task. However, they attributed this drop in performance to fewer training examples and to the challenge of obtaining properly labeled examples, rather than to a failure of the methodology or to some inherent difficulty of the classification task. Finally, they applied detectors to 291 malicious executables discovered after they gathered the original data set, and boosted decision trees achieved a true-positive rate of 0.98 for a desired false-positive rate of 0.05. According to the researchers, this result is particularly important, for it suggests that their methodology could be used as the basis for an operational system for detecting previously undiscovered malicious software.

We observe that their data set was relatively balanced. We expect malware to be a smaller set compared to the much larger universe of useful software in the wild. Another aspect that needs to be considered is how their methods would stand up to adversarial inputs.

8.11.5 Attack Generation

The generation of natural language (NLG) is an important area of research. There are many uses for this in cybersecurity, the primary use being either to generate training data such as "benign" and spam email, or "benign" user activity. The survey [165] provides a discussion of many of the current methods available.

NLG techniques were used by researchers in semi-automatic generation of email masquerade attacks [27]. The goal of their research is to improve defenses against phishing attacks by gathering insights into how humans analyze emails, and by generating challenging data sets for evaluating detectors. They used recursive transition networks (RTNs) and the DADA engine to generate emails that followed the styles of Hillary Clinton and Sarah Palin. A human study was conducted to find out whether the attack emails generated were effective or not. They found that participants could not reliably distinguish between real emails written by Hillary Clinton and Sarah Palin and emails generated using RTNs.

In [382], researchers presented an RNN-based model for generating spear phishing attacks on Twitter. Current RNN-based text generation suffers from two problems: incoherence and repetition. The character limit on tweets ensures that tweets are neither too repetitive, nor too incoherent. RNNs were also for generating fake online reviews in [511]. Here some postprocessing was used to solve the issues with RNN-based text generation. RNNs have also been used to learn the structure of malicious URLs and then generate new URLs that preserve this structure [26].

Another application is to provide users with an automatically generated description of an application and its security, in a way that the user can understand and take appropriate action. The paper [498] describes some work towards this goal. In general, the problem of generating well formatted, useful and accurate text from other sources is an important one with considerable ongoing research. A literature analysis of how artificial intelligence can be used in crime appears in [240].

Recently, the use of deep learning for language tasks has exploded. Most NLP conferences are now dominated by this line of research. The interested reader is encouraged to check out the NLP conferences.

Chapter 9

Big Data Techniques and Security

As more and more sensors and software monitoring tools are deployed by companies and storage costs decline concomitantly, the amount of data collected and stored is growing rapidly. It is estimated that humans are creating over 2 quintillion bytes of data, according to an IBM report.[1] As data increases exponentially, new techniques are required to handle these vast volumes of data. This is the realm of big data. Fraud detection is one of the major use cases for big data analytics [68]. Phone and credit card companies have been doing automatic fraud detection for a long time on large data sets. However, they used custom infrastructures for fraud detection, which were not adopted outside the settings for which they were originally designed for several reasons. Big data technologies are enabling economic infrastructures for cybersecurity. Examples include the WINE platform [129] and BotCloud [157], which uses MapReduce.

The goal of this chapter is to give an overview of the challenges and tools for handling big data, and to discuss the security challenges that arise in the process of using those tools as well as some of security applications that are enabled with their use. There are many books on big data, both from the infrastructure perspective and the analytics perspective. The survey article [235] describes at a high level many of the issues and methodologies in big data analytics. See also [78].

9.1 Key terms

There is still an ongoing debate on when a data set is referred to as big data, although [415] defines it as consisting of "extensive data sets ... that require a scalable architecture for efficient storage, manipulation and analysis." We can confidently assert that if your data set fits in the RAM memory of a single machine, and can be processed and analyzed in-memory, then it is not big data. Of course this is a moving target, since what does not fit in memory today may well fit tomorrow. The NIST report also emphasizes this, making a distinction between vertical scaling, which is making a machine faster and better, and horizontal scaling, where we add more machines to the mix (in other words, a cluster). The NIST report refers to this as the main component of the big data paradigm. However, we believe that the main determinant is whether the amount of data itself is pushing the scaling, not really whether the scaling is horizontal or vertical. For example, if all you are

[1] `https://www.livevault.com/2-5-quintillion-bytes-of-data-are-created-every-day/` - Accessed 23 July 2018.

doing is using multiple machines to speed up a complex algorithm on a small data set, it does not qualify as big data, even though you are using a cluster.

Big data is characterized by the three Vs: volume, velocity and variety. Volume refers to the size of the data. Velocity refers to the rate at which it is being generated and stored. Variety refers to the heterogeneity of the data. To this list of three Vs, two more are usually added: veracity and value. Veracity refers to the reliability of the data and value refers to the importance of the data. In [494], researchers recommend their nine V perspective: with the three Vs of volume, velocity, and variety coming from the data domain; three Vs of visibility, verdict and value coming from the Business Intelligence domain; and three Vs of validity, veracity, and variability are inspired from statistics.[2] Variability refers to both data complexity and variation, which is the change in data over time. Complexity refers to how many different types of attributes exist for the data instances. Changes include, for example, a change in the structure, the flow rate, or a wide drift in values of the attributes. Visibility refers to the insights provided by the data and the metadata. Verdict refers to the kinds of decisions that need to be made based on the data. They also claim that big data = machine learning + cloud computing.

When we observe the data on the world wide web, we observe that it is mostly unstructured data of different types: text, audio, images, and video. It is estimated that structured data in the form of relational databases or tables is only a small subset of the total data available [163]. In between structured and unstructured data is semistructured data, e.g., XML files.

The high volume and velocity of data means that individual machines are inadequate for the tasks of data handling and processing, and clusters of computers must be employed. Clustering software is used to transform the cluster of machines into a pooled resource that acts like one large machine with easy scalability and high availability.[3] A cluster should be easily scalable by adding more machines as needed without having to change the characteristics of the machines that are already in the cluster. With many machines in the cluster, even if each individual machine is highly reliable, there are bound to be failures. Therefore, a cluster should easily tolerate failures of individual machines or storage components and provide high throughput.

The data processing cycle of big data is similar to that of the data mining algorithms with key additional considerations in the data ingestion, which is a significant challenge because of the high volume and velocity of the data.

9.2 Ingesting the Data

Data ingestion is the process of adding raw data to the system, which was always present, but which is far more complex with big data. This can be a fairly complex operation, especially if the data is far from the desired format for processing and analysis. The quality of the data also plays a key role in the data ingestion process. Additional considerations with big data include: how to distribute the data when horizontal scaling is essential, and how to change the algorithm so that it can deal with this data distribution.

Many technologies and frameworks are now available to facilitate the data ingestion process. For example, Apache Sqoop can take data from relational databases into a big data system; Apache Flume and Apache Chukwa are good for aggregation and importing of

[2] They also claim that big data = machine learning + cloud computing, but we feel that this is an oversimplification, since most of the data in big data is unstructured, think text, images, audio and video clips, and not ready for analytics.

[3] `https://www.digitalocean.com/community/tutorials/an-introduction-to-big-data-concepts-and-terminology`

application and server logs; and the Gobblin framework can ease aggregation and normalization of the output of these tools towards the end of the ingestion process.

Of course, data cleaning, consistency checking and correction of errors are necessary steps, regardless of whether it is a big data system or not.

9.3 Persistent Storage

After the data is ingested, it is given to a persistent storage manager. The persistent storage manager must be able to handle the high volume and velocity of data and also ensure that it is made available in large volumes quickly and reliably. The reliability aspect is extremely important, since in a big data system, there are many processing and storage components, and the more the components in a system, the higher the chance of some component or the other failing per unit time.

Broadly speaking, there are two approaches to reliability: replication and encoding. Replication involves making copies of sensors, data, or, in general, any component or unit of a system that can fail and whose correct functioning is critical to performance and/or safety. NASA is famous for its triple modular redundancy, which means using three units (e.g., sensors), instead of one, and taking "majority" vote of their outputs. Encoding refers to redundancy in the form of error-correcting codes. For example, computing parity bits on a binary file can help to detect and correct a certain number of errors. Encoding can be more efficient than replication; the trade-off is that replication is applicable more generally, whereas encoding applies to more limited situations such as data.

A distributed file system is typically used in a big data system and frameworks include: Apache's Hadoop Distributed File System (HDFS), Ceph and GlusterFS. For structured data access, one can use distributed databases, e.g., NoSQL databases. Some of these databases can deal with heterogeneous data, and they are designed with the same fault tolerant considerations.

For reliability, HDFS used only replication, original plus two copies, in the beginning, but with Hadoop Version 3, encoding was also introduced.

A word about NoSQL databases is in order here. The relational data model was, and still remains, a great way to store data, while providing the well-known ACID properties: atomicity, consistency, isolation and durability. However, it has two issues: impedance-mismatch [373] and single-machine design.

Impedance-mismatch refers to the differences between the relational model and in-memory data structures. The relational model is based on the algebra of relations. Each relation is a set of tuples, where each tuple is a set of name-value pairs. The values in these tuples need to be simple, they cannot contain structures such as nested-records or lists. In-memory data structures can be nested and complex. This mismatch leads to a lot of time spent on, and frustration in, coding the mapping or transformation between the relational database and the in-memory data structures.

With cluster computing becoming the norm in handling big data, the relational database model suffered another blow. The term NoSQL database was coined for the more flexible, so-called "aggregate" data models that have become popular with cluster-computing. However, this term hides a lot of different data models, which are all collectively referred to as NoSQL databases. The main ones are: Key-Value Stores (e.g., Riak), Document Databases (e.g., MongoDB), Column-oriented Databases (e.g., Cassandra), and Graph Databases (e.g., Neo4).[4] They all have different characteristics and functionality and serve different application scenarios.

[4]The Graph Databases are an attempt to bring back the relationships that are lost in the aggregate data models.

A key challenge in any big data application is that at some point the solution architects have to decide the mechanisms and formats to use to store the data (e.g., flat data file, semi-structured files, NoSQL databases, or even a more conventional relational database), and this decision will to some extent influence the type of analysis and operations that can be executed efficiently [162].

We close this section with the CAP theorem, which was conjectured by Eric Brewer in the context of web services [60] and proved in [168]. It states that no distributed database can simultaneously provide all three properties of: atomic consistency (linearizability), availability (every request received by a nonfailing node will elicit a response) and tolerance to partition (i.e., the service is still available even if some nodes or links go down). Note that atomic consistency is a stronger requirement than the consistency property in ACID list above. It can be considered as roughly equivalent to consistency and isolation.

9.4 Computing and Analyzing

The computation layer is typically quite diverse since the requirements and approaches depend on the application and the analyses to be conducted. Data may be partitioned across nodes, each chunk of data may be processed in parallel, and the whole process may be iterated many times using a variety of tools.

These steps are referred to individually as splitting or partitioning, mapping, shuffling, reducing, and assembling. This is the workflow in Apache's MapReduce. The computation may be done in batch mode or real-time depending on the requirements of the application. Real-time processing is also called stream processing. Tricks are often used in real-time processing to reduce disk I/O as much as possible.

Apache Storm, Apache Flink, and Apache Spark are different computational frameworks for real-time or near real-time processing. However, there are many other tools that can be plugged into these frameworks. For example, Apache Hive is a data warehouse interface for Hadoop, Apache Pig is a high level query interface, while SQL-like data manipulations can be performed with tools such as Apache Drill, Apache Impala, Apache Spark SQL, and Presto. For machine learning, tools such as SystemML, Apache Mahout, and Apache Spark's MLlib are available. For analytics programming, both R and Python are popular choices.

The MapReduce framework had its beginnings in functional programming languages. Backus' Turing Award paper proposed the language called FP [25], which stands for Functional Programming Language and Environment, in which there was an apply-to-all operator and a reduce operator. The FP language and framework was proposed as a useful tool to not just develop, but also reason about, purely functional programs. These programs do not contain any assignment statement, functions are first-class objects, i.e., they can be passed freely as parameters and returned as results of other functions, and functions are side-effect free. The semantics is a reduction semantics as opposed to a state transition semantics, which is typical of imperative languages.[5]

Below is an example FP "program," basically a set of function definitions, for cartesian product of two sets `cart2`, represented as sequences. The auxiliary function un is for the union of two sets, both represented as a sequence. Since sequence is the only composite data structure, the sets are input to the program in a sequence. So we get the familiar problem of lots of irritating silly angular brackets like so: $\langle\langle 1,2,3\rangle,\langle 4,5,6\rangle\rangle$ and much of the cartesian product of n sets function is devoted to "flattening" a sequence of sequences.

[5] A reduction semantics specifies transformations on expressions. In a state transition semantics, there is the concept of a state, which abstracts a machine, and each command, definition and programming language construct is interpreted as a transition between states.

To understand the function definitions, we need to understand that the @ symbol denotes function composition, the ! denotes the insert-left functional of FP and binds tightly so that !f@g is the same as (!f)@g. The ampersand denotes the apply-to-all functional form. The functions apndl and apndr, append from the left and right respectively, so that *apndr* : $\langle\langle 1,2,3\rangle,\langle 4,5,6\rangle\rangle = \langle 1,2,3,\langle 4,5,6\rangle\rangle$ and *apndl* : $\langle\langle 1,2,3\rangle,\langle 4,5,6\rangle\rangle = \langle\langle 1,2,3\rangle,4,5,6\rangle$.

```
{un !apndl@apndr}
{cart2 !un@&distl@distr}
```

FP never really took off in the business world, perhaps because of its quirky syntax and one composite data structure, the sequence, but Google researchers realized that some of the operators that were part of the FP paradigm were more generally useful in parallel and distributed computational settings, and thus MapReduce was born. In the MapReduce framework, there are usually several iterations of map and reduce tasks. In each iteration, the Map function distributes work to different nodes and the reduce function aggregates the results.

Map and reduce operations work on $(key, value)$ pairs. The map operation works on every input instance, which is a key-value pair, and outputs an arbitrary number of key-value pairs. It is similar to FP's apply-to-all functional, which takes an operator as one argument, in Backus' notation $\alpha\ f$, where f is a unary function from numbers to numbers, say. For example: $\alpha\ sq$, where sq is the squaring function on numbers, can be applied to the sequence $\langle 1,2,3\rangle$ to yield the sequence $\langle 1,4,9\rangle$ as output.

The reduce operation applies an "aggregating" operation, hence the term reduce, to all the values corresponding to the same intermediate key and outputs key/value pairs. It is similar to FP's insert (notation '/' or '!') functional that takes a binary operator as argument. For example, $/+ : \langle 1,2,3,4\rangle = 10$ and $/* : \langle 1,2,3,4\rangle = 24$. The insert functional is recursive, i.e., $/+ : \langle 1,2,3,4\rangle$ can be understood as $+ : \langle 1, /+ : \langle 2,3,4\rangle\rangle$.

The difference from MapReduce is that FP does not have a *direct* concept of key-value pair.[6] Together with a distributed file system, all other aspects of execution, such as fault-tolerance and scheduling, are handled by the run-time system in a transparent manner for convenient cluster computing. The MapReduce framework hides tedious details of load balancing, data exchange and synchronization from the programmer.

We now give a short introduction to Apache Spark and the programming language Scala. Apache Spark is a general-purpose cluster computing environment. Interactions with earlier versions of Spark took place via what is known as a SparkContext *sc*. For example, the command[7]

```
./bin/spark-shell --master local[3]
```

returns a SparkContext that we can use for inputting data, running programs, etc. In the above command, the –master option is very important. With just one parameter, one can specify the resources to be used: the master URL for a distributed cluster, or local to run locally with one thread, or local[N] to run locally with N threads. Spark uses a directed acyclic graph structure for data, called RDD, for resilient distributed data. All commands take an RDD graph and return another RDD graph.

We continue with our example from the same source. The following command reads a file and then applies a filter operation to it. Spark allows anonymous functions in the style of lambda calculus and functional programming. For example, in the filter command below,

---

[6]Of course, FP is a Turing-complete programming languages, so all computable functions are expressible, but that does not imply that the simulation would be convenient or efficient.

[7]`https://spark.apache.org/docs/2.3.1/`

Table 9.1: Some Frameworks for Big Data Analytics

Framework	Summary
Lucene	Lucene is a scalable library for information retrieval from text data
Nutch	Nutch is an open-source search engine file system
Sqoop	Tool for bulk data transfer between Hadoop and structured datastores, e.g., relational databases
Cloudera	Software platform for data management and processing
Hortonworks	Similar set of tools as Cloudera

we check for each line in the file, whether it contains the string "Spark," but this function has no name. This filter operation transforms the data set. We can perform further actions on this transformed data. Spark freely allows chaining of data transformations and actions, as the last command shows.

```
scala> val textFile = spark.read.textFile("README.md")
textFile: org.apache.spark.sql.Dataset[String] = [value: string]

scala> val linesWithSpark = textFile.filter(line => line.contains("Spark"))
linesWithSpark: org.apache.spark.sql.Dataset[String] = [value: string]

scala> textFile.filter(line => line.contains("Spark")).count()
// How many lines contain "Spark"?
res3: Long = 15
```

Spark has several components, which increase its expressive power and flexibility. For example, there is Spark Streaming, for real-time processing of streaming data. There is Spark SQL, which combines relational processing and functional programming. Data can be queried via SQL queries or the Hive Query Language. There is GraphX, the Spark API for manipulating graphs. There is MLlib, for machine learning. In earlier versions of Spark, one needed a different context for each Spark API. Recently, Spark has introduced SparkSession to replace SparkContext and give a uniform interface for Spark APIs. For more details, the reader can refer to Apache Spark documentation online or one of many Spark Tutorials online.[8]

We summarize some frameworks for big data analytics in Table 9.1, which are not mentioned above. Besides Spark, there are a number of data processing engines, e.g., Microsoft Dryad, Storm, Tez, Flink, and CIEL, capable of supporting MapReduce-like processing [494].

9.5 Techniques for Handling Big Data

Finding similar objects is a fundamental problem in data mining, and the volume of data in big data applications necessitates fast solutions. We now briefly discuss two very important techniques for this problem: min hashing and locality sensitive hashing [269].

These ideas can be adapted for many situations, but it is convenient to start with text documents. Suppose we wish to find similar documents in a corpus. By taking n-grams of characters (also called shingles), we can convert the documents to sets. How big should n be? It should not be too small, nor too big. Its value depends on the length of typical

[8] https://www.edureka.co/blog/spark-tutorial/#What_Is_Apache_Spark

documents in the corpus and the size of the character set. The main consideration in picking the value of n is that the probability of any given n-gram occurring in any given document is low.

Now finding similar sets amounts to computing their intersections or their Jaccard similarity. However, the sets of n-grams we obtain can be quite large. To reduce the complexity of computing the intersection, we wish to compute similarity-preserving signatures for the sets. It is helpful to visualize sets as characteristic vectors on a common vocabulary of n-grams and a collection of sets becomes a characteristic matrix. Each set is now a column of this matrix.

Now, we take a permutation of the rows and define the minhash value of each set as the first row in which the column representing the set has a 1. The nice property of this minhash value is that the probability of two sets having the same minhash value under a random permutation equals their Jaccard similarity. This seems like magic, but see [269] for an explanation of this property. In practice, we simulate a random permutation by a random hash function that maps row numbers to buckets, where the number of buckets equals the number of rows.

In many situations, we do not want to compute pairwise similarity of all sets. Rather, we wish to find the most similar pairs or all pairs whose similarity exceeds a certain threshold. In such situations, locality-sensitive hashing is more efficient than minhashing. The idea is to hash the objects many times, so that similar items are hashed to the same bucket with higher probability than dissimilar items. Items that are hashed to the same bucket are then checked for similarity. We can construct such a set of hash functions by dividing the minhash signature matrix for the sets into b blocks of r rows each, and then independently hashing the r rows into r buckets using a good hash function.

In principle, pairwise similarity of documents can be computed using the MapReduce framework. With some additional pruning tricks, including stopword elimination and document frequency pruning, which gets rid of highly frequent terms, it can also be made efficient. For example, in [138], the authors use the inverted index data structure and MapReduce framework to compute pairwise symmetric similarity function of 900,000 documents on a 20-node cluster, where each node contains two single-core processors (2.4/2.8 GHz), 4 GB RAM and 100 GB disk, in roughly 130 minutes, or a little over 2 hours. The computation is arranged elegantly in two map-shuffle-reduce cycles. The first cycle maps each document d_i to $(term, (d_i, weight(term, d_i)))$, where the term is the key and the value is a pair. The shuffle operation groups the values by the same key or term. The reduce operation multiplies the weights for the same term in document pairs d_i, d_j, which become the keys for the next cycle. The values are the products of the weights. The shuffle operation then groups the products corresponding to the same key, which is a document pair. The reduce operation in the second cycle then sums these products, in other words, computing the inner product similarity.

MapReduce and Data Mining/Machine Learning Algorithms. Algorithms have been developed for many of the data mining and machine algorithms we discussed in Chapters 5 and 6. For example, K-means, EM clustering and CLIQUE, Canopy clustering, hierarchical, density-based and co-clustering have all been parallelized, see [388] and references cited therein. There are also parallel algorithms for frequent itemset mining, graph mining, tree ensembles, probabilistic latent semantic indexing, LDA and Hidden Markov model [388].

Scalable Algorithms for Graph Mining. The problem of finding patterns in large graphs has many applications, e.g., cybersecurity, social network analysis and fraud detection, to name a few of them. The basic primitives for mining of large graphs include: storage management and indexing, structure analysis, eigensolver, and graph layout/compression. Structure analysis includes such problems as computing the diameter of the graph (maximum of the shortest path lengths between any two nodes), finding connected components, and shortest paths between any two pairs of nodes. An eigensolver can rapidly find for us the first few eigenvalues and eigenvectors of the adjacency matrix, which are needed for solving problems such as finding the near-cliques, the number of triangles, etc. Graph layout and compression are needed for large, real-world graphs. At first glance, it seems that finding clique-like structures and compressing them is a good idea. However, it has been shown that this approach is the "wrong paradigm" since "real-world graphs have no good cuts," see [143] and references cited therein for pointers to graph mining algorithms. Kang and Faloutsos [228] have introduced the SLASHBURN technique, which gives better compression, faster execution times and avoids the "no good cuts" problem.

MapReduce for Text Mining. There are several works on text clustering with MapReduce. The field is ripe for a comparison of the various algorithms. Nagwani [325] proposed summarizing text collection using topic modeling and clustering based on MapReduce. Lin and Dyer [276] wrote a monograph that includes several primitives for mining of large text collections using MapReduce. These primitives include: inverted index, EM and EM-like algorithms, and algorithms for graphs derived from text data.

Limitations of MapReduce. Problems requiring a shared global state such as online learning and Monte Carlo simulations are challenging for the MapReduce model [276]. In online learning, the parameters of the model represent a shared global state and the algorithm updates these using every training instance. In Monte Carlo simulation, frequency statistics are computed over samples drawn from random variables that simulate the behavior of a model. The need for a shared global state in MapReduce has led to innovations such as Google's Bigtable [75], a distributed persistent map, and Amazon's Dynamo [113], a distributed key-value store. The open source versions of these are HBase and Cassandra, respectively.

9.6 Visualizing

Due to the type of information being processed in big data systems, recognizing trends or changes in data over time is often more important than the values themselves. Visualizing data is one of the most useful ways to spot trends and make sense of a large number of data points.

Fundamentally, the only way to visualize big data is to visualize a reduced view of the data. This can be accomplished by viewing a window on the data, such as all records that match a given query, or by visualizing a statistic computed on the data. The first approach is really not big data visualization – it is simply data visualization on a subset of the data. While there are computational issues on how one defines and extracts this subset, the visualization is no different than for smaller data sets.

The statistics one visualizes tend to be of two types. One may visualize a statistic related to a desired inference, such as the number of packets per second on a network to investigate trends in bandwidth usage, or the number of clients connecting to servers to look for denial-of-service attacks, or the average temperature at each geographic point to investigate global warming. In order to get a better idea of the data set as a whole, the statistic may be an aggregation of the data into clusters, followed by a visualization of the clusters.

Figure 9.1 illustrates one way to visualize big data. Here we are looking at 24 hours of flows on the Los Alamos network. Plotted are the kernel estimators of the log of the number of bytes, with the top curve corresponding to hour 0 and the bottom to hour 23 (11 pm). This type of *waterfall* plot can be computed every hour, with a new curve added at the bottom, and all the other curves moving up. This illustrates the method of plotting summary statistics of the data, in a manner that allows comparison across one of the variables, in this case time.

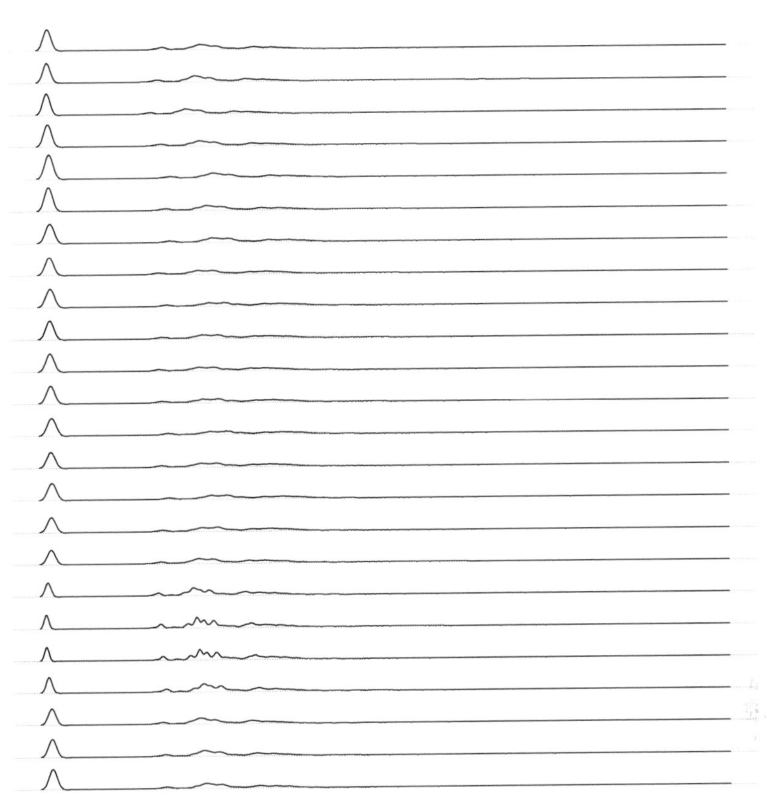

Figure 9.1: Waterfall depiction of the log of the number of bytes over 24 hours on day 3 of the LANL flows data.

Real-time processing is frequently used to visualize application and server metrics. The data changes frequently and large deltas in the metrics typically indicate significant impacts on the health of the systems or organization. In these cases, projects like Prometheus[9] can be useful for processing the data streams as a time-series database and visualizing that information.

One popular way of visualizing data is with the Elastic Stack, formerly known as the ELK Stack. Composed of Logstash for data collection, Elasticsearch for indexing data, and Kibana for visualization, the Elastic Stack can be used with big data systems to visually interface with the results of calculations or raw metrics. A similar stack can be achieved using Apache Solr for indexing and a Kibana fork called Banana for visualization. The stack created by these is called Silk.

Another visualization technology typically used for interactive data science work is a data "notebook." These projects allow for interactive exploration and visualization of the data in a format conducive to sharing, presenting, or collaborating. Popular examples of this type of visualization interface are Jupyter Notebook and Apache Zeppelin.

We should also mention ORA [69], which is a toolkit for analyzing and visualizing network data. The networks can be social networks, multimode networks, or networks on maps. ORA has been designed for dynamic and high-dimensional networks. It can import data from a variety of formats, e.g., CSV, TSV, UCINET, or Twitter JSON, and can export to Google Earth, or KML files. It also tries to reduce user training time and effort by

[9] `https://prometheus.io/`

automating common workflows. The batch mode of ORA has been used for networks with a million nodes.

For scientific data visualization, there is VisIt,[10] which is an "open source, interactive, scalable, visualization, animation and analysis tool. Users can interactively visualize and analyze data ranging in scale from small (< 10 core) desktop-sized projects to large ($> 100,000$ core) leadership-class computing facility simulation campaigns. Users can quickly generate visualizations, animate them through time, manipulate them with a variety of operators and mathematical expressions, and save the resulting images and animations for presentations." Wu et al. [497] have designed VisIt-OSPRay, a high-performance, hybrid-parallel rendering system in VisIt, using OSPRay (a CPU ray tracing framework), IceT (a distributed computing framework built on Java), and PIDX for parallel I/O.

For visualization tools specific to cybersecurity data, we refer the reader to the paper by Zhang et al. [521]. In this paper, they present a tool for visualizing traffic causality, which helps in anomaly detection. There is also the VizSec workshop proceedings for digging deeper.

9.7 Streaming Data

Streaming data was introduced in Section 6.12, in which a streaming version of the kernel estimator was given. Streaming data occurs in many cybersecurity applications, for example when one is monitoring the packets on a high bandwidth network, the activities of many users such as in a social network, or in video analysis.

The method described above utilized the recursive version of the mean calculation, which allows for streaming kernel estimators (as discussed), streaming mixture models ([433]), and many other similar methods. These methods typically implement an exponential window on the data, rather than computing the exact values that would have been computed through traditional methods with access to all the data, although the mean calculation itself can be performed exactly.

The basic idea of many streaming algorithms is to compare a new observation to existing models, determine whether this is an outlier or fits the current model. If an outlier, the algorithm must decide what to do; should it report the outlier, drop the outlier from consideration, expand the model to encompass the outlier, or temporarily retain the outlier for future processing? If the observation is not an outlier, the model provides an assessment of the observation (is it spam or ham, for example) and then the algorithm must decide whether to use the observation to update the model, or simply report the analysis and proceed to the next observation. For example, [506] utilizes a streaming kernel estimator to determine whether false data has been injected into a smart grid. This is similar to the method of [417] which uses a sliding window on the data to produce a local (in time) kernel estimator.

A nice discussion of the state-of-the-art in streaming clustering is provided in [391]. The basic idea is to maintain a representation of the clusters which allows new observations to be absorbed into the cluster, modifying the parameters, and allowing for the formation of new clusters (from outliers) merging of clusters into a single cluster, and splitting of clusters that have become too dispersed. Some algorithms retain various "hypotheses" about new potential clusters, or several different clusterings to use for comparison purposes, and to hedge against drift caused by temporary changes or adversarial attacks. Some are purely streaming, some utilize sliding windows (sometimes more than one size window, in order to have different granularity of time), while others retain a set of recent observations for off-line processing.

[10] `http://users.sdsc.edu/~amit/scivis-tutorial/`

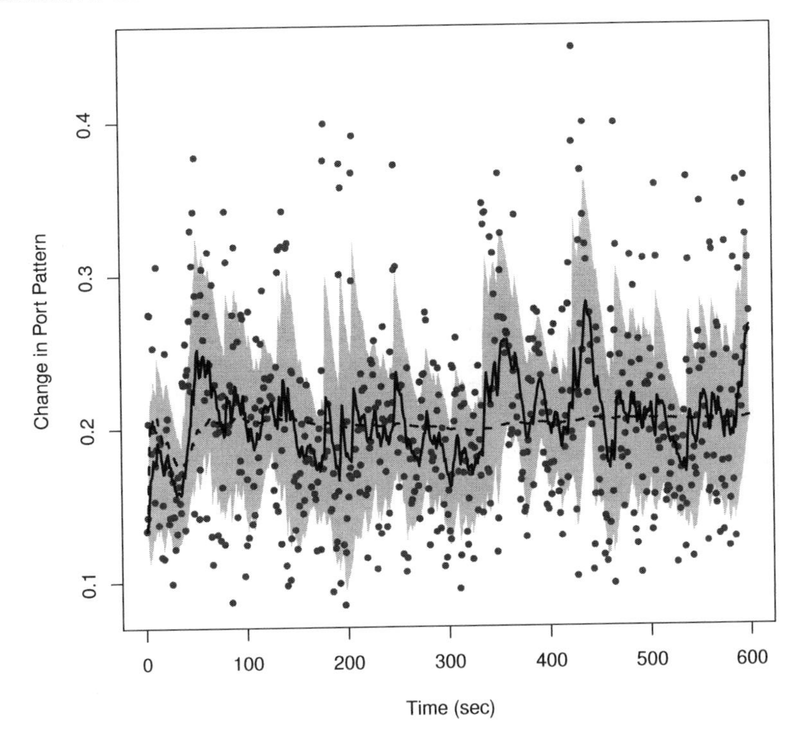

Figure 9.2: Streaming mean and variance calculations computed on the LANL flows data. The x-axis corresponds to time in seconds. On the y-axis the proportion of connections to each "common" port during that second is compared to the proportions from the previous second using the sum of the absolute differences. The points correspond to these values. The solid curve is the streaming, exponentially weighted sample mean, the dashed curve is the true mean of all the data to date, and the gray region corresponds to the region of one standard deviation around the streaming average, computed in the same manner.

Using a recursive method like an exponentially weighted average, as was discussed in Section 6.12 suggests a problem: one has to choose the weight, but this is equivalent to defining how far back one is looking in the data; larger values allow for detecting events that are "low and slow", while smaller values allow for fast response but will miss these slower attacks. One could utilize several methods, each with its own value of the parameter. These methods that utilize one or more statistics, computed in a streaming manner, with a threshold to detect change points, fall under the broad topic of control charts. One utilizes the changes to detect deviations from normal (change points) and tries to then "control" the generating process (say, in a factory) to bring the system back down in the benign range. The same ideas are used here to detect changes indicative of attacks.

Figure 9.2 depicts a streaming mean and variance calculation, using Equation (6.13) with $\theta = 0.9$. A similar formula is used to construct the variance. For each second, the proportion of times a connection is made to each of the 241 "common" TCP ports is computed, and this is compared to the vector for the previous second using the ℓ_1 norm: sum of the absolute differences. The points correspond to these values; the dashed curve corresponds to the true mean of these values computed on all the data to date; the dashed curve and the gray region correspond to the streaming mean and one standard deviation from this mean, computed from the streaming variance.

As can be seen, the streaming mean is a local estimate of the mean, and tracks the short-term trends and changes in the data. This method can be used to estimate local statistics for the purposes of investigating outliers. It requires only the previous value of the statistics to be retained, and is updated using each new observation. Although this example is univariate, the recursive equations extend easily to multivariate data.

In [216] a hybrid method is discussed which implements a streaming change point detection algorithm with a spectral method (Fourier transform or Wavelet) to detect periodicities and handle seasonality. This method allows them to set a threshold that can detect the smaller, subtle changes, while utilizing the spectral analyzer to filter the false alarms. The spectral analyzer can also be used to detect certain types of attacks that often have a periodic nature; they describe an SSH dictionary attack, in which large numbers of authorization requests are sent, embedded in benign traffic, in a periodic manner that can be detected via the spectral approach – it shows up as high energy in the appropriate frequency bin.

The methods discussed in [216] also apply to dynamic graphs, such as those that occur in network traffic, where the graph corresponds to clients and servers (vertices) and connections/flows (edges) that occur in time. Features (graph invariants) are computed and control chart methods then apply to detect changes. The authors illustrate this using a random graph model in which the vertices are grouped into communities, and the method is used to detect changes in this community structure.

Stream Processing Systems. Early systems for stream processing, e.g., TinyDB, STREAM, and TelegraphCQ, were built by database researchers, see [141] and references cited therein. More recent systems, e.g., Storm, Spark Streaming and Flink, have been influenced by big data frameworks such as Hadoop and attempt to combine low-latency with scalability. Spark, Flink and Google Cloud Dataflow are examples of systems that unify batch and streaming models in a single tightly coupled architecture. However, they take different approaches to unification. Batch mode is the primary model of Spark and stream is implemented via micro-batches. Flink and Google Cloud Dataflow have stream mode as the primary model and batch, which is essentially a bounded stream, is treated as a special case of streaming.

9.8 Big Data Security

Many organizations are unable to handle the storage and processing of huge volumes of data they create. For example, in a single day, a large hospital with an active emergency department may create hundreds of test results including MRI scans, X-ray images, CAT scans, EEG and ECG results. Creating in-house infrastructure and expertise for handling such volumes of data may be too expensive and also distracting from the main mission of the organization. So the organization may turn to outsourcing this gargantuan job, which usually means a cloud service provider.

Cloud computing makes some key promises of elastic computing, pay-as-you-use, and economies of scale. However, cloud computing brings up a fresh new set of challenges: security and privacy, apart from loss of control, vendor lock-in, reliability of the provider and downtime. Moreover, the cloud computing paradigm rests on "public" infrastructures composed of shared, heterogeneous hardware, operating systems and analytical software. Most security solutions have been designed for semi-isolated, small-scale systems [416]. The growth in streaming technologies has created the need for rapid development and deployment of security and privacy solutions designed for cloud computing infrastructures.

The following list, adapted from [416], captures the essence of new requirements created by big data paradigm.

- Big data is collected from more diverse sources. The sources of big data typically include: sensors including mobile devices and social networks, beyond the usual sources. The diversity of sources increases the combinations of attack vectors and multiplies the attack surfaces.
- Data search and selection may lead to security or privacy policy issues when the results of these steps are combined. The volume of data that is searched and filtered may lead to inadvertent re-identification.
- Privacy-preserving mechanisms may not be sufficient for the big data pipeline. A lot of research in privacy-preserving data mining and machine learning considers only individual steps. The big data processing pipeline may consist of a huge number of steps that may be individually privacy preserving, but their net effect may not preserve privacy.
- Big data has bigger potential to create targets that are more valuable.[11] Attackers will scale up their attacks, since the targets themselves are more valuable. Hence, defensive techniques will need scaling as well.
- Information assurance and disaster recovery needs for big data systems. The volume of big data and the scale of the processing architectures have not been studied in the past. Traditional methods of backing up data, testing, verification and provenance will be severely put to the test and will likely be found wanting.
- Consent traceability for identifiable information. The path that data takes through big data computing infrastructures and cloud-computing provider chains may not properly preserve the consent of the data owners, or lack of it.
- Emerging risks in open data and analytics. As analysts and researchers leverage third-party big data, there could be problems of data identification, metadata tagging, data aggregation, data linkage, and data poisoning.

9.8.1 Implications of Big Data Characteristics on Security and Privacy

Big data characteristics such as volume, variety, veracity, velocity and validity have important implications for security and privacy.

- Volume implies either distributed computation or cloud computing, and often multi-tiered storage media. Distributed computing frameworks are not designed for security or privacy. For example, securing the MapReduce framework requires securing the mappers and reducer, and protecting the data in the presence of untrusted mappers or reducer. The threat model for network-based, distributed, multitier systems includes: confidentiality, integrity, availability, consistency, and provenance. Attacks to consider include: collusion, roll-back and recordkeeping disputes [13, 416]. Securing the data and transaction logs in an auto-tiering environment is challenging. How to maintain availability in the presence of unauthorized accesses will be very important. The sheer amount of data means that data integrity attacks are much harder to find (needle in haystack problem) and also are time-consuming. Finally, volume implies that scalable and granular algorithms are needed for access control and confidentiality. Attribute-based encryption seems promising from this regard, but scalability is still an issue.

[11] Of course, the value depends on the significance of the data.

- Variety implies heterogeneity of data and computing resources. NoSQL databases are still not sufficiently mature in terms of security. Robust solutions to NoSQL injection attacks are needed [13]. With so many different hardware platforms, operating systems and applications, there are a lot more moving parts and vulnerabilities.

- Veracity implies mechanisms for data provenance have to be traceable and applicable across the big data processing pipeline, since attackers can plant sensors and devices, or poison the data.

- Velocity means streaming and we have already touched upon the problem with streaming data above.

- Validity means logical consistency of the data. This requires end-point input validation and filtering as well as scalable algorithms for dealing with missing/corrupted data.

9.8.2 Mechanisms for Big Data Security Goals

We now examine the mechanisms available for big data security goals of data confidentiality, provenance, and system health monitoring. In 2016, researchers identified security and privacy issues in big data as an area of concern [413]. According to their analysis of the research literature, the number of papers being published in the field of security and privacy for big data is declining and represents a tiny percentage of the total research output.

1. Data confidentiality: We consider that cloud computing architectures for processing big data will necessitate that, increasingly, computations will be performed on encrypted data. In 2002, Song et al. [398] showed how to perform search over encrypted data with encrypted keywords. Improvements were made by Curtmola et al. in 2006 [102]. Recently, functional encryption [47] and homormorphic encryption [52] have been proposed, which allow more general computations on encrypted data.

   In functional encryption, the decryption key allows the user to learn only a function of the data and nothing else. In a fully homomorphic encryption scheme, we can compute arbitrary functions over encrypted data without the decryption key, i.e., given encryptions $E(m_1)$, ..., $E(m_k)$ of messages m_1, ..., m_k, we can efficiently compute an encryption of $f(m_1, ..., m_k)$ for any efficiently computable function f. Constructing a fully functional encryption scheme is still an open problem. A fully homomorphic scheme is available, but its availability does not seem to be sufficient for this task. Moreover, the scalability of existing solutions is still an open question.

   Key management for big data systems is also more complex because of the growth in cloud provider-consumer relationships, greater demands, and heterogeneity of infrastructures.

2. Provenance: Digital signatures can provide a measure of source authentication and methods for message integrity (e.g., the TLS protocol) can be used for detecting any tampering of the message in transit. End point input validation, mentioned above, needs to be both syntactic and semantic. Because the meaning may depend on the context as well as the contents of the message/data, general solutions for semantic validation are unlikely. Provenance also requires authenticated computations on the data. For this purpose, trusted platform modules or cryptographic mechanisms or both of these techniques can be used. Granular audits are necessary for provenance. Techniques for retention and disposition of data and rights management are also needed. This area still needs a lot more work.

3. System health monitoring: There are two broad aspects of system health monitoring: system availability and system immunity. For system availability, mechanisms are needed that can resist large-scale denial-of-service attacks. For system immunity, we need forensics, abuse detection, security breach detection, and big data analytics on logs, cyber-physical events and intelligent agents. This is still a developing field.

   For big data forensics, Xu et al. [503] propose a data-reduction approach that exploits dependencies among system events to reduce the number of audit log entries. Some solutions have been proposed for the various infrastructural components of a big data system. For example, accountable MapReduce was introduced in [501]. Here, auditor nodes perform accountability tests to identify suspicious nodes. MapReduce has been adapted to run on multiple clusters in [471], the resulting framework is called G-Hadoop. A security framework for G-Hadoop was introduced in [522].

Appendix A

Linear Algebra Basics

Linear algebra is the field of mathematics that studies sets of points in $\mathbb{R}^d$, with operations that allow forming linear combinations of points, and provides tools for the analysis of the geometric structure of point sets.[1] A large proportion of machine learning and statistical methods rely on linear algebra, either fundamentally or as a mechanism for efficient calculation, and so it is important for cybersecurity analysts to have a working knowledge of linear algebra. This appendix provides the basic definitions and ideas that are used in the text. Good references for the material in this chapter are [24, 243, 411].

A.1 Vectors

In computer science, a vector is a one-dimensional array. We write $(x_1, x_2, \ldots, x_n)$ for the vector of n numbers.[2] In mathematics, vectors come with structure. We can add two vectors, provided they are the same length, and multiply a vector by a scalar (a single number):

$$(x_1, \ldots, x_n) + (y_1, \ldots, y_n) = (x_1 + y_1, \ldots, x_n + y_n) \tag{A.1}$$

$$a * (x_1, \ldots, x_n) + (y_1, \ldots, y_n) = (a * x_1, \ldots, a * x_n). \tag{A.2}$$

We will adopt the convention that a vector $x \in \mathbb{R}^d$ will be written in lowercase, and we will refer to the ith element of x by x_i. A (real-valued) *vector space* is a collection of points in $\mathbb{R}^d$ with the above operations of addition and scalar multiplication.

We can define multiplication of vectors in a number of ways. One way is the *dot product*, also called the *inner product*. This is defined as:

$$x \cdot y = x^T y = \sum_{i=1}^{n} x_i y_i. \tag{A.3}$$

The exponent T in the above equation denotes transpose, which will be defined when we define matrices in the next section. For now, the sum is the definition to keep in mind. Using the inner product, we can define the norm, or length, of a vector to be:

$$\|x\|_2 = \sqrt{x \cdot x} = \sqrt{x_1^2 + x_2^2 + \cdots + x_n^2}. \tag{A.4}$$

The subscript 2 on the left-hand side of the formula corresponds to the fact that this is the ℓ_2, or *Euclidean norm*. There are other norms that one could define for vectors, such as

[1] Restricting to real points is not necessary, but in cybersecurity nearly all applications of linear algebra are applied to real valued (or integer valued) points.

[2] We will only be considering numeric vectors in this appendix.

the ℓ_1 norm, which is the sum of the absolute values, but we will only consider the ℓ_2 norm here. Hence, from now on, we will drop the subscript. With this, the following formula is easy to derive:

$$x \cdot y = \|x\| \|y\| \cos(\angle xy) \tag{A.5}$$

where $\angle xy$ denotes the angle between the vectors x and y. A *unit vector* is a vector of norm 1.

In $\mathbb{R}^d$ the familiar axes can be written in vector notation, as $e_i = (0, \dots, 0, 1, 0, \dots, 0)$, where the one is in position i. Any real-valued vector can be written as a linear combination of the e_i:

$$v = (v_1, \dots, v_n) = v_1 e_1 + \dots v_n e_n.$$

We say the $\{e_i\}$ *span* the vector space $\mathbb{R}^d$. Furthermore, we cannot reduce this set by removing one or more of the e_i and still span the space, and we cannot write any of the e_i as a linear combination of the others. A set of vectors with these properties is called a *basis* for the vector space $\mathbb{R}^d$.

A.2 Matrices

A *matrix* is a two-dimensional array:

$$A = (a_{ij}) = \begin{pmatrix} a_{11} & a_{12} & \cdots & a_{1n} \\ a_{21} & a_{22} & \cdots & a_{2n} \\ \cdots & \cdots & \cdots & \cdots \\ a_{m-11} & a_{m-12} & \cdots & a_{m-1n} \\ a_{m1} & a_{n2} & \cdots & a_{mn} \end{pmatrix}. \tag{A.6}$$

We will sometimes write $A = A_{m \times n}$ to emphasize that the size of the matrix is m rows by n columns.[3] We can add matrices term-by-term, and multiply them by a scalar:

$$(a_{ij}) + (b_{ij}) = (a_{ij} + b_{ij}) \qquad cA = (ca_{ij}) \tag{A.7}$$

We can also multiply two matrices, provided the number of columns of the first is equal to the number of rows of the second. The product of an $m \times n$ A and an $n \times p$ B matrix is an $m \times p$ matrix, with elements:

$$(AB)_{ij} = \sum_{s=1}^{n} a_{is} b_{sj}. \tag{A.8}$$

The product of a matrix with a vector, Av is defined as in Equation (A.8). This is why mathematicians like to represent vectors as column vectors, or $n \times 1$ matrices.

An $m \times n$ matrix A can be thought of as a linear operator: it is a linear map from $\mathbb{R}^m$ to $\mathbb{R}^n$. It is linear, in that $A(av + bw) = aAv + bAw$ for vectors v, w and scalars a, b.

Note that a matrix is made up of vectors in two ways: it is an array of row vectors (the rows) and an array of column vectors (the columns). We can define the transpose of the matrix by swapping rows and columns: $A^T = (a_{ij})^T = (a_{ji})$. If A is $m \times n$, then A^T is $n \times m$. A matrix is *square* if the number of rows equals the number of columns: $m = n$. A square matrix is *symmetric* if $A^T = A$. It is a real matrix if all the entries are real valued. All the matrices that we will be concerned with in this appendix will be real matrices.

A matrix will be called *diagonal* if all the entries in the matrix except those along the diagonal – a_{ii} – are zero. A diagonal matrix may have some zeros on the diagonal; saying it

[3]You will also see A^{mn} and $A^{m \times n}$ in the literature.

is diagonal only guarantees that if there are any nonzero values, they occur on the diagonal. The diagonal square matrix all of whose diagonal entries are equal to 1 is called the *identity matrix*. It is written I_n, or just I if the dimension is clear. Note that $IA = AI = A$ for any square matrix (of the same dimension as I).

An inverse of a matrix A is a matrix A^{-1} with the property that

$$AA^{-1} = A^{-1}A = I.$$

Not all matrices have inverses; trivially, the matrix of all zeros does not. The *determinant* can be used to decide if a matrix has an inverse. There are many formulas for computing the determinant, most of which are fairly complicated notationally. Any linear algebra package will have code to compute the determinant. In the next section, a method for computing the determinant will be discussed, as well as another way to tell if a matrix is invertible.

In the section above, we wrote the vectors as row vectors (this is for convenience), but mathematicians think of a vector of n entries as an $n \times 1$ matrix: a column vector. Thus we can write the dot product of vectors v and w in matrix notation (as we did in Equation (A.3)) as $v^T w$.

A.2.1 Eigenvectors and Eigenvalues

An eigenvector of a square matrix is a vector that is scaled by the matrix without changing the direction in which it points (except possibly by flipping it to point in the opposite direction). That is, an eigenvector is a unit vector[4] v that solves the equation:

$$Av = \lambda v$$

for some $\lambda \in \mathbb{R}$. The number λ is the *eigenvalue* associated with v, and can be thought of as how much the linear operator defined by A scales in the direction of v. This fact is used, along with the spectral theorem described below, in Section 4.10.

A useful fact is that the determinant of a square matrix is the product of its eigenvalues. A consequence of this is that a matrix is invertible if and only if none of its eigenvalues is 0.

It is easy to see that for a diagonal matrix, the e_i are eigenvectors and the diagonal entries of the matrix are the eigenvalues associated with them. It is generally not this easy to see what the eigenvectors and eigenvalues of a given matrix are, but there are plenty of linear algorithm packages and libraries, and any program or language such as MATLAB, R, Python, etc. for data analysis will have functions to compute eigenvectors and eigenvalues.

For a real symmetric square $n \times n$ matrix A, all the eigenvectors and eigenvalues are real, and there are exactly n distinct eigenvectors (the eigenvalues may not be distinct – consider the identity matrix, all of whose eigenvalues are obviously 1). For nonsymmetric matrices there may not be n eigenvectors, and the eigenvectors and eigenvalues may be complex.

The spectral theorem relates a matrix A to its eigenvectors and eigenvalues. It states that for any real symmetric $n \times n$ matrix, there exists a matrix U such that

$$A = U\Lambda U^T, \tag{A.9}$$

where U is the $n \times n$ matrix whose columns correspond to the eigenvectors of A and Λ is the diagonal matrix whose entries are the eigenvalues of A. When one refers to this decomposition, it is assumed that the vectors/values have been ordered so that the entries of Λ are in decreasing order. Further, the eigenvectors (columns of U) are *orthogonal*. Two

[4]If the matrix is not symmetric, it is possible that the zero vector satisfies the equation and is thus an eigenvector which is obviously not a unit vector. We will ignore this case, since we will only be considering symmetric matrices. Any other vector that satisfies the equation can be scaled by its norm, and thus turned into a unit vector.

vectors v, w are orthogonal if their inner product is 0: $v^T w = 0$. Note that geometrically, this means that the angle between them is 90 degrees – think of the coordinate axes e_i. The set of eigenvalues of the matrix A is called the *spectrum*, although when a *spectral method* for data analysis is discussed it generally utilizes both the eigenvalues and eigenvectors.

The spectral theorem gives a way to compute the inverse of a matrix:

$$A^{-1} = U\Lambda^{-1}U^T. \tag{A.10}$$

That is, replace all the eigenvalues λ_i with $1/\lambda_i$ in Λ. This makes sense if the eigenvalues are all nonzero. Computing eigenvalues and eigenvectors is usually no easier than computing the inverse directly, but if one needs them anyway, this provides a quick way to obtain the inverse of the matrix once one has the spectrum (including the eigenvectors).

Equation (A.10) also provides a method for computing a *generalized inverse* of the matrix. This is "almost" an inverse, and can often be used in calculations in which an inverse appears, for matrices which are not invertible. The only difference is that in Equation (A.10), the 0 eigenvalues are left alone, only the nonzero ones are inverted. In practice, code will implement a version of this called the Moore-Penrose inverse, which is technically a *pseudoinverse*, using the singular value decomposition described below.

A.2.2 The Singular Value Decomposition

The spectral theorem is for symmetric (hence square) matrices. What about nonsquare matrices, such as the term-document matrix found in text analysis or nonsymmetric matrices such as the adjacency matrix of a directed graph (see Appendix B)? In this case, we use the singular value decomposition (SVD). Given any real valued matrix $A_{m \times n}$ (the theorem is more general than this, but we are only concerned with real matrices in this book), it can be decomposed into a product of three matrices:

$$A = UDV^T. \tag{A.11}$$

Here U is $m \times m$ matrix of *left singular vectors*, which correspond to the eigenvectors of AA^T, D is the diagonal $m \times n$ matrix[5] of singular values, and V is the matrix of *right singular vectors*, the $n \times n$ matrix of the eigenvectors of $A^T A$. Once again, one orders these according to decreasing singular values. It turns out, not only are all the entries of the vectors and the singular values real, but the singular values are also nonnegative. As noted, the singular value decomposition is useful in the analysis of text documents (see Section 7.5) and directed graphs.

Two recent references for the material in this section are [412] and [46]. These two books discuss the ideas and techniques of linear algebra with a focus on the topic's utility for data analysis. The book [412] is particularly concerned with applying the methods to machine learning, in particular neural networks and deep learning.

[5] In a rectangular matrix where $m \neq n$ the term "diagonal" refers to the entries with the same index: a_{ii}, just as in the square case.

Appendix B

Graphs

Aside from graphs constructed from data for the purpose of manifold learning (as discussed in Section 5.7) or topological data analysis (as discussed in Section 6.9), graphs often occur in cybersecurity as data objects themselves. Some applications are in email analysis ([353]) and on the analysis of network flows.

To be precise, a graph is a pair $G = (V, E)$, where V is the set of nodes or vertices, and E is the set of edges – pairs of vertices, and we usually write uv for the edge from vertex u to vertex v. The edges can be directed, in which case the edges are ordered pairs and G is called a *directed graph* or *digraph*, and $uv = (u, v) \neq (v, u) = vu$, or they can be undirected. A graph is *simple* if it does not contain multiple edges – there is at most one (directed or undirected, as appropriate) edge between any pair of vertices – and there are no loops – edges from a vertex to itself. The number of vertices in G, $|V|$, is called the *order*, and the number of edges, $|E|$ is called the *size*. Sometimes, particularly if one is dealing with more than one graph, the notation for the vertex and edge sets may be decorated with the graph, such as $V(G)$ or V_G. This extra notation will be unnecessary in this book.

Graphs appear in cybersecurity in a number of places. The web is a graph, with pages as nodes, and edges corresponding to hyperlinks between pages. TCP/IP sessions also can be thought of in terms of graphs, with the computers corresponding to the vertices. In this case, the graph can be directed (there is an edge from a client to a server) or undirected (there was a communication between the two computers).

A *weighted graph* is one in which the edges have weights, generally positive numbers. These can correspond to the number of connections between computers, the amount of information flow (such as number of packets or bytes), the strength of the relationship in a social network, or a measure of the certainty that the edge is "real" for applications in which edges are not perfectly observed.

The degree of a vertex is the number of edges incident to the vertex. In the case of directed graphs, there are three kinds of degree: out degree, in degree, and total degree which is the sum of the other two. These are defined as one would expect. The degree sequence is the vector of degrees sorted from smallest to largest. This, like the order and size of a graph, is an *invariant*. That is, it is constant under relabeling of the vertices: no matter how one presents a graph, these measures are the same.

A *path* is a collection of vertices $\{v_1, v_2, \ldots, v_n\}$ for which $v_i v_j \in E$ for $1 \leq i, j \leq n$. In a directed graph A vertex v is *reachable* from u if there is a path from u to v. A connected component of a graph is a set of vertices all of which are reachable from any given vertex in the set. In the case of directed graphs, this is called strongly connected if the paths are required to follow the directions of the edges; otherwise, if the edge directions are ignored, it is called weakly connected. In this case it is equivalent to the concept of "connected" for an undirected graph.

The adjacency matrix is the binary matrix $S = (a_{i,j})$ which has a 1 in position $a_{i,j}$ if and only if there is an edge $v_i v_j \in E$. In a simple graph, the diagonal of the adjacency matrix is always 0 – no self-loops. In a weighted graph, one may substitute the weight for 1, in which case the matrix is referred to as the *weighted adjacency matrix*.

As mentioned above, in the analysis of network flows or network connections, there is a natural directed graph in which the vertices correspond to computers, and the directed edges go from client to server. The direction of the edges may or may not be important, depending on the application. For example, it may be sufficient to know that two computers have communicated, possibly sharing malicious code or corporate secrets, without consideration of which computer initiated the communication. Since these connections occur in time, one has a dynamic, or temporal, graph – the graph is changing in time as new edges are created (and, possibly, old edges may disappear, for example when the connection has closed or a fixed amount of time has elapsed). This can be represented in a purely stochastic manner (as in [263]) or constructed by considering all connections in a window (as in [353]). Given a time series of graphs, one wishes to investigate the graph structure and detect changes and anomalies.

B.1 Graph Invariants

A typical method of analysis for a time series of graphs is to extract a graph invariant (such as the number of edges, the number of connected components, the largest eigenvalue of the adjacency matrix, etc.) and track this invariant over time. In [353] a *scan statistic* is computed; this is a local measure such as the number of edges in the neighborhood of a vertex. By standardizing this measure with a windowed mean and standard deviation, one can identify vertices whose behavior changes, indicating for example an increase in "chatter" by the vertex and their immediate communicants. On the other hand, the embedding methods discussed in the text can be used to produce a representation of the nodes in $\mathbb{R}^d$ for further analysis.

For an example, consider the Los Alamos network flows data described in [441]. One can construct the graph of connections between computers for a given period of time, and investigate how this changes in time. Figure B.1 shows the number of vertices for two time periods. The solid curve depicts graphs defined on all the flows within each day, while the dotted curve depicts hourly graphs, in which all flows within one hour are used. In all cases, graphs between time periods utilize different data – there is no overlap in time for these graphs. One can clearly see the weekly patterns in these data.

Figure 5.7 (Chapter 5) depicts four hourly graphs from day 3 of the LANL data. In each case, the graph consists of:

- Vertices corresponding to computers.
- A directed edge between computers c and d if there was a flow with source c and destination d within the hour.

The number of connections is not retained in this analysis; however, it is straightforward to add the number of connections as weights on the edges.

B.2 The Laplacian

Given the degrees of the vertices of an undirected graph, one can form the matrix D, the diagonal matrix whose diagonal entries correspond to these degrees. Given the adjacency

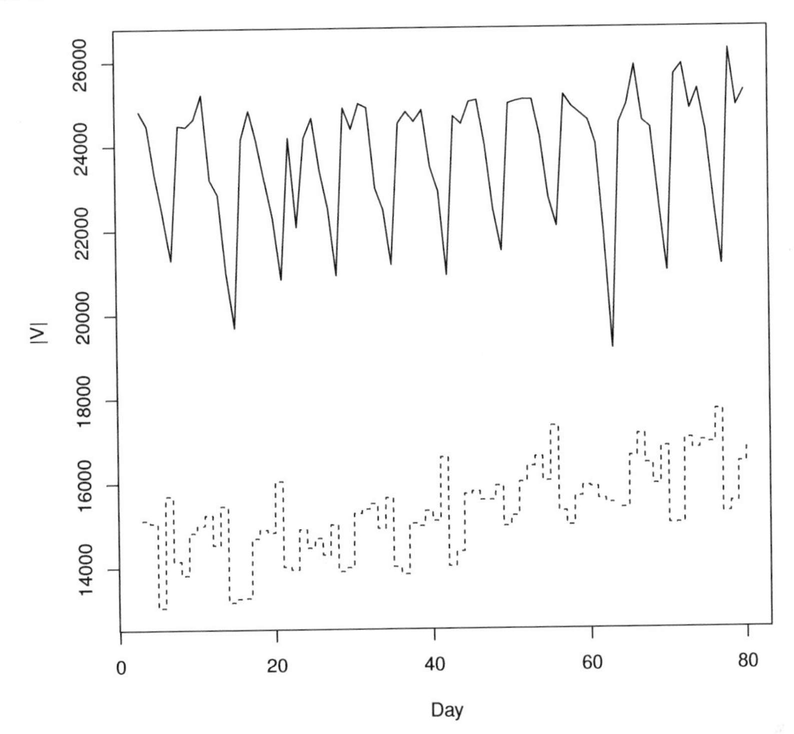

Figure B.1: The number of vertices (computers) in the network flows data. The solid curve corresponds to daily graphs, while the dotted curves correspond to hourly graphs.

matrix A one can form the Laplacian matrix: $L = D - A$. This matrix is positive semi-definite, meaning that all the eigenvalues are nonnegative. Equivalently, this means that

$$xAx^T \geq 0,$$

for all unit vectors x. It is easy to see that the number of connected components is equal to the number of zero eigenvalues of the adjacency matrix. The normalized Laplacian is written as

$$\mathcal{L} = D^{\frac{1}{2}} L D^{\frac{1}{2}}.$$

The Laplacian is used in spectral clustering and spectral embedding, as well as other graph theoretic applications. Chapter 5 discusses several of these, see also [23, 288, 330, 463] for discussions of both the Laplacian embedding method and the adjacency spectral embedding method.

Appendix C

Probability

Most people have an intuitive idea of what "randomness" means. If you flip a coin high in the air (and you are not a magician), the side that comes up when it lands is unpredictable – it is "random". In practice, it is impossible to distinguish between a truly random event and one which is simply too complex to predict; some might believe that if we knew a perfect model of the coin and the forces applied to it, including the drag and motion caused by the air, and a perfect model of the surface that the coin lands on, it would be possible to accurately determine the outcome of the toss. Whether this is true or not is irrelevant – we don't have the ability to do this in arbitrary – or even reasonably realistic – environments, and so a coin toss is "random".

A philosopher studying computer security might suggest that many of the things that we study in this book are not truly random, they are the result of deterministic processes on a physical network. We believe this to be short-sighted. There are many complex processes that affect cyber systems, and many of these are arguably random. Power fluctuations can cause certain systems to fail or to act unreliably. Random events can affect how and when users interact with systems. There is a random component to the time for a vulnerability to be detected, and who detects it first – the security professionals or the hackers. From a practical standpoint network traffic, user actions, software vulnerabilities (and whether/when they are discovered, and by whom) are all random processes, and these are best modeled and analyzed through probability and statistics.

C.1 Probability

Probability is the study of randomness, nondeterministic error and uncertainty. A probability model assigns a probability to an event. In the discrete world, where there are a finite number of distinct events, a probability model assigns a number in $[0, 1]$ to each event. This is the probability of the event. If multiple events are independent, exactly one of the events can occur, then the probabilities sum to one. More formally, we define probability as a function $P : \mathcal{P}(\Omega) \to [0, 1]$. Ω is the sample space, and $\mathcal{P}(\Omega)$ is the *power set* – the set of all subsets – and so P assigns a probability to outcomes (subsets of Ω). We require:

1. $P(\emptyset) = 0$.
2. $P(\Omega) = 1$.
3. If $A \cap B = \emptyset$, then $P(A \cup B) = P(A) + P(B)$.
4. If $A \subset B$, then $P(A) \leq P(B)$ and $P(B \setminus A) = P(B) - P(A)$.

5. Writing A^c for $\Omega \setminus A$, a consequence of #2 and #3 is that $P(A^c) = 1 - P(A)$.

Probability models can be discrete: heads or tails are the only possible outcomes of a coin flip.[1] Or they could be continuous: one usually models height, weight, velocity, the time between packets and many other physical properties as being continuous. In this case, it doesn't make sense to define the probability of a single number – the probability that the time between two packets will be exactly 1.294551221578... is 0. Rather, we define the model in terms of the probabilities of intervals. We'll see this in more detail below.

A random variable is a measurement that is random – it is measuring a random phenomenon. This is a somewhat circular and imprecise definition – a book on probability would make it precise – but it serves our purposes. Most of the basic probability deals with sets – events that may or may not occur – and a random variable can be defined in terms of functions on this "sample space", but we don't need this level of formalism.

C.1.1 Conditional Probability and Bayes' Rule

The joint probability of A and B is the probability of both A and B happening. For example, consider flipping a coin and rolling a die. These events are *independent*, meaning that the outcome of the coin does not affect the outcome of the die, and vice versa. Assuming that both are fair, the probability of getting a heads is 1/2 and the probability of rolling a 3 is 1/6. The joint probability of getting both a heads and a 3 is 1/12.

The conditional probability is a way of computing how an event depends on the outcome of another event. It is defined as:

$$P(A|B) = \frac{P(A \cap B)}{P(B)}. \tag{C.1}$$

Note that if the events are independent, then $P(A \cap B) = P(A)P(B)$ and the conditional probability in Equation (C.1) reduces to $P(A)$, which matches our intuition – if the two events don't "depend on each other", then whether or not B has occurred has no effect on the probability of A occurring.

The *law of total probability* relates the probability of A to the conditional probabilities of A given other random variables. Assume that $B_1, \ldots, B_n$ is a partition of the sample space: disjoint events whose union is Ω, the space of all possible outcomes. Then

$$P(A) = \sum_{i=1}^{n} P(A|B_i)P(B_i). \tag{C.2}$$

Bayes' rule allows us to flip the conditioning.

$$P(B|A) = \frac{P(A|B)P(B)}{P(A)}. \tag{C.3}$$

C.1.2 Base Rate Fallacy

Suppose you have a very good detector of email spam. Your algorithm correctly marks spam with an error rate of 1 in 1000. That is, for every 1000 spam emails, you expect one of them to get through. Suppose that it is also very good at correctly marking a nonspam email, with an error rate of 1 in 1000. So out of 1000 legitimate emails, you expect only one of them to be incorrectly labeled as spam. Finally, suppose that 1 email in 500 is spam.

[1]Ignoring the coin landing on its edge or falling through a hole in the space-time continuum.

Now, an email comes in and your algorithm says it is spam. What is the probability that it really is spam? It is tempting to say that it is 0.999. After all, the detector is only wrong about spam once in a thousand, right? Well, no.

Let S be the event that the classifier says an email is spam. Let's organize this:

$$\begin{aligned} P(S|\text{spam}) &= 0.999 \\ P(S|\text{legit}) &= 0.001 \\ P(\text{spam}) &= 0.002 \\ P(\text{legit}) &= 0.998 \end{aligned}$$

$P(S)$ is the probability that the classifier calls an email spam, and is:

$$\begin{aligned} P(S) &= P(S|\text{spam})P(\text{spam}) + P(S|\text{legit})P(\text{legit}) \\ &= 0.999 * 0.002 + 0.001 * 0.998 \\ &= 0.002996 \end{aligned}$$

Bayes' rule says:

$$\begin{aligned} P(\text{spam}|S) &= \frac{P(S|\text{spam})P(\text{spam})}{P(S)} \\ &= \frac{0.999 * 0.002}{0.002996} \\ &= 0.6668892. \end{aligned}$$

So, when your really good classifier says an email is spam, about 1/3 of the time it's wrong! This is the *base rate fallacy*, whose name comes from the fact that the important rate is not the rate at which emails arrive, or the rate at which spam arrives, but rather the rate at which the event of interest occurs.

Here's how to think about this. Suppose you get 1000 emails. On average, you'd expect two to be spam, and your classifier would most likely so classify them. However, you'd expect one of the legitimate emails to also be flagged as spam. Hence, of the three emails flagged as spam, one (or one third) is legitimate. This kind of analysis is likely much more important to you, the user of email, than all these 1-in-a-thousand error rates, and it is why computer security algorithm developers, dealing with many thousands or even millions of objects (emails, flows, logs, packets) worry quite a lot about false alarm rates – the rate at which legitimate objects are misclassified.

C.1.3 Expected Values and Moments

The most commonly used property of a random variable is the mean, or first moment. The mean is a measure of *location*: it represents a type of "center" of the data. Let X be a discrete random variable. The expected value, or mean, of the random variable is defined as:

$$E(X) = \sum_{i=1}^{n} p(x_i)x_i. \tag{C.4}$$

We can generalize this to powers of X: the kth moment of X is $E(X^k)$. In particular, the variance of X is $E((X - E(X))^2)$, i.e., the second moment of the random variable once it has been centered at 0 by subtracting the mean. This is a measure of how "variable" the random variable is, hence the name. The standard deviation is the square root of the variance, and is in the same units as X. Thus, the variance is a measure of "spread": how far the data typically is from the center, as defined by the mean.

Another measure of the "spread" of a univariate random variable is the inter-quartile range. The p-quartile is the value for which $100 * p\%$ of the data is less than or equal to the value. The first quartile is the 0.25-quantile, the third quartile is the 0.75-quantile. So one quarter of the data is less than the first quantile, three quarters of the data is less than the fourth quantile. The inter-quartile range is the difference of the third and first quartiles.

There are other definitions of "center" that are useful. For example, the median is the "middle" point of univariate (one dimensional) data; it is the point for which half the data is larger and half is smaller, i.e., the 0.5-quantile. There is no multivariate median – more accurately, there are many possible multivariate medians, referred to as *statistical depth*, each with different properties, and each satisfying a subset of the properties of the univariate median that statisticians have identified as being desirable. See [360, 369, 530, 531] for discussion of these issues and many examples of statistical depth functions.

C.1.4 Distribution Functions and Densities

Some measurements are naturally modeled as being continuously valued, for example taking values in $[0, 1]$ or $[0, \infty]$. To define what we mean by these, we introduce two new concepts, the *cumulative distribution function* (CDF) and the *probability density function* (PDF).

The CDF is the function $CDF(t)$ defined as the probability of obtaining a value at most t from a sample for the random variable X. For discrete random variables, we can compute this as:

$$P[X \leq t] = \sum_{x \leq t} P(X = x). \tag{C.5}$$

Thus the CDF is exactly as named: it is the cumulative probability up to t. For continuous variables it is still perfectly well defined, it is the probability of the event $X \leq t$, and in fact many continuous distributions are defined by providing the form of their CDF, although as we shall see most are defined in terms of the closely related PDF.

When we define a discrete random variable by stating the probabilities of each of the possible values, what we are really doing is defining the *probability mass function* of the random variable. This is the function that is defined only on the possible values of the random variable, and returns the probability of that value.

The probability density function is the continuous version of the probability mass function. We can't have a function $f(x)$ that returns the probability that $X = x$, because in the continuous case this is zero.[2] Instead, we could define it in terms of limits of shrinking intervals around x, or equivalently we can define it as:

$$P[X \leq x] = \int_{t=-\infty}^{x} f(t)dt. \tag{C.6}$$

That is (assuming sufficient smoothness conditions) we can think of the probability density as the derivative of the cumulative density function. Note that although CDFs

[2]Technically, it is zero for all but a finite – set of measure zero – number of values. For the vast majority of cases of concern in computer security, this set is empty.

are always defined, PDFs are not. The way to see this is to think of the PDF as the derivative of the CDF, and so if the CDF is not differentiable, there will be no PDF for that distribution. This is worth knowing, but not of much practical importance for computer security applications. For our purposes, the distributions we will be dealing with will always have a well-defined probability density function.

While we cannot view the probability density function as a method for obtaining the probability at a point, we can use it to obtain a measure of how "likely" the point is to "have been drawn from" the distribution. Given a probability density function $f(x)$, the likelihood of a point x_i is defined to be $f(x_i)$. More generally, given points $x_1, \ldots, x_n$, the likelihood is the joint distribution $f(x_1, \ldots, x_n)$. If the points are drawn independently, then the distribution factors as a product, and the likelihood is:

$$L(x_1, \ldots, x_n) = \prod_{i=1}^{n} f(x_i). \tag{C.7}$$

Since products can be awkward, particularly when maximizing them (which will come up later in this chapter) one often deals with the log-likelihood,

$$\ell(x_1, \ldots, x_n) = \log(L(x_1, \ldots, x_n)) = \log(\prod_{i=1}^{n} f(x_i)) = \sum_{i=1}^{n} \log(f(x_i)). \tag{C.8}$$

C.2 Models

C.2.1 Bernoulli and Binomial

The Bernoulli random variable is the "coin toss" distribution. Suppose you are flipping a coin, and you believe that the chances that it will come up "heads" or "tails" are equal. If we denote the probability of heads as p, then this is a Bernoulli random variable with probability p. The Bernoulli(p) mass function is:

$$\begin{aligned}\text{Prob}(x = 1) &= p\\ \text{Prob}(x = 0) &= 1 - p\end{aligned}$$

A coin with $p = 1/2$ is called a *fair coin.* In common parlance, one will often hear the phrase "the coin is random" to mean that the coin is fair, with the implied (and sometimes explicitly stated) consequence that if $p \neq 1/2$, the coin is "not random". It is important to keep in mind that "random" does not mean "equally probable".

Suppose we flip a coin n times.[3] What is the probability that the first time we see a "heads" is at flip j? This is easy to compute, simply by analyzing the situation: the only way to obtain the event "first heads is at flip j" is by getting $j-1$ tails first, and then a head at flip j. Hence the probability is $(1-p)^{j-1}p$, provided that $n \geq j$. If $n < j$ the question is not well posed: we are really asking what is the probability of flipping n times without a "heads", and this is $(1-p)^n$.

The Bernoulli distribution is for a binary variable: true/false, heads/tails. If we associate the value 0 to tails and 1 to heads, the coin becomes a method for generating a binary sequence – a sequence of 0s and 1s. Suppose we flip the coin 100 times and record how many heads we observe. This is equivalent to adding the binary variable. This results in a new distribution: the binomial. It is the first distribution one generally learns about for

[3]Unless we explicitly state a dependence in an experiment like this, it should be assumed that the draws of the random variable ("flips" in this case) are independent.

variables that take on integer values, in this case nonnegative values. In our coin tossing example, it is the count of the number of times a coin comes up "heads" in n tosses. Mathematically, it is the sum of independent Bernoulli variables.

Recall that $n! = 1 * 2 * 3 \cdots * n$, $0! = 1$ by convention, and that

$$\binom{n}{k} = \frac{n!}{k!(n-k)!}.$$

Then a random variable from the binomial distribution, written $X \sim B(n,p)$ has probability mass function

$$P(X = k) = \binom{n}{k} p^k (1-p)^{n-k}, \quad \text{(C.9)}$$

for $k = 0, 1, \ldots, n$ and 0 otherwise.

C.2.2 Multinomial

As we've seen, the Bernoulli and binomial distributions are ways to model a coin toss. Sticking with the gambling theme, what about dice? Consider a single die. A fair one has equal probability for each face, or $P(X = k) = 1/6$ for $k = 1, 2, \ldots, 6$. Suppose we roll the die n times (or equivalently, roll n identical dice).[4] The multinomial distribution models the number of times each face occurs. To be completely general, let's not assume the dice are fair, or that they are standard dice. Instead, assume they have K distinct faces, labeled 1 through K, and suppose the probability of rolling a k is p_k. Write $(x_1, \ldots, x_K)$ for the event that we observe the ith face x_i times in our n rolls (or n dice). Assume $x_i \geq 0$ and $\sum x_i = n$. Then

$$P(x_1, \ldots, x_K) = \frac{n!}{x_1! \cdots x_K!} p_1^{x_1} \cdots p_K^{x_K}, \quad \text{(C.10)}$$

otherwise it is 0.

C.2.3 Uniform

When a lay-person says something is "random", they are often thinking "equiprobable" – that is, that all possible events have equal probability. The continuous distribution associated with this is the uniform distribution. The uniform density on the interval $[a, b]$ is

$$f(x) = \frac{1}{b-a} \quad \text{(C.11)}$$

or 0 if $x \notin [a, b]$.

We can expand this in many ways. In higher dimensions, we can consider squares, cubes and hyper cubes, with constant density over the entire area/volume. Similarly, we can (in principal) pick an object of any shape and define the uniform distribution to be constant for points in the object, and zero outside. In practice, it is not always trivial to produce a uniform sample in an object – think about how you might generate data uniformly distributed on the sphere, the boundary of a 3-dimensional ball. While it seems obvious that one should move to polar coordinates, and then sample the angles uniformly, this doesn't work. For more discussion, and several correct algorithms, see `http://corysimon.github.io/articles/uniformdistn-on-sphere/`.[5]

Suppose that X is a continuous random variable with distribution function F_X. Then $F_X(X)$ is uniform on $[0, 1]$. The inverse of this is also true, and can be used to generate

[4]One usually sees the multinomial distribution described in terms of colored balls in an urn or bag; in this case, the colors correspond to the sides of the die, and the proportion of balls of color k is p_k.

[5]Visited on Dec 10, 2017.

random variables from a given distribution: if U is a uniform random variable on $[0, 1]$, then $F_X^{-1}(U)$ (F_X^{-1} denotes the inverse of the distribution function F_X) has distribution F_X.

Of course, the above only works for the case where the distribution function is invertible. It assumes that one can draw uniform random variables, but fortunately every software package that generates random variables has a uniform random number generator, and every computer language does as well or has a function to generate random positive integers, which can be turned into a uniform number generator by dividing by the maximum integer.[6]

[6]Technically, this is false for two reasons. First, random number generators in software are pseudo-random, that is they make use of clever (but deterministic) mathematics to generate numbers that are indistinguishable, for most practical purposes, from random numbers. Second, since computers can only represent a finite number of values, they can never produce continuous random variables. For our purposes we can safely ignore these issues, except perhaps for certain cryptographic purposes, where the quality of random number generators may become critical.

Bibliography

[1] Ayman El Aassal, Luis Moraes, Shahryar Baki, Avisha Das, and Rakesh M. Verma. Anti-phishing pilot at ACM IWSPA 2018: Evaluating performance with new metrics for unbalanced datasets. In *Proc. 1st Anti-Phishing Shared Pilot at ACM IWSPA 2018*, volume 2124. http://ceur-ws.org, 2018.

[2] Onur Aciiçmez, Shay Gueron, and Jean-Pierre Seifert. New branch prediction vulnerabilities in openssl and necessary software countermeasures. In *IMA Conference on Cryptography and Coding*, pages 185–203. Springer, 2007.

[3] Niall Adams and Nicholas Heard. *Data analysis for network cyber-security.* World Scientific Publishing, 2014.

[4] Andreas Adolfsson, Margareta Ackerman, and Naomi C. Brownstein. To cluster, or not to cluster: An analysis of clusterability methods. *Pattern Recognition*, 88:13–26, 2019.

[5] Charu Aggarwal. *Outlier analysis.* Springer, 2013.

[6] Charu C. Aggarwal, Xiangnan Kong, Quanquan Gu, Jiawei Han, and Philip S. Yu. Active learning: A survey. In *Data Classification: Algorithms and Applications*, pages 571–606. Chapman & Hall/CRC Data Mining and Knowledge Discovery Series, 2014.

[7] Alan Agresti. *Categorical data analysis.* Wiley, 1990.

[8] Helena Ahonen, Oskari Heinonen, Mika Klemettinen, and A. Inkeri Verkamo. Applying data mining techniques for descriptive phrase extraction in digital document collections. In *ADL*, pages 2–11. IEEE, 1998.

[9] Leman Akoglu, Hanghang Tong, Jilles Vreeken, and Christos Faloutsos. Fast and reliable anomaly detection in categorical data. In *CIKM*, pages 415–424, 2012.

[10] Maruan Al-Shedivat, Andrew Gordon Wilson, Yunus Saatchi, Zhiting Hu, and Eric P. Xing. Learning scalable deep kernels with recurrent structure. *Journal of Machine Learning Research*, 18:82:1–82:37, 2017.

[11] Ahmed Aleroud and Lina Zhou. Phishing environments, techniques, and countermeasures: A survey. *Computers & Security*, 68:160–196, 2017.

[12] Mehdi Allahyari, Seyedamin Pouriyeh, Mehdi Assefi, Saied Safaei, Elizabeth D Trippe, Juan B Gutierrez, and Krys Kochut. A brief survey of text mining: Classification, clustering and extraction techniques. *arXiv preprint arXiv:1707.02919*, 2017.

[13] Cloud Security Alliance. Top ten big data security and privacy challenges. Technical report, Cloud Security Alliance, 2012.

[14] Aijun An, Shakil M. Khan, and Xiangji Huang. Objective and subjective algorithms for grouping association rules. In *ICDM*, pages 477–480, 2003.

[15] R. Anand, K. G. Mehrotra, C. K. Mohan, and S. Ranka. An improved algorithm for neural network classification of imbalanced training sets. *IEEE Transactions on Neural Networks*, 4(6):962–969, 1993.

[16] Ion Androutsopoulos, John Koutsias, Konstantinos V. Chandrinos, and Constantine D. Spyropoulos. An experimental comparison of naive Bayesian and keyword-based anti-spam filtering with personal e-mail messages. In *Proceedings of the 23rd Annual International ACM SIGIR Conference on Research and Development in Information Retrieval*, SIGIR '00, pages 160–167, 2000.

[17] Daniele Apiletti, Elena Baralis, Tania Cerquitelli, and Vincenzo D'Elia. Characterizing network traffic by means of the netmine framework. *Computer Networks*, 53(6):774–789, 2009.

[18] Giovanni Apruzzese, Michele Colajanni, Luca Ferretti, Alessandro Guido, and Mirco Marchetti. On the effectiveness of machine and deep learning for cyber security. In *2018 10th International Conference on Cyber Conflict (CyCon)*, pages 371–390. IEEE, 2018.

[19] David Arthur and Sergei Vassilvitskii. How slow is the k-means method? In *ACM SOCG*, pages 144–153, 2006.

[20] David Arthur and Sergei Vassilvitskii. K-means++: The advantages of careful seeding. In *SODA*, pages 1027–1035, 2007.

[21] Sunil Arya, David M Mount, Nathan S Netanyahu, Ruth Silverman, and Angela Y Wu. An optimal algorithm for approximate nearest neighbor searching fixed dimensions. *JACM*, 45(6):891–923, 1998.

[22] Anish Athalye, Nicholas Carlini, and David A. Wagner. Obfuscated gradients give a false sense of security: Circumventing defenses to adversarial examples. In *Proceedings of the 35th International Conference on Machine Learning, ICML 2018, Stockholmsmässan, Stockholm, Sweden, July 10-15, 2018*, pages 274–283, 2018.

[23] Avanti Athreya, Donniell E Fishkind, Keith Levin, Vince Lyzinski, Youngser Park, Yichen Qin, Daniel L Sussman, Minh Tang, Joshua T Vogelstein, and Carey E. Priebe. Statistical inference on random dot product graphs: a survey. *arXiv preprint arXiv:1709.05454*, 2017.

[24] Sheldon Jay Axler. *Linear algebra done right*, volume 2. Springer, 1997.

[25] John W. Backus. Can programming be liberated from the von Neumann style? A functional style and its algebra of programs. *Commun. ACM*, 21(8):613–641, 1978.

[26] Alejandro Correa Bahnsen, Ivan Torroledo, David Camacho, and Sergio Villegas. Deepphish: Simulating malicious AI. In *2018 APWG Symposium on Electronic Crime Research (eCrime)*, pages 1–8, 2018.

[27] Shahryar Baki, Rakesh M. Verma, Arjun Mukherjee, and Omprakash Gnawali. Scaling and effectiveness of email masquerade attacks: Exploiting natural language generation. In *ASIACCS*, pages 469–482. ACM, 2017.

[28] Marco Barreno, Blaine Nelson, Anthony D. Joseph, and J. D. Tygar. The security of machine learning. *Machine Learning*, 81(2):121–148, 2010.

[29] Rodrigo C. Barros, André C.P.L.F de Carvalho, and Alex A. Freitas. *Automatic design of decision-tree induction algorithms.* Springer, 2015.

[30] Ronnie Bathoorn, Arne Koopman, and Arno Siebes. Reducing the frequent pattern set. In *IEEE ICDM Workshops*, pages 55–59, 2006.

[31] Ulrich Bayer, Paolo Comparetti, Clemens Hlauschek, Christopher Kruegel, and Engin Kirda. Scalable, behavior-based malware clustering. In *NDSS*, 2009.

[32] Viacheslav Belenko, Valery Chernenko, Maxim Kalinin, and Vasiliy Krundyshev. Evaluation of gan applicability for intrusion detection in self-organizing networks of cyber physical systems. In *2018 International Russian Automation Conference (RusAutoCon)*, pages 1–7. IEEE, 2018.

[33] Mikhail Belkin and Partha Niyogi. Laplacian eigenmaps and spectral techniques for embedding and clustering. In *Advances in neural information processing systems*, pages 585–591, 2002.

[34] Mikhail Belkin and Partha Niyogi. Laplacian eigenmaps for dimensionality reduction and data representation. *Neural Computation*, 15(6):1373–1396, 2003.

[35] David A Belsley, Edwin Kuh, and Roy E Welsch. *Regression diagnostics: Identifying influential data and sources of collinearity*, volume 571. John Wiley & Sons, 2005.

[36] Fernando Benites and Elena P. Sapozhnikova. Evaluation of hierarchical interestingness measures for mining pairwise generalized association rules. *IEEE Trans. Knowl. Data Eng.*, 26(12):3012–3025, 2014.

[37] Seif-Eddine Benkabou, Khalid Benabdeslem, and Bruno Canitia. L2-type regularization-based unsupervised anomaly detection from temporal data. In *IJCNN*, pages 2354–2361, 2017.

[38] Alina Beygelzimer, Sham Kakade, and John C. Langford. Cover trees for nearest neighbor. In *ICML*, pages 97–104. ACM, 2006.

[39] Peter J. Bickel and Kjell A. Doksum. *Mathematical statistics: Basic ideas and selected topics*, volume 2. CRC Press, 2015.

[40] Battista Biggio and Fabio Roli. Wild patterns: Ten years after the rise of adversarial machine learning. *Pattern Recognition*, 84:317–331, 2018.

[41] Saroj Kr. Biswas, Monali Bordoloi, and Jacob Shreya. A graph based keyword extraction model using collective node weight. *Expert Systems with Applications*, 97:51–59, 2018.

[42] Julien Blanchard, Fabrice Guillet, Régis Gras, and Henri Briand. Using information-theoretic measures to assess association rule interestingness. In *ICDM*, pages 66–73, 2005.

[43] Enrico Blanzieri and Anton Bryl. A survey of learning-based techniques of email spam filtering. *AI Review*, 29(1):63–92, 2008.

[44] David M. Blei, Andrew Y. Ng, and Michael I. Jordan. Latent dirichlet allocation. *JMLR*, 3(Jan):993–1022, 2003.

[45] Avrim Blum, Dawn Xiaodong Song, and Shobha Venkataraman. Detection of interactive stepping stones: Algorithms and confidence bounds. In *RAID*, pages 258–277, 2004.

[46] Przemyslaw Bogacki. *Linear algebra: Concepts and applications.* MAA Press, Providence, RI, 2019.

[47] Dan Boneh, Amit Sahai, and Brent Waters. Functional encryption: Definitions and challenges. In Yuval Ishai, editor, *Theory of Cryptography*, pages 253–273. Springer, 2011.

[48] Shyam Boriah, Varun Chandola, and Vipin Kumar. Similarity measures for categorical data: A comparative evaluation. In *SIAM Conference on Data Mining*, pages 243–254, 2008.

[49] Subhasree Bose and Saptarsi Goswami. *RSentiment: Analyse Sentiment of English Sentences*, 2017. R package version 2.2.1.

[50] George E.P. Box. Science and statistics. *Journal of the American Statistical Association*, 71(356):791–799, 1976.

[51] George E.P. Box, W.G. Hunter, and J.S. Hunter. *Statistics for experimenters.* Wiley, 1978.

[52] Zvika Brakerski, Craig Gentry, and Vinod Vaikuntanathan. Fully homomorphic encryption without bootstrapping. Cryptology ePrint Archive, Report 2011/277, 2011.

[53] Adrian M.P. Brasoveanu, Giuseppe Rizzo, Philipp Kuntschik, Albert Weichselbraun, and Lyndon Nixon. Framing named entity linking error types. In *LREC*, 2018.

[54] Pavel Brazdil, Christophe G. Giraud-Carrier, Carlos Soares, and Ricardo Vilalta. *Metalearning - Applications to data mining.* Cognitive Technologies. Springer, 2009.

[55] Pavel Brazdil, Ricardo Vilalta, Christophe G. Giraud-Carrier, and Carlos Soares. Metalearning. In *Encyclopedia of Machine Learning and Data Mining*, pages 818–823. Springer, 2017.

[56] Leo Breiman. Bagging predictors. *Machine Learning*, 24(2):123–140, 1996.

[57] Leo Breiman. Random forests. *Machine Learning*, 45(1):5–32, 2001.

[58] Leo Breiman. Statistical modeling: The two cultures (with comments and a rejoinder by the author). *Statistical Science*, 16(3):199–231, 2001.

[59] Leo Breiman, Jerome Friedman, R. A. Olshen, and Charles J Stone. *Classification and regression trees.* Chapman & Hall/CRC, 1984.

[60] Eric A. Brewer. Towards robust distributed systems (abstract). In *Proceedings of the Nineteenth Annual ACM Symposium on Principles of Distributed Computing*, PODC '00, pages 7–, New York, NY, USA, 2000. ACM.

[61] Andrei Z. Broder, Steven C. Glassman, Mark S. Manasse, and Geoffrey Zweig. Syntactic clustering of the web. *Computer Networks*, 29(8-13):1157–1166, 1997.

[62] Anna L. Buczak and Erhan Guven. A survey of data mining and machine learning methods for cyber security intrusion detection. *IEEE Communications Surveys and Tutorials*, 18(2):1153–1176, 2016.

[63] Thomas Bühler and Matthias Hein. Spectral clustering based on the graph p-laplacian. In *Proceedings of the 26th Annual International Conference on Machine Learning, ICML 2009, Montreal, Quebec, Canada, June 14-18, 2009*, pages 81–88, 2009.

[64] Kenneth P. Burnham and David R. Anderson. *Model selection and multimodel inference: a practical information-theoretic approach.* Springer Science & Business Media, 2003.

[65] Steven Burrows and Seyed MM Tahaghoghi. Source code authorship attribution using n-grams. In *Proceedings of the Twelfth Australasian Document Computing Symposium, Melbourne, Australia, RMIT University*, pages 32–39. Citeseer, 2007.

[66] Ricardo Campos, Vítor Mangaravite, Arian Pasquali, Alípio Mário Jorge, Célia Nunes, and Adam Jatowt. Yake! collection-independent automatic keyword extractor. In *ECIR*, pages 806–810. Springer, 2018.

[67] Feng Cao, Martin Estert, Weining Qian, and Aoying Zhou. Density-based clustering over an evolving data stream with noise. In *Proceedings of the 2006 SIAM International Conference on Data Mining*, pages 328–339. SIAM, 2006.

[68] Alvaro A. Cárdenas, Pratyusa K. Manadhata, and Sreeranga P. Rajan. Big data analytics for security. *IEEE Security & Privacy*, 11(6):74–76, 2013.

[69] Kathleen M. Carley. ORA: A toolkit for dynamic network analysis and visualization. In *Encyclopedia of Social Network Analysis and Mining, 2nd Edition.* Springer, 2018.

[70] Gunnar Carlsson. Topology and data. *Bull. American Mathematical Society*, 46(2):255–308, 2009.

[71] Wenyaw Chan, George Cybenko, Murat Kantarcioglu, Ernst L. Leiss, Thamar Solorio, Bhavani Thuraisingham, and Rakesh M. Verma. Panel: Essential data analytics knowledge for cybersecurity professionals and students. In *Proceedings IWSPA*, pages 55–57, New York, NY, USA, 2015. ACM.

[72] Varun Chandola, Arindam Banerjee, and Vipin Kumar. Anomaly detection: A survey. *ACM Comput. Surv.*, 41(3):15:1–15:58, 2009.

[73] Madhusudhanan Chandrasekaran, Krishnan Narayanan, and Shambhu Upadhyaya. Phishing email detection based on structural properties. In *Proc. NY Academy of Science*, pages 2–8, 2006.

[74] Chih-Chung Chang and Chih-Jen Lin. LIBSVM: A library for support vector machines. *Transactions on Intelligent Systems and Technology*, 2(3):27, 2011.

[75] Fay Chang, Jeffrey Dean, Sanjay Ghemawat, Wilson C. Hsieh, Deborah A. Wallach, Michael Burrows, Tushar Chandra, Andrew Fikes, and Robert Gruber. Bigtable: A distributed storage system for structured data. In *7th Symposium on Operating Systems Design and Implementation (OSDI '06), November 6-8, Seattle, WA, USA*, pages 205–218, 2006.

[76] Olivier Chapelle, Bernhard Schölkopf, and Alexander Zien. *Semi-supervised learning.* MIT Press, Cambridge, 2006.

[77] Devendra Singh Chaplot and Ruslan Salakhutdinov. Knowledge-based word sense disambiguation using topic models. *arXiv preprint arXiv:1801.01900*, 2018.

[78] C.L. Philip Chen and Chun-Yang Zhang. Data-intensive applications, challenges, techniques and technologies: A survey on big data. *Information Sciences*, 275:314–347, 2014.

[79] Kuan-Ta Chen, Hsing-Kuo Kenneth Pao, and Hong-Chung Chang. Game bot identification based on manifold learning. In *SIGCOMM Workshop on Network and System Support for Games*, pages 21–26. ACM, 2008.

[80] Ping Chen, Rakesh M. Verma, Janet C. Meininger, and Wenyaw Chan. Semantic analysis of association rules. In *FLAIRS Conference*, pages 270–275, 2008.

[81] Ting Chen, Lu An Tang, Yizhou Sun, Zhengzhang Chen, and Kai Zhang. Entity embedding-based anomaly detection for heterogeneous categorical events. In *IJCAI*, pages 1396–1403, 2016.

[82] William W.S. Chen. *Statistical methods in computer security*. CRC Press, 2004.

[83] Yun-Nung Chen and Alexander Rudnicky. Two-stage stochastic email synthesizer. In *INLG*, pages 99–102, 2014.

[84] Yun-Nung Chen and Alexander Rudnicky. Two-stage stochastic natural language generation for email synthesis by modeling sender style and topic structure. In *INLG*, pages 152–156, 2014.

[85] Hong Cheng, Xifeng Yan, Jiawei Han, and Philip S. Yu. Direct discriminative pattern mining for effective classification. In *IEEE ICDE*, pages 169–178, 2008.

[86] Boris Chernis and Rakesh M. Verma. Machine learning methods for software vulnerability detection. In *IWSPA@CODASPY*, pages 31–39. ACM, 2018.

[87] Davide Chicco. Ten quick tips for machine learning in computational biology. *BioData Mining*, 10(1):35, 2017.

[88] Kang Leng Chiew, Kelvin Sheng Chek Yong, and Choon Lin Tan. A survey of phishing attacks: Their types, vectors and technical approaches. *Expert Systems with Applications*, 106:1–20, 2018.

[89] Daniel Y.T. Chino, Alceu Ferraz Costa, Agma J. M. Traina, and Christos Faloutsos. VolTime: Unsupervised anomaly detection on users' online activity volume. In *SDM*, pages 108–116, 2017.

[90] Ashish Chiplunkar, Michael Kapralov, Sanjeev Khanna, Aida Mousavifar, and Yuval Peres. Testing graph clusterability: Algorithms and lower bounds. In *59th IEEE Annual Symposium on Foundations of Computer Science, FOCS 2018, Paris, France, October 7-9, 2018*, pages 497–508, 2018.

[91] Jiyeon Choo, Rachsuda Jiamthapthaksin, Chun-sheng Chen, Oner Ulvi Celepcikay, Christian Giusti, and Christoph F. Eick. MOSAIC: A proximity graph approach for agglomerative clustering. In Il Yeal Song, Johann Eder, and Tho Manh Nguyen, editors, *Data Warehousing and Knowledge Discovery*, pages 231–240, Berlin, Heidelberg, 2007. Springer Berlin Heidelberg.

[92] Chenhui Chu and Rui Wang. A survey of domain adaptation for neural machine translation. In *Proceedings of the 27th International Conference on Computational Linguistics*, pages 1304–1319, Santa Fe, New Mexico, USA, August 2018. Association for Computational Linguistics.

[93] David A. Cieslak, T. Ryan Hoens, Nitesh V. Chawla, and W. Philip Kegelmeyer. Hellinger distance decision trees are robust and skew-insensitive. *DMKD Journal*, 24(1):136–158, 2012.

[94] Moustapha Cisse, Yossi Adi, Natalia Neverova, and Joseph Keshet. Houdini: Fooling deep structured prediction models. *arXiv preprint arXiv:1707.05373*, 2017.

[95] William S. Cleveland. Lowess: A program for smoothing scatterplots by robust locally weighted regression. *The American Statistician*, 35(1):54–54, 1981.

[96] William S. Cleveland. *Visualizing data.* Hobart Press, 1993.

[97] William W. Cohen. Fast effective rule induction. In *Proc. ICML*, pages 115–123, 1995.

[98] Ronan Collobert and Jason Weston. A unified architecture for natural language processing: Deep neural networks with multitask learning. In *ICML*, pages 160–167. ACM, 2008.

[99] Ronan Collobert, Jason Weston, Léon Bottou, Michael Karlen, Koray Kavukcuoglu, and Pavel Kuksa. Natural language processing (almost) from scratch. *JMLR*, 12(Aug):2493–2537, 2011.

[100] Victor Costan and Srinivas Devadas. Intel SGX explained. *IACR Cryptology ePrint Archive*, 2016(086):1–118, 2016.

[101] T. F. Cox and M. A. A. Cox. *Multidimensional scaling.* Chapman & Hall, 2001.

[102] Reza Curtmola, Juan Garay, Seny Kamara, and Rafail Ostrovsky. Searchable symmetric encryption: improved definitions and efficient constructions. *Journal of Computer Security*, 19(5):895–934, 2011. Preliminary version in ACM CCS 2006.

[103] Avisha Das, Shahryar Baki, Ayman El Aassal, and Rakesh M. Verma. Systematization of Knowledge (SoK): Reexamining phishing and spear phishing detection research from the security perspective, 2018. Under Review.

[104] Avisha Das and Rakesh M. Verma. Automated email generation for targeted attacks using natural language. In *TA-COS@LREC*, 2018.

[105] Kaustav Das, Jeff Schneider, and Daniel B. Neill. Anomaly pattern detection in categorical datasets. In *SIGKDD*, 2008.

[106] Dipankar Dasgupta, Senhua Yu, and Fernando Nino. Recent advances in artificial immune systems: Models and applications. *Applied Soft Computing*, 11(2):1574–1587, 2011.

[107] Sanjoy Dasgupta and Anupam Gupta. An elementary proof of a theorem of Johnson and Lindenstrauss. *Random Structures & Algorithms*, 22(1):60–65, 2003.

[108] Neil Daswani, Christoph Kern, and Anita Kesavan. *Foundations of security - what every programmer needs to know.* Apress, 2007.

[109] Rajesh N. Davé. Characterization and detection of noise in clustering. *Pattern Recognition Letters*, 12(11):657–664, 1991.

[110] Rajesh N. Davé and Raghuram Krishnapuram. Robust clustering methods: A unified view. *IEEE Trans. Fuzzy Systems*, 5(2):270–293, 1997.

[111] C. Davies and R. Ganesan. BApasswd: A new proactive password checker. In *National Computer Security Conference*, pages 1–15, 1993.

[112] David Douglas DeBarr. *Spam, phishing, and fraud detection using random projections, adversarial learning, and semi-supervised learning.* PhD thesis, George Mason University, 2013.

[113] Giuseppe DeCandia, Deniz Hastorun, Madan Jampani, Gunavardhan Kakulapati, Avinash Lakshman, Alex Pilchin, Swaminathan Sivasubramanian, Peter Vosshall, and Werner Vogels. Dynamo: Amazon's highly available key-value store. In *Proceedings of the 21st ACM Symposium on Operating Systems Principles 2007, SOSP 2007, Stevenson, Washington, USA, October 14-17, 2007*, pages 205–220, 2007.

[114] Jacob Devlin, Ming-Wei Chang, Kenton Lee, and Kristina Toutanova. BERT: Pre-training of deep bidirectional transformers for language understanding. In *2019 Annual Conference of the North American Chapter for Computational Linguistics, NAACL-HLT*, pages 4171–4186, Minneapolis, USA, 2019.

[115] Luc Devroye, László Györfi, and Gábor Lugosi. *A probabilistic theory of pattern recognition.* Springer, 1996.

[116] Dua Dheeru and Efi Karra Taniskidou. UCI machine learning repository, 2017. `http://archive.ics.uci.edu/ml`.

[117] Eric Diehl. *Ten laws for security.* Springer, 2016.

[118] Selma Dilek, Hüseyin Çakır, and Mustafa Aydin. Applications of artificial intelligence techniques to combating cyber crimes: A review. *arXiv preprint arXiv:1502.03552*, 2015.

[119] Gregory Ditzler, Manuel Roveri, Cesare Alippi, and Robi Polikar. Learning in non-stationary environments: A survey. *IEEE Comp. Int. Mag.*, 10(4):12–25, 2015.

[120] David Doermann. The indexing and retrieval of document images: A survey. *Computer Vision and Image Understanding*, 70(3):287–298, 1998.

[121] Carl Doersch. Tutorial on variational autoencoders. *CoRR*, abs/1606.05908, 2016.

[122] Qi Dong, Shaogang Gong, and Xiatian Zhu. Imbalanced deep learning by minority class incremental rectification. *CoRR*, abs/1804.10851, 2018.

[123] David Donoho. 50 years of data science. In *Princeton NJ, Tukey Centennial Workshop*, 2015.

[124] David L. Donoho. High-dimensional data analysis: The curses and blessings of dimensionality. In *AMS Conference On Math Challenges Of The 21st Century*, 2000.

[125] Zuochao Dou, Issa Khalil, Abdallah Khreishah, Ala I. Al-Fuqaha, and Mohsen Guizani. Systematization of knowledge (sok): A systematic review of software-based web phishing detection. *IEEE Communications Surveys and Tutorials*, 19(4):2797–2819, 2017.

[126] Sumeet Dua and Xian Du. *Data mining and machine learning in cybersecurity.* CRC Press, 2011.

[127] Lian Duan, Lida Xu, Feng Guo, Jun Lee, and Baopin Yan. A local-density based spatial clustering algorithm with noise. *Inf. Syst.*, 32(7):978–986, 2007.

[128] Richard O. Duda, Peter E. Hart, and David G. Stork. *Pattern classification.* John Wiley & Sons, 2012.

[129] Tudor Dumitras and Darren Shou. Toward a standard benchmark for computer security research: the worldwide intelligence network environment (WINE). In *First Workshop on Building Analysis Datasets and Gathering Experience Returns for Security, BADGERS@EuroSys 2011*, pages 89–96, 2011.

[130] John Durkin and John Durkin. *Expert systems: design and development.* Macmillan New York, 1994.

[131] Jennifer G. Dy and Carla E. Brodley. Feature selection for unsupervised learning. *Journal of Machine Learning Research*, 5:845–889, 2004.

[132] Sean R. Eddy. Hidden markov models. *Current Opinion in Structural Biology*, 6(3):361–365, 1996.

[133] Bradley Efron and Carl Morris. Data analysis using Stein's estimator and its generalizations. *Journal of the American Statistical Association*, 70(350):311–319, 1975.

[134] Bradley Efron and Carl Morris. Stein's paradox in statistics. *Scientific American*, 236(5):119–127, 1977.

[135] Manuel Egele, Theodoor Scholte, Engin Kirda, and Christopher Kruegel. A survey on automated dynamic malware-analysis techniques and tools. *ACM Comput. Surv.*, 44(2):6:1–6:42, 2012.

[136] Gal Egozi and Rakesh M. Verma. Phishing email detection using robust NLP techniques. In *2018 IEEE International Conference on Data Mining Workshops, ICDM Workshops, Singapore, Singapore, November 17-20, 2018*, pages 7–12, 2018.

[137] Ahmed Salah El-Din and Neamat El Gayar. New feature splitting criteria for co-training using genetic algorithm optimization. In *Multiple Classifier Systems, 9th International Workshop, MCS 2010, Cairo, Egypt, April 7-9, 2010. Proceedings*, pages 22–32, 2010.

[138] Tamer Elsayed, Jimmy Lin, and Douglas W Oard. Pairwise document similarity in large collections with MapReduce. In *Proceedings of the 46th Annual Meeting of the Association for Computational Linguistics on Human Language Technologies: Short Papers*, pages 265–268. Association for Computational Linguistics, 2008.

[139] TH Emerson, CC Olson, and T Doster. A study of the effect of alternative similarity measures on the performance of graph-based anomaly detection algorithms. In *Algorithms and Technologies for Multispectral, Hyperspectral, and Ultraspectral Imagery XXIV*, volume 10644, page 106440F. International Society for Optics and Photonics, 2018.

[140] Charles Epstein, Gunnar Carlsson, and Herbert Edelsbrunner. Topological data analysis. *Inverse Problems*, 27(12):1–2, 2011.

[141] Kyumars Sheykh Esmaili. Reflections on almost two decades of research into stream processing: Tutorial. In *Proceedings of the 11th ACM International Conference on Distributed and Event-based Systems*, DEBS '17, pages 21–23, New York, NY, USA, 2017. ACM.

[142] Martin Ester, Hans-Peter Kriegel, Jörg Sander, and Xiaowei Xu. A density-based algorithm for discovering clusters a density-based algorithm for discovering clusters in large spatial databases with noise. In *SIGKDD*, pages 226–231, 1996.

[143] Christos Faloutsos and U. Kang. Managing and mining large graphs: Patterns and algorithms. In *Proceedings of the 2012 ACM SIGMOD International Conference on Management of Data*, SIGMOD '12, pages 585–588, New York, NY, USA, 2012. ACM.

[144] Yong Fang, Cheng Zhang, Cheng Huang, Liang Liu, and Yue Yang. Phishing email detection using improved RCNN model with multilevel vectors and attention mechanism. *IEEE Access*, 7:56329–56340, 2019.

[145] Houtan Faridi, Srivathsan Srinivasagopalan, and Rakesh M. Verma. Performance evaluation of features and clustering algorithms for malware. In *2018 IEEE International Conference on Data Mining Workshops, ICDM Workshops, Singapore, Singapore, November 17-20, 2018*, pages 13–22, 2018.

[146] Houtan Faridi, Srivathsan Srinivasagopalan, and Rakesh M. Verma. Parameter tuning and confidence limits of malware clustering. In *Proceedings of the Ninth ACM Conference on Data and Application Security and Privacy, CODASPY 2019, Richardson, TX, USA, March 25-27, 2019*, pages 169–171, 2019.

[147] Houtan Faridi and Rakesh M. Verma. Analysis of malware datasets, 2018.

[148] Gowhar Farooq. Politics of fake news: How whatsapp became a potent propaganda tool in india. *Media Watch*, 9(1):106–117, 2017.

[149] Christiane Fellbaum, editor. *WordNet an electronic lexical database.* MIT Press, 1998.

[150] Manuel Fernández-Delgado, Eva Cernadas, Senén Barro, and Dinani Amorim. Do we need hundreds of classifiers to solve real world classification problems? *JMLR*, 15:3133–3181, 2014.

[151] Ian Fette, Norman Sadeh, and Anthony Tomasic. Learning to detect phishing emails. In *WWW*, pages 649–656, 2007.

[152] Charles J. Fillmore. Frame semantics and the nature of language. *Annals of the New York Academy of Sciences*, 280(1):20–32, 1976.

[153] Jenny Rose Finkel, Trond Grenager, and Christopher D. Manning. Incorporating non-local information into information extraction systems by Gibbs sampling. In *ACL*, pages 363–370, 2005.

[154] Stephanie Forrest, Steven A. Hofmeyr, Anil Somayaji, and Thomas A. Longstaff. A sense of self for Unix processes. In *IEEE Symposium on Security and Privacy*, pages 120–128, 1996.

[155] Eric Fosler-Lussier. Markov models and hidden markov models: A brief tutorial. Technical Report TR-98-041, International Computer Science Institute, 1998.

[156] Chris Fraley and Adrian E Raftery. Model-based clustering, discriminant analysis, and density estimation. *Journal of the American statistical Association*, 97(458):611–631, 2002.

[157] Jerome Francois, Shaonan Wang, Walter Bronzi, Radu State, and Thomas Engel. Botcloud: Detecting botnets using MapReduce. In *Information Forensics and Security Workshop*, pages 1–6. IEEE, 2011.

[158] Yoav Freund, Robert Schapire, and Naoki Abe. A short introduction to boosting. *Journal-Japanese Society For Artificial Intelligence*, 14(771-780):1612, 1999.

[159] Yoav Freund, Robert E Schapire, et al. Experiments with a new boosting algorithm. In *ICML*, volume 96, pages 148–156, 1996.

[160] Brendan J. Frey and Delbert Dueck. Clustering by passing messages between data points. *Science*, 315(5814):972–976, 2007.

[161] Jerome Friedman. A variable span scatterplot smoother. Technical Report 5, Stanford University, 1984.

[162] Edgar Gabriel. Private communication, May 2019.

[163] Amir Gandomi and Murtaza Haider. Beyond the hype: Big data concepts, methods, and analytics. *International Journal of Information Management*, 35(2):137–144, 2015.

[164] Joseph Gardiner and Shishir Nagaraja. On the security of machine learning in malware c&c detection: A survey. *ACM Comput. Surv.*, 49(3):59:1–59:39, 2016.

[165] Albert Gatt and Emiel Krahmer. Survey of the state of the art in natural language generation: Core tasks, applications and evaluation. *Journal of AI Research*, 61:65–170, 2018.

[166] S. Gaudin. The omega files, 2000. `https://www.computerworld.com.au/article/69055/omega_files/`.

[167] Matthew Gerber and Joyce Yue Chai. Beyond NomBank: A study of implicit arguments for nominal predicates. In *ACL 2010, Proceedings of the 48th Annual Meeting of the Association for Computational Linguistics, July 11-16, 2010, Uppsala, Sweden*, pages 1583–1592, 2010.

[168] Seth Gilbert and Nancy Lynch. Brewer's conjecture and the feasibility of consistent, available, partition-tolerant web services. *SIGACT News*, 33(2):51–59, June 2002.

[169] Kendall E Giles, Michael W. Trosset, David J. Marchette, and Carey E. Priebe. Iterative denoising. *Computational Statistics*, 23(4):497–517, 2008.

[170] Kevin Gimpel, Nathan Schneider, Brendan O'Connor, Dipanjan Das, Daniel Mills, Jacob Eisenstein, Michael Heilman, Dani Yogatama, Jeffrey Flanigan, and Noah A. Smith. Part-of-speech tagging for twitter: Annotation, features, and experiments. Technical report, Carnegie-Mellon Univ Pittsburgh Pa School of Computer Science, 2010.

[171] Yoav Goldberg. *Neural network methods for natural language processing*. Synthesis Lectures on Human Language Technologies. Morgan & Claypool, 2017.

[172] Ian Goodfellow, Yoshua Bengio, and Aaron Courville. *Deep learning*, volume 1. MIT Press, 2016.

[173] Ian J. Goodfellow, Jonathon Shlens, and Christian Szegedy. Explaining and harnessing adversarial examples. *arXiv preprint arXiv:1412.6572*, 2014.

[174] Johannes Götzfried, Moritz Eckert, Sebastian Schinzel, and Tilo Müller. Cache attacks on intel SGX. In *European Workshop on Systems Security*, pages 2:1–2:6, 2017.

[175] John C. Gower and Garmt B. Dijksterhuis. *Procrustes problems.* Oxford University Press, 2004.

[176] Christopher N. Gutierrez, Taegyu Kim, Raffaele Della Corte, Jeffrey Avery, Saurabh Bagchi, Dan Goldwasser, and Marcello Cinque. Learning from the ones that got away: Detecting new forms of phishing attacks. *IEEE TDSC*, 2018.

[177] M.A. Hall and G. Holmes. Benchmarking attribute selection techniques for discrete class data mining. *IEEE Transactions on Knowledge and Data Engineering*, 15(6):1437–1447, Nov 2003.

[178] Joonas Hämäläinen, Susanne Jauhiainen, and Tommi Kärkkäinen. Comparison of internal clustering validation indices for prototype-based clustering. *Algorithms*, 10(3):105, 2017.

[179] Felix Hamborg, Corinna Breitinger, Moritz Schubotz, Soeren Lachnit, and Bela Gipp. Extraction of main event descriptors from news articles by answering the journalistic five w and one h questions. In *ACM/IEEE JCDL*, 2018.

[180] Greg Hamerly and Charles Elkan. Learning the k in k-means. In *Advances in Neural Information Processing Systems 16 [Neural Information Processing Systems, NIPS 2003, December 8-13, 2003, Vancouver and Whistler, British Columbia, Canada]*, pages 281–288, 2003.

[181] James Douglas Hamilton. *Time series analysis*, volume 2. Princeton University Press, 1994.

[182] Jiawei Han, Jian Pei, and Micheline Kamber. *Data mining: Concepts and techniques.* Elsevier, 2011.

[183] Jiawei Han, Jian Pei, and Yiwen Yin. Mining frequent patterns without candidate generation. In *ACM SIGMOD Record*, volume 29, pages 1–12. ACM, 2000.

[184] YuFei Han and Yun Shen. Accurate spear phishing campaign attribution and early detection. In *Proceedings of the 31st Annual ACM Symposium on Applied Computing*, SAC '16, pages 2079–2086, New York, NY, USA, 2016. ACM.

[185] David J. Hand et al. Classifier technology and the illusion of progress. *Statistical Science*, 21(1):1–14, 2006.

[186] Jacob A Harer, Louis Y Kim, Rebecca L Russell, Onur Ozdemir, Leonard R Kosta, Akshay Rangamani, Lei H Hamilton, Gabriel I Centeno, Jonathan R Key, Paul M Ellingwood, et al. Automated software vulnerability detection with machine learning. *arXiv preprint arXiv:1803.04497*, 2018.

[187] Trevor Hastie, Saharon Rosset, Ji Zhu, and Hui Zou. Multi-class Adaboost. *Statistics and its Interface*, 2(3):349–360, 2009.

[188] Trevor Hastie and Robert Tibshirani. *Generalized additive models.* Wiley Online Library, 1990.

[189] Trevor Hastie, Robert Tibshirani, and Jerome Friedman. *The elements of statistical learning: Data mining, inference, and prediction.* Springer, second edition, 2011.

[190] Kathryn Hempstalk, Eibe Frank, and Ian H Witten. One-class classification by combining density and class probability estimation. In *ECML PKDD*, pages 505–519. Springer, 2008.

[191] Geoffrey E Hinton, Simon Osindero, and Yee-Whye Teh. A fast learning algorithm for deep belief nets. *Neural computation*, 18(7):1527–1554, 2006.

[192] Grant Ho, Aashish Sharma, Mobin Javed, Vern Paxson, and David A. Wagner. Detecting credential spearphishing in enterprise settings. In *USENIX Security*, pages 469–485, 2017.

[193] Thomas Hofmann. Probabilistic latent semantic analysis. In *UAI*, pages 289–296, 1999.

[194] Steven A. Hofmeyr and Stephanie Forrest. Architecture for an artificial immune system. *Evolutionary computation*, 8(4):443–473, 2000.

[195] Steven A. Hofmeyr, Stephanie Forrest, and Anil Somayaji. Intrusion detection using sequences of system calls. *Journal of Computer Security*, 6(3):151–180, 1998.

[196] David I Holmes and Richard S Forsyth. The federalist revisited: New directions in authorship attribution. *Literary and Linguistic Computing*, 10(2):111–127, 1995.

[197] Elizabeth E Holmes, Eric J. Ward, and Kellie Wills. MARSS: Multivariate autoregressive state-space models for analyzing time-series data. *R Journal*, 4(1), 2012.

[198] Susan Holmes. Statistical proof? The problem of irreproducibility. *Bulletin American Mathematical Society*, 55(1):31–55, 2018.

[199] Kurt Hornik, Maxwell Stinchcombe, and Halbert White. Universal approximation of an unknown mapping and its derivatives using multilayer feedforward networks. *Neural networks*, 3(5):551–560, 1990.

[200] Tobias Horsmann, Nicolai Erbs, and Torsten Zesch. Fast or accurate? - A comparative evaluation of pos tagging models. In *Proceedings of the International Conference of the German Society for Computational Linguistics and Language Technology, GSCL 2015, University of Duisburg-Essen, Germany, 30th September - 2nd October 2015*, pages 22–30, 2015.

[201] Han-Wei Hsiao, Huey-Min Sun, and Wei-Cheng Fan. Detecting stepping-stone intrusion using association rule mining. *Security and Communication Networks*, 6(10):1225–1235, 2013.

[202] Chu-Ren Huang, Petr Šimon, Shu-Kai Hsieh, and Laurent Prévot. Rethinking chinese word segmentation: tokenization, character classification, or wordbreak identification. In *ACL (poster and demonstrations)*, pages 69–72. ACL, 2007.

[203] Peter J. Huber. *Robust statistical procedures*, volume 68. SIAM, 1996.

[204] Peter J. Huber. Robust statistics. In *International Encyclopedia of Statistical Science*, pages 1248–1251. Springer, 2011.

[205] Anette Hulth. Improved automatic keyword extraction given more linguistic knowledge. In *EMNLP*, pages 216–223. ACL, 2003.

[206] R. Hyndman, A.B. Koehler, J.K. Ord, and R.D. Snyder. *Forecasting with Exponential Smoothing: The State Space Approach.* Springer, Berlin, 2008.

[207] Rob J. Hyndman, G. Athanasopoulos, S. Razbash, D. Schmidt, Z. Zhou, Y. Khan, C. Bergmeir, and E. Wang. forecast: Forecasting functions for time series and linear models. *R package version*, 6(6):7, 2015.

[208] Jay Jacobs and Bob Rudis. *Data-driven security: Analysis, visualization and dashboards.* John Wiley & Sons, 2014.

[209] Aditya Jain, Gandhar Kulkarni, and Vraj Shah. Natural language processing. *International Journal of Computer Sciences and Engineering*, 6(1), 2018.

[210] Anil K. Jain. Data clustering: 50 years beyond k-means. *Pattern Recognition Letters*, 31(8):651–666, 2010.

[211] Anil K. Jain, Robert P.W. Duin, and Jianchang Mao. Statistical pattern recognition: A review. *Transactions on Pattern Analysis and Machine Intelligence*, 22(1):4–37, 2000.

[212] Anil K. Jain, Alexander P. Topchy, Martin H. C. Law, and Joachim M. Buhmann. Landscape of clustering algorithms. In *17th International Conference on Pattern Recognition, ICPR 2004, Cambridge, UK, August 23-26, 2004.*, pages 260–263, 2004.

[213] Gareth James, Daniela Witten, Trevor Hastie, and Robert Tibshirani. *An introduction to statistical learning*, volume 6. Springer, 2013.

[214] Lancelot F James, Carey E. Priebe, and David J. Marchette. Consistent estimation of mixture complexity. *The Annals of Statistics*, pages 1281–1296, 2001.

[215] Ahmad Javaid, Quamar Niyaz, Weiqing Sun, and Mansoor Alam. A deep learning approach for network intrusion detection system. In *Proceedings of the 9th EAI International Conference on Bio-inspired Information and Communications Technologies (formerly BIONETICS)*, pages 21–26. ICST (Institute for Computer Sciences, Social Informatics and Telecommunications Engineering), 2016.

[216] Daniel R. Jeske, Nathaniel T. Stevens, Alexander G. Tartakovsky, and James D. Wilson. Statistical methods for network surveillance. *Applied Stochastic Models in Business and Industry*, 34(4):425–445, 2018.

[217] Karel Ježek and Josef Steinberger. Automatic text summarization (the state of the art 2007 and new challenges). In *Proceedings of Znalosti*, pages 1–12, 2008.

[218] Xue-peng Jia and Xiao-feng Rong. A self-training method for detection of phishing websites. In *Data Mining and Big Data - Third International Conference, DMBD 2018, Shanghai, China, June 17-22, 2018, Proceedings*, pages 414–425, 2018.

[219] Thorsten Joachims. Transductive inference for text classification using support vector machines. In *Proc. ICML*, pages 200–209, 1999.

[220] William B. Johnson and Joram Lindenstrauss. Extensions of Lipschitz mappings into a Hilbert space. *Contemporary mathematics*, 26(189-206):1, 1984.

[221] M. Chris Jones. Simple boundary correction for kernel density estimation. *Statistics and Computing*, 3(3):135–146, 1993.

[222] Anthony D. Joseph, Blaine Nelson, Benjamin I. P. Rubinstein, and J. D. Tygar. *Adversarial machine learning.* Cambridge University Press, 2019.

[223] V. Joshi. *Unsupervised anomaly detection in numerical datasets.* PhD thesis, University of Cincinnati, 2015.

[224] Antoine Joux. A new index calculus algorithm with complexity L(1/4+o(1)) in small characteristic. In Tanja Lange, Kristin Lauter, and Petr Lisoněk, editors, *Selected Areas in Cryptography – SAC 2013*, pages 355–379. Springer, 2014.

[225] Roberto J. Bayardo Jr. and Rakesh Agrawal. Mining the most interesting rules. In *KDD*, pages 145–154, 1999.

[226] Daniel Jurafsky and James H. Martin. *Speech and language processing (2nd Edition).* Prentice-Hall, Inc., Upper Saddle River, NJ, USA, 2009.

[227] Paul C. Kainen. Utilizing geometric anomalies of high dimension: When complexity makes computation easier. In Kevin Warwick and Moroslav Kárný, editors, *Computer-Intensive Methods in Control and Signal Processing.* Birkhäuser, Boston, 1997.

[228] U. Kang and Christos Faloutsos. Beyond 'caveman communities': Hubs and spokes for graph compression and mining. In *11th IEEE International Conference on Data Mining, ICDM 2011, Vancouver, BC, Canada, December 11-14, 2011*, pages 300–309, 2011.

[229] Murat Kantarcioglu and Rakesh M. Verma, editors. *IWSPA '18: Proceedings of the Fourth ACM International Workshop on Security and Privacy Analytics*, New York, NY, USA, 2018. ACM.

[230] Leonard Kaufman and Peter J. Rousseeuw. *Finding groups in data: An introduction to cluster analysis*, volume 344. John Wiley & Sons, 2009.

[231] John D. Kelleher, Brian Mac Namee, and Aoife D'Arcy. *Fundamentals of machine learning for predictive data analytics: algorithms, worked examples, and case studies.* MIT Press, 2015.

[232] Alexander D. Kent. Comprehensive, Multi-Source Cyber-Security Events. Los Alamos National Laboratory, 2015.

[233] Alexander D. Kent. Cybersecurity Data Sources for Dynamic Network Research. In *Dynamic Networks in Cybersecurity.* Imperial College Press, 2015.

[234] Alexander Khalitmonenko and Oleg Kupreev. DDOS attacks in Q1 2017, 2017. https://securelist.com/ddos-attacks-in-q1-2017/78285/ - Accessed July 18, 2017.

[235] Shehroz S. Khan and Michael G. Madden. A survey of recent trends in one class classification. In *Irish Conference on AI and Cognitive Science*, pages 188–197. Springer, 2009.

[236] Shehroz S. Khan and Michael G. Madden. One-class classification: Taxonomy of study and review of techniques. *The Knowledge Engineering Review*, 29(3):345–374, 2014.

[237] P. Khanna, C. Rebeiro, and A. Hazra. XFC: A framework for eXploitable fault characterization in block ciphers. In *DAC*, pages 1–6, 2017.

[238] Alison Kidd. *Knowledge acquisition for expert systems: A practical handbook.* Springer Science & Business Media, 2012.

[239] Won-Young Kim, Young-Koo Lee, and Jiawei Han. CCMine: Efficient mining of confidence-closed correlated patterns. In *PAKDD*, volume 3056 of *LNCS*, pages 569–579. Springer, 2004.

[240] Thomas King, Nikita Aggarwal, Mariarosaria Taddeo, and Luciano Floridi. Artificial intelligence crime:An interdisciplinary analysis of foreseeable threats and solutions, 2018. `http://dx.doi.org/10.2139/ssrn.3183238`.

[241] Paul Kingsbury and Martha Palmer. From TreeBank to PropBank. In *Proceedings of the Third International Conference on Language Resources and Evaluation, LREC 2002, May 29-31, 2002, Las Palmas, Canary Islands, Spain*, 2002.

[242] Ryan Kiros, Yukun Zhu, Ruslan Salakhutdinov, Richard S. Zemel, Antonio Torralba, Raquel Urtasun, and Sanja Fidler. Skip-thought vectors. *CoRR*, abs/1506.06726, 2015.

[243] Philip N. Klein. *Coding the matrix: Linear algebra through applications to computer science*. Newtonian Press, 2013.

[244] Paul Kocher, Daniel Genkin, Daniel Gruss, Werner Haas, Mike Hamburg, Moritz Lipp, Stefan Mangard, Thomas Prescher, Michael Schwarz, and Yuval Yarom. Spectre attacks: Exploiting speculative execution. *CoRR*, abs/1801.01203, 2018.

[245] J. Zico Kolter and Marcus A. Maloof. Dynamic weighted majority: An ensemble method for drifting concepts. *JMLR*, 8(Dec):2755–2790, 2007.

[246] Jeremy Z. Kolter and Marcus A. Maloof. Dynamic weighted majority: A new ensemble method for tracking concept drift. In *ICDM*, pages 123–130. IEEE, 2003.

[247] Jeremy Z. Kolter and Marcus A. Maloof. Learning to detect and classify malicious executables in the wild. *JMLR*, 6:2721–2744, 2006.

[248] Alexander Kott and Michael Ownby. Toward a research agenda in adversarial reasoning: Computational approaches to anticipating the opponent's intent and actions. *arXiv preprint arXiv:1512.07943*, 2015.

[249] Dimitrios Kotzias, Misha Denil, Nando De Freitas, and Padhraic Smyth. From group to individual labels using deep features. In *Proceedings of the 21th ACM SIGKDD International Conference on Knowledge Discovery and Data Mining*, pages 597–606. ACM, 2015.

[250] Ferenc Kovács, Csaba Legány, and Attila Babos. Cluster validity measurement techniques. In *6th International Symposium of Hungarian Researchers on Computational Intelligence*. Citeseer, 2005.

[251] Bartosz Krawczyk, Michał Woźniak, and Bogusław Cyganek. Clustering-based ensembles for one-class classification. *Information Sciences*, 264:182–195, 2014.

[252] Nikolaus Kriegeskorte. Deep neural networks: A new framework for modeling biological vision and brain information processing. *Annual review of vision science*, 1:417–446, 2015.

[253] Alex Krizhevsky, Ilya Sutskever, and Geoffrey E Hinton. Imagenet classification with deep convolutional neural networks. In *Advances in neural information processing systems*, pages 1097–1105, 2012.

[254] Ram Shankar Siva Kumar, Andrew Wicker, and Matt Swann. Practical machine learning for cloud intrusion detection: Challenges and the way forward. In *Proceedings of the 10th ACM Workshop on Artificial Intelligence and Security*, pages 81–90. ACM, 2017.

[255] Ludmila I. Kuncheva and Christopher J. Whitaker. Measures of diversity in classifier ensembles and their relationship with the ensemble accuracy. *Machine learning*, 51(2):181–207, 2003.

[256] Ludmila I. Kuncheva, Christopher J. Whitaker, C.A. Shipp, and Robert P.W. Duin. Limits on the majority vote accuracy in classifier fusion. *Pattern Analysis & Applications*, 6(1):22–31, 2003.

[257] Siwei Lai, Liheng Xu, Kang Liu, and Jun Zhao. Recurrent convolutional neural networks for text classification. In *AAAI Conference on Artificial Intelligence*, 2015.

[258] Louisa Lam and Ching Y. Suen. Application of majority voting to pattern recognition: An analysis of its behavior and performance. *IEEE Trans. Systems, Man, and Cybernetics, Part A*, 27(5):553–568, 1997.

[259] Thomas K. Landauer, Danielle S. McNamara, Simon Dennis, and Wlater Kintsch, editors. *Handbook of latent semantic analysis.* Lawrence Erlbaum Associates, Mahwah, NJ, 2007.

[260] Terran Lane and Carla E. Brodley. Approaches to online learning and concept drift for user identification in computer security. In *ACM SIGKDD*, pages 259–263, 1998.

[261] Robert L. Launer and Graham N. Wilkinson. *Robustness in statistics.* Academic Press, 2014.

[262] Yann LeCun, Yoshua Bengio, and Geoffrey Hinton. Deep learning. *Nature*, 521(7553):436, 2015.

[263] Nam H Lee and Carey E. Priebe. A latent process model for time series of attributed random graphs. *Statistical Inference for Stochastic Processes*, 14(3):231–253, 2011.

[264] Sangho Lee, Ming-Wei Shih, Prasun Gera, Taesoo Kim, Hyesoon Kim, and Marcus Peinado. Inferring fine-grained control flow inside SGX enclaves with branch shadowing. In *USENIX Security*, pages 16–18, 2017.

[265] Wenke Lee. Applying data mining to intrusion detection: The quest for automation, efficiency, and credibility. *SIGKDD Explor. Newsl.*, 4(2):35–42, December 2002.

[266] Wenke Lee and Salvatore J. Stolfo. Data mining approaches for intrusion detection. In *USENIX Security*, 1998.

[267] Wenke Lee, Salvatore J. Stolfo, and Philip K. Chan. Learning patterns from Unix process execution traces for intrusion detection. In *AAAI Workshop on AI Approaches to Fraud Detection and Risk Management*, pages 50–56, 1997.

[268] Wendy G. Lehnert and Martin H. Ringle. *Strategies for natural language processing.* Psychology Press, 2014.

[269] Jure Leskovec, Anand Rajaraman, and Jeffrey D. Ullman. *Mining of massive datasets, 2nd Ed.* Cambridge University Press, 2014.

[270] Keith Levin and Vince Lyzinski. Laplacian eigenmaps from sparse, noisy similarity measurements. *IEEE Transactions on Signal Processing*, 65(8):1988–2003, 2017.

[271] Peng Li, Limin Liu, Debin Gao, and Michael K. Reiter. On challenges in evaluating malware clustering. In *RAID*, pages 238–255. Springer, 2010.

[272] Yuancheng Li, Rui Xiao, Jingang Feng, and Liujun Zhao. A semi-supervised learning approach for detection of phishing webpages. *Optik*, 124(23):6027–6033, 2013.

[273] Zhiyao Liang and Rakesh M. Verma. Improving techniques for proving undecidability of checking cryptographic protocols. In *ARES*, pages 1067–1074, 2008.

[274] Zhiyao Liang and Rakesh M. Verma. Correcting and improving the NP proof for cryptographic protocol insecurity. In *ICISS*, pages 101–116, 2009.

[275] Shu-Hsien Liao. Expert system methodologies and applications - a decade review from 1995 to 2004. *Expert systems with applications*, 28(1):93–103, 2005.

[276] Jimmy J. Lin and Chris Dyer. *Data-Intensive Text Processing with MapReduce*. Synthesis Lectures on Human Language Technologies. Morgan & Claypool Publishers, 2010.

[277] Moritz Lipp, Michael Schwarz, Daniel Gruss, Thomas Prescher, Werner Haas, Stefan Mangard, Paul Kocher, Daniel Genkin, Yuval Yarom, and Mike Hamburg. Meltdown. *CoRR*, abs/1801.01207, 2018.

[278] Bing Liu, Wynne Hsu, and Yiming Ma. Integrating classification and association rule mining. In *Proc. SIGKDD*, pages 80–86, 1998.

[279] Fangfei Liu, Yuval Yarom, Qian Ge, Gernot Heiser, and Ruby B Lee. Last-level cache side-channel attacks are practical. In *IEEE Symposium on Security and Privacy Symposium*, pages 605–622, 2015.

[280] Fanghui Liu, Xiaolin Huang, Chen Gong, Jie Yang, and Li Li. Learning Data-adaptive Nonparametric Kernels. *arXiv e-prints*, page arXiv:1808.10724, Aug 2018.

[281] Qiang Liu, Pan Li, Wentao Zhao, Wei Cai, Shui Yu, and Victor CM Leung. A survey on security threats and defensive techniques of machine learning: A data driven view. *IEEE access*, 6:12103–12117, 2018.

[282] Julie Beth Lovins. Development of a stemming algorithm. *Mech. Translat. & Comp. Linguistics*, 11(1-2):22–31, 1968.

[283] Daniel Lowd and Christopher Meek. Adversarial learning. In *SIGKDD*, pages 641–647. ACM, 2005.

[284] Gavin Lowe. Breaking and fixing the Needham-Schroeder public-key protocol using FDR. *Software - Concepts and Tools*, 17(3):93–102, 1996.

[285] Hans Peter Luhn. The automatic creation of literature abstracts. *IBM Journal of Research and Development*, 2(2):159–165, 1958.

[286] Sebastian Lühr and Mihai Lazarescu. Incremental clustering of dynamic data streams using connectivity based representative points. *Data Knowl. Eng.*, 68(1):1–27, 2009.

[287] Bin Luo, Richard C Wilson, and Edwin R Hancock. Spectral embedding of graphs. *Pattern Recognition*, 36(10):2213–2230, 2003.

[288] Vince Lyzinski, Daniel L Sussman, Minh Tang, Avanti Athreya, Carey E. Priebe, et al. Perfect clustering for stochastic blockmodel graphs via adjacency spectral embedding. *Electronic Journal of Statistics*, 8(2):2905–2922, 2014.

[289] Justin Ma, Lawrence K. Saul, Stefan Savage, and Geoffrey M. Voelker. Beyond blacklists: Learning to detect malicious web sites from suspicious URLs. In *SIGKDD*, pages 1245–1254. ACM, 2009.

[290] Laurens van der Maaten and Geoffrey Hinton. Visualizing data using t-SNE. *Journal of Machine Learning Research*, 9(Nov):2579–2605, 2008.

[291] Matthew V. Mahoney and Philip K. Chan. An analysis of the 1999 DARPA/Lincoln laboratory evaluation data for network anomaly detection. In *RAID*, pages 220–237. Springer, 2003.

[292] Markus Maier, Ulrike von Luxburg, and Matthias Hein. How the result of graph clustering methods depends on the construction of the graph. *arXiv e-prints*, page arXiv:1102.2075, Feb 2011.

[293] Marcus A. Maloof. *Machine learning and data mining for computer security: Methods and applications*. Springer, 2006.

[294] Michael Mampaey, Jilles Vreeken, and Nikolaj Tatti. Summarizing data succinctly with the most informative itemsets. *TKDD*, 6(4):16:1–16:42, 2012.

[295] Heikki Mannila, Hannu Toivonen, and A. Inkeri Verkamo. Discovery of frequent episodes in event sequences. *Data Min. Knowl. Discov.*, 1(3):259–289, 1997.

[296] Christopher D. Manning, Prabhakar Raghavan, and Hinrich Schütze. *Introduction to Information Retrieval*. Cambridge Press, 2008.

[297] Mohammad Hossein Manshaei, Quanyan Zhu, Tansu Alpcan, Tamer Bacşar, and Jean-Pierre Hubaux. Game theory meets network security and privacy. *ACM Comput. Surv.*, 45(3):25:1–25:39, July 2013.

[298] Samuel Marchal, Kalle Saari, Nidhi Singh, and N Asokan. Know your phish: Novel techniques for detecting phishing sites and their targets. In *ICDCS*, pages 323–333. IEEE, 2016.

[299] David J. Marchette. *Computer intrusion detection and network monitoring: A statistical viewpoint*. Springer Science & Business Media, 2001.

[300] David J. Marchette. *Random graphs for statistical pattern recognition*, volume 565. John Wiley & Sons, 2005.

[301] Daniel L. Marino, Chathurika S. Wickramasinghe, and Milos Manic. An adversarial approach for explainable AI in intrusion detection systems. In *IECON 2018-44th Annual Conference of the IEEE Industrial Electronics Society*, pages 3237–3243. IEEE, 2018.

[302] Luis Marujo, Wang Ling, Isabel Trancoso, Chris Dyer, Alan W Black, Anatole Gershman, David Martins de Matos, Joao Neto, and Jaime Carbonell. Automatic keyword extraction on Twitter. In *ACL - IJCNLP (Volume 2: Short Papers)*, volume 2, pages 637–643, 2015.

[303] Mmalerato Masombuka, Marthie Grobler, and Bruce Watson. Towards an artificial intelligence framework to actively defend cyberspace. In *European Conference on Cyber Warfare and Security*, pages 589–XIII. Academic Conferences International Limited, 2018.

[304] David Sergio Matusevich, Carlos Ordonez, and Veerabhadran Baladandayuthapani. A fast convergence clustering algorithm merging MCMC and EM methods. In *CIKM*, pages 1525–1528, 2013.

[305] Ujjwal Maulik and Sanghamitra Bandyopadhyay. Performance evaluation of some clustering algorithms and validity indices. *Transactions on Pattern Analysis and Machine Intelligence*, 24(12):1650–1654, 2002.

[306] Pamela McCorduck. *Machines who think: A personal inquiry into the history and prospects of artificial intelligence.* AK Peters/CRC Press, 2009.

[307] D. Kevin McGrath, Andrew Kalafut, and Minaxi Gupta. Phishing infrastructure fluxes all the way. *IEEE Security & Privacy*, 7(5):21–28, 2009.

[308] Geoffrey McLachlan and Thriyambakam Krishnan. *The EM algorithm and extensions*, volume 382. John Wiley & Sons, 2007.

[309] Geoffrey McLachlan and David Peel. *Finite mixture models.* John Wiley & Sons, 2004.

[310] Leigh Metcalf and William Casey. *Cybersecurity and applied mathematics.* Syngress, 2016.

[311] Vangelis Metsis, Ion Androutsopoulos, and Georgios Paliouras. Spam filtering with naive Bayes-which naive Bayes? In *CEAS*, volume 17, pages 28–69, 2006.

[312] Tony A. Meyer and Brendon Whateley. Spambayes: Effective open-source, Bayesian based, email classification system. In *CEAS 2004*, 2004.

[313] Adam Meyers, Ruth Reeves, Catherine Macleod, Rachel Szekely, Veronika Zielinska, Brian Young, and Ralph Grishman. The NomBank project: An interim report. In *Proceedings of the Workshop Frontiers in Corpus Annotation, HLT-NAACL 2004, Boston, MA, USA, May 6, 2004*, 2004.

[314] Rada Mihalcea. Graph-based ranking algorithms for sentence extraction, applied to text summarization. In *ACL (posters and demonstration sessions)*, page 20. ACL, 2004.

[315] Rada Mihalcea and Ehsanul Faruque. Senselearner: Minimally supervised word sense disambiguation for all words in open text. In *Proceedings of the Third International Workshop on the Evaluation of Systems for the Semantic Analysis of Text, SENSEVAL@ACL 2004, Barcelona, Spain, July 25-26, 2004*, 2004.

[316] Rada Mihalcea and Paul Tarau. Textrank: Bringing order into text. In *EMNLP*, 2004.

[317] Tomas Mikolov, Kai Chen, Greg Corrado, and Jeffrey Dean. Efficient estimation of word representations in vector space. *CoRR*, abs/1301.3781, 2013.

[318] Marvin Minsky and Seymour A Papert. *Perceptrons: An introduction to computational geometry.* MIT Press, 2017.

[319] Yisroel Mirsky, Naor Kalbo, Yuval Elovici, and Asaf Shabtai. Vesper: Using echo analysis to detect man-in-the-middle attacks in LANs. *IEEE Transactions on Information Forensics and Security*, 14(6):1638–1653, 2019.

[320] Rami Mohammad, Lee McCluskey, and Fadi Thabtah. Phishing web sites data set. `http://eprints.hud.ac.uk/id/eprint/24330/`, 2012.

[321] Stefano Monti, Pablo Tamayo, Jill Mesirov, and Todd Golub. Consensus clustering: a resampling-based method for class discovery and visualization of gene expression microarray data. *Machine learning*, 52(1-2):91–118, 2003.

[322] Arjun Mukherjee. Detecting deceptive opinion spam using linguistics, behavioral and statistical modeling. In *ACL - IJCNLP 2015, Tutorial Abstracts*, pages 21–22, 2015.

[323] Srinivas Mukkamala and Andrew H. Sung. A comparative study of techniques for intrusion detection. In *IEEE ICTAI*, page 570, 2003.

[324] David Nadeau and Satoshi Sekine. A survey of named entity recognition and classification. *Lingvisticae Investigationes*, 30(1):3–26, 2007.

[325] N. K. Nagwani. Summarizing large text collection using topic modeling and clustering based on MapReduce framework. *Journal of Big Data*, 2(1):6, Jun 2015.

[326] Roberto Navigli. Word sense disambiguation: A survey. *ACM Comput. Surv.*, 41(2):10:1–10:69, February 2009.

[327] Richard E. Neapolitan. *Probabilistic reasoning in expert systems: Theory and algorithms*. CreateSpace Independent Publishing Platform, 2012.

[328] Graham Neubig, Taro Watanabe, Eiichiro Sumita, Shinsuke Mori, and Tatsuya Kawahara. An unsupervised model for joint phrase alignment and extraction. In *ACL:HLT-Volume 1*, pages 632–641. Association for Computational Linguistics, 2011.

[329] Andrew Y. Ng and Michael I. Jordan. On discriminative vs. generative classifiers: A comparison of logistic regression and naive Bayes. In *Advances in Neural Information Processing Systems 14 [Neural Information Processing Systems: Natural and Synthetic, NIPS 2001, December 3-8, 2001, Vancouver, British Columbia, Canada]*, pages 841–848, 2001.

[330] Andrew Y. Ng, Michael I. Jordan, and Yair Weiss. On spectral clustering: Analysis and an algorithm. In *Advances in Neural Information Processing Systems*, pages 849–856, 2002.

[331] Anh Nguyen, Jason Yosinski, and Jeff Clune. Deep neural networks are easily fooled: High confidence predictions for unrecognizable images. In *IEEE CVPR*, pages 427–436, 2015.

[332] Kamal P. Nigam. Using unlabeled data to improve text classification. Technical report, Carnegie-Mellon University Pittsburgh PA School of Computer Science, 2001.

[333] Stephen Northcutt and Judy Novak. *Network intrusion detection*. Sams Publishing, 2002.

[334] Andrew P. Norton and Yanjun Qi. Adversarial-playground: A visualization suite showing how adversarial examples fool deep learning. In *2017 IEEE Symposium on Visualization for Cyber Security (VizSec)*, pages 1–4. IEEE, 2017.

[335] Colin C. Olson, K. Peter Judd, and Jonathan M. Nichols. Manifold learning techniques for unsupervised anomaly detection. *Expert Syst. Appl.*, 91:374–385, 2018.

[336] Myle Ott, Yejin Choi, Claire Cardie, and Jeffrey T Hancock. Finding deceptive opinion spam by any stretch of the imagination. In *ACL: HLT Volume 1*, pages 309–319, 2011.

[337] Houman Owhadi and Gene Ryan Yoo. Kernel flows: from learning kernels from data into the abyss. *Journal of Computational Physics*, 389:22–47, 2019.

[338] Alexis Palmer and Caroline Sporleder. Evaluating framenet-style semantic parsing: The role of coverage gaps in framenet. In *COLING 2010, 23rd International Conference on Computational Linguistics, Posters Volume, 23-27 August 2010, Beijing, China*, pages 928–936, 2010.

[339] Hsing-Kuo Pao, Ching-Hao Mao, Hahn-Ming Lee, Chi-Dong Chen, and Christos Faloutsos. An intrinsic graphical signature based on alert correlation analysis for intrusion detection. In *Technologies and Applications of AI*, pages 102–109. IEEE, 2010.

[340] Nicolas Papernot, Patrick McDaniel, Somesh Jha, Matt Fredrikson, Z Berkay Celik, and Ananthram Swami. The limitations of deep learning in adversarial settings. In *Security and Privacy (EuroS&P), 2016 IEEE European Symposium on*, pages 372–387. IEEE, 2016.

[341] Dhaval Patel, Miral Patel, and Yogesh Dangar. A survey of different stemming algorithm. *International Journal of Advanced Engineering and Research Development*, 2(6), 2015.

[342] Neal Patwari, Alfred O Hero III, and Adam Pacholski. Manifold learning visualization of network traffic data. In *SIGCOMM workshop on Mining network data*, pages 191–196. ACM, 2005.

[343] Yudi Pawitan. *In all likelihood: statistical modelling and inference using likelihood.* Oxford University Press, 2001.

[344] Dan Pelleg and Andrew W. Moore. X-means: Extending k-means with efficient estimation of the number of clusters. In *Proceedings of the Seventeenth International Conference on Machine Learning (ICML 2000), Stanford University, Stanford, CA, USA, June 29 - July 2, 2000*, pages 727–734, 2000.

[345] Jeffrey Pennington, Richard Socher, and Christopher D. Manning. GloVe: Global vectors for word representation. In *Empirical Methods in Natural Language Processing (EMNLP)*, pages 1532–1543, 2014.

[346] Roberto Perdisci, Andrea Lanzi, and Wenke Lee. Mcboost: Boosting scalability in malware collection and analysis using statistical classification of executables. In *ACSAC*, pages 301–310, 2008.

[347] Radia Perlman. *Interconnections: bridges and routers.* Addison-Wesley, 1992.

[348] Matthew E. Peters, Mark Neumann, Mohit Iyyer, Matt Gardner, Christopher Clark, Kenton Lee, and Luke Zettlemoyer. Deep contextualized word representations. In *Proc. of NAACL*, 2018.

[349] Charles P Pfleeger and Shari Lawrence Pfleeger. *Security in computing.* Prentice Hall, 2002.

[350] Martin F Porter. An algorithm for suffix stripping. *Program*, 14(3):130–137, 1980.

[351] Adarsh Prasad, Alexandru Niculescu-Mizil, and Pradeep Ravikumar. On separability of loss functions, and revisiting discriminative vs generative models. In *Advances in Neural Information Processing Systems 30: Annual Conference on Neural Information Processing Systems 2017, 4-9 December 2017, Long Beach, CA, USA*, pages 7053–7062, 2017.

[352] Mila Dalla Preda, Mihai Christodorescu, Somesh Jha, and Saumya K. Debray. A semantics-based approach to malware detection. *ACM TOPLAS*, 30(5):25:1–25:54, 2008.

[353] Carey E. Priebe, John M. Conroy, David J. Marchette, and Youngser Park. Scan statistics on Enron graphs. *Computational & Mathematical Organization Theory*, 11(3), 2005.

[354] Carey E. Priebe and David J. Marchette. Adaptive mixture density estimation. *Pattern Recognition*, 26(5):771–785, 1993.

[355] Carey E. Priebe and David J. Marchette. Alternating kernel and mixture density estimates. *Computational Statistics & Data Analysis*, 35(1):43–65, 2000.

[356] Carey E. Priebe, David J. Marchette, and Dennis M. Healy, Jr. Integrated sensing and processing for statistical pattern recognition. In Daniel Rockmore and Dennis M. Healy, Jr., editors, *Modern Signal Processing*. Cambridge University Press, 2003.

[357] Carey E. Priebe, David J. Marchette, Youngser Park, Edward J. Wegman, Jeffrey L. Solka, Diego A. Socolinsky, Damianos Karakos, Ken W. Church, Roland Guglielmi, Ronald R. Coifman, et al. Iterative denoising for cross-corpus discovery. In *COMPSTAT 2004 - Proceedings in Computational Statistics*, pages 381–392. Springer, 2004.

[358] R Core Team. *R: A Language and Environment for Statistical Computing*. R Foundation for Statistical Computing, 2015.

[359] Alec Radford, Jeffrey Wu, Rewon Child, David Luan, Dario Amodei, and Ilya Sutskever. Language models are unsupervised multitask learners. *OpenAI Blog*, 1:8, 2019.

[360] Eynat Rafalin and Diane L. Souvaine. Computational geometry and statistical depth measures. In *Theory and applications of recent robust methods*, pages 283–295. Springer, 2004.

[361] Erhard Rahm and Hong Hai Do. Data cleaning: Problems and current approaches. `https://www.betterevaluation.org/sites/default/files/data_cleaning.pdf`, 2000.

[362] Venkatesh Ramanathan and Harry Wechsler. phishGILLNET—phishing detection methodology using probabilistic latent semantic analysis, adaboost, and co-training. *EURASIP Journal on Information Security*, 2012(1), 2012.

[363] Tom Richens. Anomalies in the wordnet verb hierarchy. In *ICCL-Volume 1*, pages 729–736. ACL, 2008.

[364] Alan Ritter, Sam Clark, Oren Etzioni, et al. Named entity recognition in tweets: an experimental study. In *Proceedings of the conference on empirical methods in natural language processing*, pages 1524–1534. Association for Computational Linguistics, 2011.

[365] Lior Rokach and Oded Maimon. *Data mining With decision trees: Theory and applications.* World Scientific Publishing Co., Inc., River Edge, NJ, USA, 2nd edition, 2014.

[366] Royi Ronen, Marian Radu, Corina Feuerstein, Elad Yom-Tov, and Mansour Ahmadi. Microsoft malware classification challenge. *CoRR*, abs/1802.10135, 2018.

[367] Stuart Rose, Dave Engel, Nick Cramer, and Wendy Cowley. Automatic keyword extraction from individual documents. In Michael W. Berry and Jacob Kogan, editors, *Text mining: Applications and theory*, pages 1–20. John Wiley & Sons, 2010.

[368] Irving N. Rothman, Rakesh M. Verma, Tom Woodell, and Blake Whitaker. Defoe's contribution to Robert Drury's journal: A stylometric analysis. In *Borck Festschrift.* AMS Press Inc., New York, 2017.

[369] Peter J. Rousseeuw and Mia Hubert. Statistical depth meets computational geometry: a short survey. *arXiv preprint arXiv:1508.03828*, 2015.

[370] David E Rumelhart, James L McClelland, and PDP Research Group. *Parallel distributed processing*, volume 1. MIT Press, 1987.

[371] Stuart J. Russell and Peter Norvig. *Artificial intelligence: a modern approach.* Pearson Education Limited, 2016.

[372] Sara Sabour, Yanshuai Cao, Fartash Faghri, and David J. Fleet. Adversarial manipulation of deep representations. *arXiv preprint arXiv:1511.05122*, 2015.

[373] Pramod J. Sadalge and Martin Fowler. *NoSQL distilled. A brief guide to the emerging world of polyglot persistence.* Addison-Wesley, 2013.

[374] Doyen Sahoo, Chenghao Liu, and Steven C. H. Hoi. Malicious URL detection using machine learning: A survey. *CoRR*, abs/1701.07179, 2017.

[375] Beatrice Santorini. Part-of-speech tagging guidelines for the Penn Treebank Project (3rd revision). Technical Report (CIS), University of Pennsylvania, 1995. Second Printing.

[376] Bernhard Scholkopf and Alexander J. Smola. *Learning with kernels: support vector machines, regularization, optimization, and beyond.* MIT Press, 2001.

[377] Erich Schubert, Jörg Sander, Martin Ester, Hans Peter Kriegel, and Xiaowei Xu. DBSCAN revisited, revisited: Why and how you should (still) use DBSCAN. *ACM Trans. Database Syst.*, 42(3):19:1–19:21, 2017.

[378] Matthew G. Schultz, Eleazar Eskin, Erez Zadok, and Salvatore J. Stolfo. Data mining methods for detection of new malicious executables. In *IEEE Symposium on Security and Privacy*, pages 38–49, 2001.

[379] Edward J. Schwartz, Thanassis Avgerinos, and David Brumley. All you ever wanted to know about dynamic taint analysis and forward symbolic execution (but might have been afraid to ask). In *IEEE Symposium on Security and Privacy*, pages 317–331, 2010.

[380] George A.F. Seber and Alan J. Lee. *Linear regression analysis*, volume 329. John Wiley & Sons, 2012.

[381] António Joaquim Serralheiro, Yariv Ephraim, and Lawrence R. Rabiner. On nonstationary hidden markov modeling of speech signals. In *EUROSPEECH*, pages 1159–1162, 1989.

[382] John Seymour and Philip Tully. Weaponizing data science for social engineering: Automated E2E spear phishing on Twitter. *Black Hat USA*, page 37, 2016.

[383] M. Zubair Shafiq, S. Momina Tabish, Fauzan Mirza, and Muddassar Farooq. Peminer: Mining structural information to detect malicious executables in realtime. In *RAID*, pages 121–141, 2009.

[384] Claude Shannon. A mathematical theory of cryptography. Memorandum MM 45-110-02, Bell Laboratories, 1945.

[385] Iman Sharafaldin, A. Habibi Lashkari, and Ali A. Ghorbani. Toward generating a new intrusion detection dataset and intrusion traffic characterization. In *ICISSP*, 2018.

[386] Razieh Sheikhpour, Mehdi Agha Sarram, Sajjad Gharaghani, and Mohammad Ali Zare Chahooki. A survey on semi-supervised feature selection methods. *Pattern Recognition*, 64:141–158, 2017.

[387] Jianbo Shi and Jitendra Malik. Normalized cuts and image segmentation. *IEEE Trans. Pattern Anal. Mach. Intell.*, 22(8):888–905, 2000.

[388] Kyuseok Shim. MapReduce algorithms for big data analysis. *Proc. VLDB Endow.*, 5(12), August 2012.

[389] Peter W. Shor. Polynomial-time algorithms for prime factorization and discrete logarithms on a quantum computer. *SIAM review*, 41(2):303–332, 1999.

[390] Muazzam Siddiqui, Morgan C. Wang, and Joohan Lee. A survey of data mining techniques for malware detection using file features. In *Annual Southeast Regional Conference*, pages 509–510, 2008.

[391] Jonathan A. Silva, Elaine R. Faria, Rodrigo C. Barros, Eduardo R. Hruschka, Andre CPLF De Carvalho, and João Gama. Data stream clustering: A survey. *ACM Computing Surveys*, 46(1):13, 2013.

[392] David Silver, Aja Huang, Chris J. Maddison, Arthur Guez, Laurent Sifre, George Van Den Driessche, Julian Schrittwieser, Ioannis Antonoglou, Veda Panneershelvam, Marc Lanctot, et al. Mastering the game of Go with deep neural networks and tree search. *Nature*, 529(7587):484–489, 2016.

[393] Bernard W. Silverman. *Density estimation for statistics and data analysis*, volume 26. CRC Press, 1986.

[394] Bongkee Sin and Jin H. Kim. Nonstationary hidden markov model. *Signal Processing*, 46(1):31–46, 1995.

[395] Ankush Singla and Elisa Bertino. How deep learning is making information security more intelligent. *IEEE Security & Privacy*, 17(3):56–65, 2019.

[396] Sami Smadi, Nauman Aslam, and Li Zhang. Detection of online phishing email using dynamic evolving neural network based on reinforcement learning. *Decision Support Systems*, 107:88–102, 2018.

[397] Robin Sommer and Vern Paxson. Outside the closed world: On using machine learning for network intrusion detection. In *IEEE Symposium on Security and Privacy*, pages 305–316, 2010.

[398] Dawn Xiaodong Song, David A. Wagner, and Adrian Perrig. Practical techniques for searches on encrypted data. In *IEEE Symposium on Security and Privacy*, pages 44–55, 2000.

[399] Shi-Jie Song, Zunguo Huang, Hua-Ping Hu, and Shi-Yao Jin. A sequential pattern mining algorithm for misuse intrusion detection. In Hai Jin, Yi Pan, Nong Xiao, and Jianhua Sun, editors, *Grid and Cooperative Computing Workshops*, pages 458–465. Springer, 2004.

[400] Abimbola Soriyan and Theresa Omodunbi. Trends in multi-document summarization system methods. *International Journal of Computer Applications*, 97(16), 2014.

[401] Eugene H. Spafford. OPUS: Preventing weak password choices. *Computers & Security*, 11(3):273–278, 1992.

[402] William Stallings. *Cryptography and network security - principles and practice (3rd. ed.)*. Prentice Hall, 2003.

[403] William Stallings. *Cryptography and network security - principles and practice (6th. ed.)*. Prentice Hall, 2014.

[404] William Stallings, Lawrie Brown, Michael D. Bauer, and Arup Kumar Bhattacharjee. *Computer security: principles and practice.* Pearson Education, 2012.

[405] Efstathios Stamatatos. A survey of modern authorship attribution methods. *JAIST*, 60(3):538–556, 2009.

[406] Katarzyna Stąpor. Evaluating and comparing classifiers: Review, some recommendations and limitations. In *International Conference on Computer Recognition Systems*, pages 12–21. Springer, 2017.

[407] Salvatore J. Stolfo, Shlomo Hershkop, Ke Wang, Olivier Nimeskern, and Chia-Wei Hu. Behavior profiling of email. In *International Conference on Intelligence and Security Informatics*, pages 74–90. Springer, 2003.

[408] Charles J. Stone. Additive regression and other nonparametric models. *The Annals of Statistics*, pages 689–705, 1985.

[409] Michael Stonebraker, Daniel Bruckner, Ihab F. Ilyas, George Beskales, Mitch Cherniack, Stanley B. Zdonik, Alexander Pagan, and Shan Xu. Data curation at scale: The data tamer system. In *CIDR*, 2013.

[410] Michael Stonebraker and Ihab F. Ilyas. Data integration: The current status and the way forward. *IEEE Data Eng. Bull.*, 41(2):3–9, 2018.

[411] Gilbert Strang. *Introduction to linear algebra.* Wellesley-Cambridge Press, Wellesley, MA, fifth edition, 2016.

[412] Gilbert Strang. *Linear algebra and learning from data.* Wellesley-Cambridge Press, Wellesley, MA, 2019.

[413] Kenneth David Strang and Zhaohao Sun. Meta-analysis of big data security and privacy: Scholarly literature gaps. In *IEEE BigData*, pages 4035–4037, 2016.

[414] Jiawei Su, Danilo Vasconcellos Vargas, and Sakurai Kouichi. One pixel attack for fooling deep neural networks. *arXiv preprint arXiv:1710.08864*, 2017.

[415] Big Data Public WG (Sec. & Priv. Subgroup). NIST big data interoperability framework: Volume 4, security and privacy. Technical Report NIST Special Publication 1500-1, NIST, 2015.

[416] Big Data Public WG (Sec. & Priv. Subgroup). NIST big data interoperability framework: Volume 4, security and privacy. Technical Report NIST Special Publication 1500-4, NIST, 2015.

[417] Sharmila Subramaniam, Themis Palpanas, Dimitris Papadopoulos, Vana Kalogeraki, and Dimitrios Gunopulos. Online outlier detection in sensor data using non-parametric models. In *VLDB*, pages 187–198, 2006.

[418] Daniel L Sussman, Minh Tang, Donniell E. Fishkind, and Carey E. Priebe. A consistent adjacency spectral embedding for stochastic blockmodel graphs. *Journal of the American Statistical Association*, 107(499):1119–1128, 2012.

[419] Peter Svenmarck, Linus Luotsinen, Mattias Nilsson, and Johan Schubert. Possibilities and challenges for artificial intelligence in military applications. In *Proceedings of the NATO Big Data and Artificial Intelligence for Military Decision Making Specialists' Meeting*, 2018.

[420] Pedro Tabacof and Eduardo Valle. Exploring the space of adversarial images. In *IJCNN*, pages 426–433. IEEE, 2016.

[421] Ashwin Tamilarasan, Srinivas Mukkamala, Andrew H. Sung, and Krishna Yendrapalli. Feature ranking and selection for intrusion detection using artificial neural networks and statistical methods. In *IJCNN*, pages 4754–4761, 2006.

[422] Ying Tan. *Artificial immune system: Applications in computer security*. John Wiley & Sons, 2016.

[423] P.N. Tang, M. Steinbach, and V. Kumar. *Introduction to data mining*. Pearson Education, 2006.

[424] Mahbod Tavallaee, Ebrahim Bagheri, Wei Lu, and Ali A. Ghorbani. A detailed analysis of the KDD CUP 99 data set. In *Computational Intelligence for Security and Defense Applications Symposium*, pages 1–6. IEEE, 2009.

[425] David M.J. Tax and Robert P.W. Duin. Uniform object generation for optimizing one-class classifiers. *JMLR*, 2(Dec):155–173, 2001.

[426] Joshua B Tenenbaum, Vin De Silva, and John C. Langford. A global geometric framework for nonlinear dimensionality reduction. *Science*, 290(5500):2319–2323, 2000.

[427] Tanmay Thakur and Rakesh M. Verma. Catching classical and hijack-based phishing attacks. In *ICISS*, pages 318–337. Springer LNCS 8880, 2014.

[428] Robert Tibshirani. Regression shrinkage and selection via the Lasso. *Journal of the Royal Statistical Society. Series B*, pages 267–288, 1996.

[429] Robert Tibshirani, Martin Wainwright, and Trevor Hastie. *Statistical learning with sparsity: The Lasso and generalizations*. Chapman & Hall/CRC, 2015.

[430] Robert Tibshirani and Guenther Walther. Cluster validation by prediction strength. *Journal of Computational and Graphical Statistics*, 14(3):511–528, 2005.

[431] Time Inc. *Computer security, understanding computers series.* Time-Life Books, 1990.

[432] Jon Timmis, Mark Neal, and John Hunt. An artificial immune system for data analysis. *Biosystems*, 55(1-3):143–150, 2000.

[433] D. Michael Titterington, Adrian F.M. Smith, and Udi E. Makov. *Statistical analysis of finite mixture distributions.* Wiley, 1985.

[434] Rocco Tripodi and Marcello Pelillo. A game-theoretic approach to word sense disambiguation. *Computational Linguistics*, 43(1):31–70, 2017.

[435] Michael W. Trosset. *An introduction to statistical inference and its applications with R.* CRC Press, 2009.

[436] Gerard V. Trunk. A problem of dimensionality: A simple example. *Transactions on Pattern Analysis and Machine Intelligence*, PAMI-1(3):306–307, 1979.

[437] Edward R. Tufte. *Envisioning information.* Graphics Press, Cheshire, Connecticut, 1990.

[438] Edward R. Tufte. *Visual explanations.* Graphics Press, Cheshire, Connecticut, 1997.

[439] Edward R. Tufte. *The visual display of quantitative information.* Graphics Press, Cheshire, Connecticut, 2001.

[440] Harshal Tupsamudre, Ajeet Kumar Singh, and Sachin Lodha. Everything is in the name - A URL based approach for phishing detection. In *Cyber Security Cryptography and Machine Learning - Third International Symposium, CSCML 2019, Beer-Sheva, Israel, June 27-28, 2019, Proceedings*, pages 231–248, 2019.

[441] Melissa J.M. Turcotte, Alexander D. Kent, and Curtis Hash. Unified host and network data set. *arXiv preprint arXiv:1708.07518*, 2017.

[442] Sun Tzu. The art of war. In *Strategic Studies*, pages 63–91. Routledge, 2008.

[443] Daniele Ucci, Leonardo Aniello, and Roberto Baldoni. Survey on the usage of machine learning techniques for malware analysis. *CoRR*, abs/1710.08189, 2017.

[444] Alper Kursat Uysal and Serkan Gunal. A novel probabilistic feature selection method for text classification. *Knowledge-Based Systems*, 36:226–235, 2012.

[445] John R. Vacca. *Computer and information security handbook.* Newnes, 2012.

[446] Gaurav Varshney, Manoj Misra, and Pradeep K. Atrey. A survey and classification of web phishing detection schemes. *Security and Comm. Networks*, 9(18):6266–6284, 2016.

[447] Ashish Vaswani, Noam Shazeer, Niki Parmar, Jakob Uszkoreit, Llion Jones, Aidan N. Gomez, Lukasz Kaiser, and Illia Polosukhin. Attention is all you need. *CoRR*, abs/1706.03762, 2017.

[448] Ashish Venugopal, Stephan Vogel, and Alex Waibel. Effective phrase translation extraction from alignment models. In *ACL-Volume 1*, pages 319–326, 2003.

[449] Rakesh M. Verma, editor. *IWSPA '15: Proceedings of the 2015 ACM International Workshop on International Workshop on Security and Privacy Analytics*, New York, NY, USA, 2015. ACM.

[450] Rakesh M. Verma and Ayman El Aassal. Comprehensive method for detecting phishing emails using correlation-based analysis and user participation. In *CODASPY*, pages 155–157, 2017.

[451] Rakesh M. Verma and Avisha Das. What's in a URL: Fast feature extraction and malicious URL detection. In *Proceedings of the 3rd ACM on International Workshop on Security And Privacy Analytics*, IWSPA '17, pages 55–63. ACM, 2017.

[452] Rakesh M. Verma and Avisha Das, editors. *Anti-Phishing Pilot at ACM IWSPA 2018: Evaluating Performance with New Metrics for Unbalanced Datasets*, volume 2124. ceur-ws.org, 2018.

[453] Rakesh M. Verma and Keith Dyer. On the character of phishing URLs: Accurate and robust statistical learning classifiers. In *CODASPY*, pages 111–122. ACM, 2015.

[454] Rakesh M. Verma and Nabil Hossain. A semantic feature selection technique with application to phishing email detection. In *ICISC*, 2013.

[455] Rakesh M. Verma, Murat Kantarcioglu, David J. Marchette, Ernst L. Leiss, and Thamar Solorio. Security analytics: Essential data analytics knowledge for cybersecurity professionals and students. *IEEE Security & Privacy*, 13(6):60–65, 2015.

[456] Rakesh M. Verma and Daniel Lee. Extractive summarization: Limits, compression, generalized model and heuristics. *Computación y Sistemas*, 21(4), 2017.

[457] Rakesh M. Verma, Narasimha Shashidhar, and Nabil Hossain. Phishing email detection the natural language way. In *Proc. 17th ESORICS*, pages 824–841, 2012.

[458] Rakesh M. Verma and Srivathsan Srinivasagopalan. Clustering for security challenges. In *Proceedings of the ACM International Workshop on Security and Privacy Analytics*, IWSPA '19, pages 1–2, New York, NY, USA, 2019. ACM.

[459] Rakesh M. Verma and Vasanthi Vuppuluri. A new approach for idiom identification using meanings and the web. In *RANLP*, pages 681–687, 2015.

[460] Rakesh M. Verma, Vasanthi Vuppuluri, An Nguyen, Arjun Mukherjee, Ghita Mammar, Shahryar Baki, and Reed Armstrong. Mining the web for collocations: IR models of term associations. In *CICLing Part I*, pages 177–194, 2016.

[461] Roman Vershynin. *High-dimensional probability: An introduction with applications in data science*, volume 47. Cambridge University Press, 2018.

[462] Ricardo Vilalta, Christophe Giraud-Carrier, Pavel Brazdil, and Carlos Soares. *Inductive Transfer*, pages 666–671. Springer US, Boston, MA, 2017.

[463] Ulrike von Luxburg. A tutorial on spectral clustering. *CoRR*, abs/0711.0189, 2007.

[464] Yevgeniy Vorobeychik and Murat Kantarcioglu. *Adversarial Machine Learning*. Synthesis Lectures on Artificial Intelligence and Machine Learning. Morgan & Claypool Publishers, 2018.

[465] Vasanthi Vuppuluri, Shahryar Baki, An Nguyen, and Rakesh M. Verma. ICE: Idiom and collocation extractor for research and education. In *EACL*, pages 108–111, 2017.

[466] Kiri Wagstaff, Claire Cardie, Seth Rogers, and Stefan Schrödl. Constrained k-means clustering with background knowledge. In *ICML*, pages 577–584, 2001.

[467] Martin J. Wainwright. *High-dimensional statistics: A non-asymptotic viewpoint*, volume 48. Cambridge University Press, 2019.

[468] Matt P. Wand and M. Chris Jones. *Kernel smoothing*. Crc Press, 1994.

[469] Cliff Wang and Zhuo Lu. Cyber deception: Overview and the road ahead. *IEEE Security & Privacy*, 16(2):80–85, 2018.

[470] Ding Wang, Zijian Zhang, Ping Wang, Jeff Yan, and Xinyi Huang. Targeted online password guessing: An underestimated threat. In *CCS*, pages 1242–1254. ACM, 2016.

[471] Lizhe Wang, Jie Tao, Rajiv Ranjan, Holger Marten, Achim Streit, Jingying Chen, and Dan Chen. G-hadoop: MapReduce across distributed data centers for data-intensive computing. *FGCS*, 29(3):739–750, 2013.

[472] S. Wang, W. Liu, J. Wu, L. Cao, Q. Meng, and P. J. Kennedy. Training deep neural networks on imbalanced data sets. In *IJCNN*, pages 4368–4374, 2016.

[473] William Yang Wang, Sameer Singh, and Jiwei Li. Deep adversarial learning for NLP. In *Proceedings of the 2019 Conference of the North American Chapter of the Association for Computational Linguistics: Human Language Technologies, NAACL-HLT 2019, Minneapolis, MN, USA, June 2, 2019, Tutorial Abstracts*, pages 1–5, 2019.

[474] Zheng Wang. Deep learning-based intrusion detection with adversaries. *IEEE Access*, 6:38367–38384, 2018.

[475] Christina Warrender, Stephanie Forrest, and Barak A. Pearlmutter. Detecting intrusions using system calls: Alternative data models. In *IEEE Symposium on Security and Privacy*, pages 133–145, 1999.

[476] Larry Wasserman. Topological data analysis. *Annual Review of Statistics and Its Application*, 5:501–532, 2018.

[477] Martin Wattenberg, Fernanda Viégas, and Ian Johnson. How to use t-SNE effectively. *Distill*, 1(10):e2, 2016.

[478] Geoffrey I. Webb. Discovering significant rules. In *SIGKDD*, pages 434–443. ACM, 2006.

[479] Edward J. Wegman. Hyperdimensional data analysis using parallel coordinates. *Journal of the American Statistical Association*, 85(411):664–675, 1990.

[480] Edward J. Wegman and David J. Marchette. On some techniques for streaming data: A case study of internet packet headers. *Journal of Computational and Graphical Statistics*, 12(4):893–914, 2003.

[481] Nico Weichbrodt, Anil Kurmus, Peter Pietzuch, and Rüdiger Kapitza. Asyncshock: Exploiting synchronisation bugs in Intel SGX enclaves. In *ESORICS*, pages 440–457. Springer, 2016.

[482] M. Weir, S. Aggarwal, B. D. Medeiros, and B. Glodek. Password cracking using probabilistic context-free grammars. In *IEEE Symposium on Security and Privacy*, pages 391–405, 2009.

[483] Roy E. Welsch and Edwin Kuh. Linear regression diagnostics. Working Paper 173, National Bureau of Economic Research, March 1977.

[484] Li Wenliang, Dougal Sutherland, Heiko Strathmann, and Arthur Gretton. Learning deep kernels for exponential family densities. In Kamalika Chaudhuri and Ruslan Salakhutdinov, editors, *Proceedings of the 36th International Conference on Machine Learning*, volume 97 of *Proceedings of Machine Learning Research*, pages 6737–6746, Long Beach, California, USA, 09–15 Jun 2019. PMLR.

[485] Colin Whittaker, Brian Ryner, and Marria Nazif. Large-scale automatic classification of phishing pages. In *NDSS*, volume 10, 2010.

[486] Dominic Widdows and Beate Dorow. Automatic extraction of idioms using graph analysis and asymmetric lexicosyntactic patterns. In *Proceedings of the ACL-SIGLEX Workshop on Deep Lexical Acquisition*, pages 48–56, 2005.

[487] W John Wilbur and Karl Sirotkin. The automatic identification of stop words. *Journal of Information Science*, 18(1):45–55, 1992.

[488] Leland Wilkinson. *The grammar of graphics.* Springer Science & Business Media, 2006.

[489] Ian H. Witten, Eibe Frank, Mark A. Hall, and Christopher J Pal. *Data mining: Practical machine learning tools and techniques.* Morgan Kaufmann, 2016.

[490] Paul A Wortman, Fatemeh Tehranipoor, and John A Chandy. An adversarial risk-based approach for network architecture security modeling and design. In *2018 International Conference on Cyber Security and Protection of Digital Services (Cyber Security)*, pages 1–8. IEEE, 2018.

[491] Charles V. Wright, Lucas Ballard, Scott E. Coull, Fabian Monrose, and Gerald M. Masson. Spot me if you can: Uncovering spoken phrases in encrypted voip conversations. In *IEEE Symposium on Security and Privacy*, pages 35–49, 2008.

[492] Charles V. Wright, Lucas Ballard, Fabian Monrose, and Gerald M. Masson. Language identification of encrypted VOIP traffic: Alejandra y roberto or alice and bob? In *USENIX Security Symposium*, volume 3, pages 43–54, 2007.

[493] Charles V. Wright, Scott E. Coull, and Fabian Monrose. Traffic morphing: An efficient defense against statistical traffic analysis. In *NDSS*, volume 9, 2009.

[494] Caesar Wu, Rajkumar Buyya, and Kotagiri Ramamohanarao. Big data analytics = machine learning + cloud computing. *CoRR*, abs/1601.03115, 2016.

[495] Dekai Wu and Pascale Fung. Improving chinese tokenization with linguistic filters on statistical lexical acquisition. In *ANLP*, pages 180–181. ACL, 1994.

[496] Han-Ching Wu and Shou-Hsuan Stephen Huang. Neural networks-based detection of stepping-stone intrusion. *Expert Syst. Appl.*, 37(2):1431–1437, 2010.

[497] Qi Wu, Will Usher, Steve Petruzza, Sidharth Kumar, Feng Wang, Ingo Wald, Valerio Pascucci, and Charles D. Hansen. VisIt-OSPRay: Toward an Exascale Volume Visualization System. In Hank Childs and Fernando Cucchietti, editors, *Eurographics Symposium on Parallel Graphics and Visualization*. The Eurographics Association, 2018.

[498] Tingmin Wu, Lihong Tang, Zhiyu Xu, Sheng Wen, Cecile Paris, Surya Nepal, Marthie Grobler, and Yang Xiang. Catering to your concerns: Automatic generation of personalised security-centric descriptions for android apps. *arXiv preprint arXiv:1805.07070*, 2018.

[499] Guang Xiang, Jason Hong, Caroline Rosé, and Lorrie Cranor. CANTINA+: A feature-rich machine learning framework for detecting phishing web sites. *ACM Trans. Inf. Syst. Secur.*, 14(2):21:1–21:28, 2011.

[500] Xiao Xiao, Yuxin Ding, Yibin Zhang, Ke Tang, and Dai Wei. Malware detection based on objective-oriented association mining. In *ICMLC*, pages 375–380, 2013.

[501] Zhifeng Xiao and Yang Xiao. Achieving accountable MapReduce in cloud computing. *FGCS*, 30:1–13, 2014.

[502] Rui Xu and Donald C Wunsch. Survey of clustering algorithms. *IEEE Trans. on Neural Networks*, 16(5), 2005.

[503] Zhang Xu, Zhenyu Wu, Zhichun Li, Kangkook Jee, Junghwan Rhee, Xusheng Xiao, Fengyuan Xu, Haining Wang, and Guofei Jiang. High fidelity data reduction for big data security dependency analyses. In *ACM CCS*, 2016.

[504] Jing-Hao Xue and D.M. Titterington. Comment on "on discriminative vs. generative classifiers: A comparison of logistic regression and naive Bayes". *Neural Processing Letters*, 28(3):169–187, 2008.

[505] Albert Yang, Chung-Kang Peng, and Ary L Goldberger. The marlowe-shakespeare authorship debate: Approaching an old problem with new methods, 2017.

[506] Lei Yang and Fengjun Li. Detecting false data injection in smart grid in-network aggregation. In *Smart Grid Communications*, pages 408–413. IEEE, 2013.

[507] Wenchuan Yang, Wen Zuo, and Baojiang Cui. Detecting malicious URLs via a keyword-based convolutional gated-recurrent-unit neural network. *IEEE Access*, 7:29891–29900, 2019.

[508] Yiming Yang. Noise reduction in a statistical approach to text categorization. In *SIGIR*, pages 256–263. ACM, 1995.

[509] Zhilin Yang, Zihang Dai, Yiming Yang, Jaime Carbonell, Ruslan Salakhutdinov, and Quoc V. Le. XLNet: Generalized Autoregressive Pretraining for Language Understanding. *arXiv e-prints*, page arXiv:1906.08237, Jun 2019.

[510] Zichao Yang, Zhiting Hu, Ruslan Salakhutdinov, and Taylor Berg-Kirkpatrick. Improved variational autoencoders for text modeling using dilated convolutions. In *Proceedings of the 34th International Conference on Machine Learning-Volume 70*, pages 3881–3890. JMLR. org, 2017.

[511] Yuanshun Yao, Bimal Viswanath, Jenna Cryan, Haitao Zheng, and Ben Y. Zhao. Automated crowdturfing attacks and defenses in online review systems. In *ACM CCS*, pages 1143–1158, 2017.

[512] David Yarowsky. Unsupervised word sense disambiguation rivaling supervised methods. In *ACL*, pages 189–196, 1995.

[513] Adwan Yasin and Abdelmunem Abuhasan. An intelligent classification model for phishing email detection. *arXiv preprint arXiv:1608.02196*, 2016.

[514] Nathan Yau. *Visualize this: The flowing data guide to design, visualization, and statistics.* John Wiley & Sons, 2011.

[515] Nathan Yau. *Data points: Visualization that means something.* John Wiley & Sons, 2013.

[516] Yanfang Ye, Tao Li, Donald Adjeroh, and S. Sitharama Iyengar. A survey on malware detection using data mining techniques. *ACM Comput. Surv.*, 50(3):41:1–41:40, 2017.

[517] Jason Yosinski, Jeff Clune, Anh Nguyen, Thomas Fuchs, and Hod Lipson. Understanding neural networks through deep visualization. *arXiv preprint arXiv:1506.06579*, 2015.

[518] I. You and K. Yim. Malware obfuscation techniques: A brief survey. In *Conference on Broadband, Wireless Computing, Communication and Applications*, pages 297–300, Nov 2010.

[519] Saman Taghavi Zargar, James Joshi, and David Tipper. A survey of defense mechanisms against distributed denial of service (ddos) flooding attacks. *IEEE Communications Surveys and Tutorials*, 15(4):2046–2069, 2013.

[520] Han Zhang, Yuanbo Guo, and Tao Li. Multifeature named entity recognition in information security based on adversarial learning. *Security and Communication Networks*, 2019, 2019.

[521] Hao Zhang, Maoyuan Sun, Danfeng (Daphne) Yao, and Chris North. Visualizing traffic causality for analyzing network anomalies. In *Proceedings of the 2015 ACM International Workshop on Security and Privacy Analytics*, IWSPA '15, pages 37–42, New York, NY, USA, 2015. ACM.

[522] Jiaqi Zhao, Lizhe Wang, Jie Tao, Jinjun Chen, Weiye Sun, Rajiv Ranjan, Joanna Kolodziej, Achim Streit, and Dimitrios Georgakopoulos. A security framework in G-Hadoop for big data computing across distributed cloud data centres. *J. Comput. Syst. Sci.*, 80(5):994–1007, 2014.

[523] Peilin Zhao and Steven C. H. Hoi. Cost-sensitive online active learning with application to malicious URL detection. In *SIGKDD*, pages 919–927. ACM, 2013.

[524] Mu Zhu and Ali Ghodsi. Automatic dimensionality selection from the scree plot via the use of profile likelihood. *Computational Statistics & Data Analysis*, 51:918–930, 2006.

[525] Quanyan Zhu and Stefan Rass. Game theory meets network security: A tutorial. In *Proceedings of the 2018 ACM SIGSAC Conference on Computer and Communications Security*, CCS '18, pages 2163–2165, New York, NY, USA, 2018. ACM.

[526] Xiaojin Zhu and Andrew B. Goldberg. Introduction to semi-supervised learning. *Synthesis Lectures on Artificial Intelligence and Machine Learning*, 3(1):1–130, 2009.

[527] Ziyun Zhu and Tudor Dumitras. Featuresmith: Automatically engineering features for malware detection by mining the security literature. In *ACM CCS*, pages 767–778, 2016.

[528] Giulio Zizzo, Chris Hankin, Sergio Maffeis, and Kevin Jones. Adversarial machine learning beyond the image domain. In *Proceedings of the 56th Annual Design Automation Conference 2019*, DAC '19, pages 176:1–176:4, New York, NY, USA, 2019. ACM.

[529] Walter Zucchini, Iain L MacDonald, and Roland Langrock. *Hidden Markov models for time series: An introduction using R*. CRC press, 2017.

[530] Yi-jun Zuo and Heng-jian Cui. Statistical depth functions and some applications. *Advances in Mathematics*, 33(1):1–25, 2004.

[531] Yijun Zuo and Robert Serfling. General notions of statistical depth function. *The Annals of statistics*, pages 461–482, 2000.

Author Index

Index